VOLUME ONE HUNDRED AND SIXTY FOUR

PROGRESS IN MOLECULAR BIOLOGY AND TRANSLATIONAL SCIENCE

Cancer Immunotherapy

VOLUME ONE HUNDRED AND SIXTY FOUR

PROGRESS IN MOLECULAR BIOLOGY AND TRANSLATIONAL SCIENCE

Cancer Immunotherapy

Edited by

DAVID B. TEPLOW
Department of Neurology
David Geffen School of Medicine at UCLA
Los Angeles, CA, United States

ELSEVIER

ACADEMIC PRESS
An imprint of Elsevier

Academic Press is an imprint of Elsevier
50 Hampshire Street, 5th Floor, Cambridge, MA 02139, United States
525 B Street, Suite 1650, San Diego, CA 92101, United States
The Boulevard, Langford Lane, Kidlington, Oxford OX5 1GB, United Kingdom
125 London Wall, London, EC2Y 5AS, United Kingdom

First edition 2019

Notices

Knowledge and best practice in this field are constantly changing. As new research and experience broaden our understanding, changes in research methods, professional practices, or medical treatment may become necessary.

Practitioners and researchers must always rely on their own experience and knowledge in evaluating and using any information, methods, compounds, or experiments described herein. In using such information or methods they should be mindful of their own safety and the safety of others, including parties for whom they have a professional responsibility.

To the fullest extent of the law, neither the Publisher nor the authors, contributors, or editors, assume any liability for any injury and/or damage to persons or property as a matter of products liability, negligence or otherwise, or from any use or operation of any methods, products, instructions, or ideas contained in the material herein.

ISBN: 978-0-12-816575-1
ISSN: 1877-1173

For information on all Academic Press publications
visit our website at https://www.elsevier.com/books-and-journals

Publisher: Zoe Kruze
Acquisition Editor: Ashlie M. Jackman
Editorial Project Manager: Joanna Collett
Production Project Manager: James Selvam
Cover Designer: Greg Harris

Typeset by SPi Global, India

Contents

Contributors ix
Preface xiii

1. From the Hellstrom paradox toward cancer cure **1**

Karl Erik Hellstrom and Ingegerd Hellstrom

1. Cancer immunology has a long history 2
2. A Th2-type inflammation promotes tumor growth 4
3. Treating tumor-bearing mice with immunomodulatory mAbs 5
4. Mechanisms responsible for the therapeutic efficacy of immunomodulating mAbs 8
5. Synergistic effect combining immunomodulatory mAbs with cisplatin 12
6. In vitro evaluation of immunomodulatory mAbs 15
7. Outlook 15

Acknowledgments 18
References 18

2. Bioinformatic methods for cancer neoantigen prediction **25**

Sebastian Boegel, John C. Castle, Julia Kodysh, Timothy O'Donnell, and Alex Rubinsteyn

1. Introduction 26
2. What do we mean by "immunogenicity"? 28
3. Neoantigen sources 28
4. Identifying neoantigens from high-throughput sequencing data 29
5. Which protein changes result in neoantigens? 35
6. Multi-step neoantigen prediction workflows 41
7. Identifying neoantigen responsive T-cell receptors 46
8. Conclusion 47

Acknowledgment 48
References 48

3. Tumor mechanisms of resistance to immune attack **61**

David J. Zahavi and Louis M. Weiner

1. Introduction 62
2. Loss of antigenicity 66
3. Loss of immunogenicity 73

4. Immunosuppressive microenvironment 81
5. Conclusion 87
References 90

4. TNFR2-expressing $CD4^{+}Foxp3^{+}$ regulatory T cells in cancer immunology and immunotherapy 101

Jiang He, Ruixin Li, Yibo Chen, Yuanjia Hu, and Xin Chen

1. Introduction 102
2. Ovarian cancer 103
3. Lung cancer 105
4. Acute myelocytic leukemia (AML) 105
5. Lymphoma 106
6. Colon cancer 106
7. Cervical intraepithelial neoplasia (CIN) and cervical cancer 107
8. Hepatocellular carcinoma (HCC) 108
9. Closing remarks 109

Acknowledgments 113
Conflict of interest 113
References 113

5. Cancer stem cell immunology and immunotherapy: Harnessing the immune system against cancer's source 119

Roberto Ruiu, Lidia Tarone, Valeria Rolih, Giuseppina Barutello, Elisabetta Bolli, Federica Riccardo, Federica Cavallo, and Laura Conti

1. Introduction 120
2. Cancer stem cells 121
3. The immunology of cancer stem cells 137
4. Conclusions and future perspectives 170

Acknowledgments 175
References 175

6. Immunotherapy of pancreatic cancer 189

Yoshiaki Sunami and Jörg Kleeff

1. Introduction 190
2. Cancer immunosurveillance and immunoediting in pancreatic cancer 190
3. Immunotherapy with natural killer cells and dendritic cells in pancreatic cancer 191
4. Classical immunotherapy with immune checkpoint inhibitors in pancreatic cancer 195

5. Immunosuppressive tumor microenvironment in pancreatic cancer 196
6. Reprogramming the microenvironment by targeting tumor-associated macrophages as a therapeutic strategy for pancreatic cancer 199
7. Reprogramming the microenvironment by targeting regulators expressed by pancreatic cancer cells or stromal cells 204
8. Adaptive T-cell therapy for pancreatic cancer 206
9. Safety considerations in CAR T-cell therapy in pancreatic cancer 208
10. Conclusion 210
References 210

7. Synergistic combination of oncolytic virotherapy with CAR T-cell therapy 217

Adam Ajina and John Maher

1. Background 218
2. Oncolytic virotherapy—An overview 219
3. The development of chimeric antigen receptor T-cell therapy 227
4. Synergies between OVs and CAR T-cell therapy 234
5. Engineered OVs with payloads conducive to CAR function 239
6. Conclusions 273
Acknowledgments 273
Ethical approval and consent to participate 274
Consent for publication 274
Availability of supporting data 274
Competing interests 274
Funding 274
Authors' contributions 274
References 274

8. Application of CAR T cells for the treatment of solid tumors 293

Jonathan F. Khan, Abdul Salam Khan, and Renier J. Brentjens

1. Introduction 294
2. CAR T cell therapy for solid tumors 297
3. Overcoming challenges to CAR T cell therapy of solid tumors 302
4. Concluding remarks 317
References 318

Index *329*

Contributors

Adam Ajina
King's College London, Division of Cancer Studies, Guy's Hospital, London, United Kingdom

Giuseppina Barutello
Department of Molecular Biotechnology and Health Sciences, Molecular Biotechnology Center, University of Torino, Torino, Italy

Sebastian Boegel
Johannes Gutenberg University, Mainz, Germany

Elisabetta Bolli
Department of Molecular Biotechnology and Health Sciences, Molecular Biotechnology Center, University of Torino, Torino, Italy

Renier J. Brentjens
Department of Pharmacology, Weill Cornell Graduate School of Medical Sciences, Weill Cornell Medicine; Molecular Pharmacology Program, Sloan Kettering Institute; Department of Medicine, Memorial Sloan Kettering Cancer Center, New York, NY, United States

John C. Castle
Agenus Inc., Lexington, MA, United States

Federica Cavallo
Department of Molecular Biotechnology and Health Sciences, Molecular Biotechnology Center, University of Torino, Torino, Italy

Xin Chen
State Key Laboratory of Quality Research in Chinese Medicine, Institute of Chinese Medical Sciences, University of Macau, Macau, China

Yibo Chen
State Key Laboratory of Quality Research in Chinese Medicine, Institute of Chinese Medical Sciences, University of Macau, Macau, China

Laura Conti
Department of Molecular Biotechnology and Health Sciences, Molecular Biotechnology Center, University of Torino, Torino, Italy

Jiang He
State Key Laboratory of Quality Research in Chinese Medicine, Institute of Chinese Medical Sciences, University of Macau, Macau; Investment Banking, Shenzhen Rhino Star Information Co. Ltd., Shenzhen, China

Ingegerd Hellstrom
Department of Pathology, University of Washington, Harborview Medical Center, Seattle, WA, United States

Karl Erik Hellstrom
Department of Pathology, University of Washington, Harborview Medical Center, Seattle, WA, United States

Yuanjia Hu
State Key Laboratory of Quality Research in Chinese Medicine, Institute of Chinese Medical Sciences, University of Macau, Macau, China

Abdul Salam Khan
Department of Medicine, Memorial Sloan Kettering Cancer Center, New York, NY, United States

Jonathan F. Khan
Department of Pharmacology, Weill Cornell Graduate School of Medical Sciences, Weill Cornell Medicine; Molecular Pharmacology Program, Sloan Kettering Institute, Memorial Sloan Kettering Cancer Center, New York, NY, United States

Jörg Kleeff
Department of Visceral, Vascular and Endocrine Surgery, Martin-Luther-University Halle-Wittenberg, University Medical Center Halle, Halle, Germany

Julia Kodysh
Icahn School of Medicine at Mount Sinai, New York, NY, United States

Ruixin Li
State Key Laboratory of Quality Research in Chinese Medicine, Institute of Chinese Medical Sciences, University of Macau, Macau, China

John Maher
King's College London, Division of Cancer Studies, Guy's Hospital; Department of Clinical Immunology and Allergy, King's College Hospital NHS Foundation Trust, London; Department of Immunology, Eastbourne Hospital, East Sussex, United Kingdom

Timothy O'Donnell
Icahn School of Medicine at Mount Sinai, New York, NY, United States

Federica Riccardo
Department of Molecular Biotechnology and Health Sciences, Molecular Biotechnology Center, University of Torino, Torino, Italy

Valeria Rolih
Department of Molecular Biotechnology and Health Sciences, Molecular Biotechnology Center, University of Torino, Torino, Italy

Alex Rubinsteyn
Icahn School of Medicine at Mount Sinai, New York, NY, United States

Roberto Ruiu
Department of Molecular Biotechnology and Health Sciences, Molecular Biotechnology Center, University of Torino, Torino, Italy

Yoshiaki Sunami
Department of Visceral, Vascular and Endocrine Surgery, Martin-Luther-University Halle-Wittenberg, University Medical Center Halle, Halle, Germany

Lidia Tarone
Department of Molecular Biotechnology and Health Sciences, Molecular Biotechnology Center, University of Torino, Torino, Italy

Louis M. Weiner
Georgetown Lombardi Comprehensive Cancer Center, Department of Oncology, Georgetown University Medical Center, Washington, DC, United States

David J. Zahavi
Tumor Biology Training Program; Georgetown Lombardi Comprehensive Cancer Center, Department of Oncology, Georgetown University Medical Center, Washington, DC, United States

Preface

The disease we know today as "cancer" was described in the earliest medical literature found thus far, the Edwin Smith Papyrus, which is thought to have been written by an Egyptian physician circa 3000 BC. Early treatment of cancer included surgery, but it was not until the mid-18th century, with the invention of anesthetics, that the modern age of surgical oncology began. This was followed, at the turn of the 20th century, by the development of radiation therapy and, subsequently, by hormone and chemical therapies. All these therapeutic approaches target cell division, either directly, by inhibiting replication, or indirectly, by blocking receptors on hormone-sensitive tumor cells. Tumor therapies that directly target cell division cause significant side effects because of their toxicity to normal cells that also must divide to maintain tissue function. Unfortunately, in many cases, DNA damage caused by radiation and chemotherapy can also induce new tumors. In contrast, cancer immunotherapy is a targeted approach that has the potential, although not yet fully realized, to eliminate tumors with fewer effects on normal cells and thus with less morbidity than that associated with radiation and chemotherapy. Immunotherapy harnesses the innate power of a patient's own cellular and humoral immune responses to identify tumor cells as foreign and to eliminate them. These responses involve myriad different cell types as effectors and regulators of tumor cell-specific cytotoxicity. The challenge faced with immunotherapeutic approaches is tumor cell resistance to immune elimination, which is complex, dynamic, and highly effective. This volume provides the reader with the latest information on the promise and problems of cancer immunotherapy.

In Chapter 1, "From the Hellstrom paradox toward cancer cure," Karl Erik Hellstrom and Ingegerd Hellstrom begin by referring to early studies showing that tumor-bearing rodents, as well as human cancer patients, mount an immunological response to their tumors, but that the ability of this response to damage the tumor is normally inhibited. They then discuss checkpoint inhibition, a therapeutic approach that seeks to block lymphocyte cell surface receptors and ligands (e.g., CTLA-4, PD-1, and PD-L1) that, when bound in their conjugate receptor-ligand pairs, repress immune responses. The importance of this general strategy, i.e., "Cancer therapy by inhibition of negative immune regulation," was recognized in 2018 with the awarding of the Nobel Prize in Physiology or Medicine to James P. Allison and Tasuku Honjo. The Hellstrom's work has focused on mouse models of cancer that allow the study of the efficacy of using monoclonal antibodies to target checkpoint agonists/antagonists. This work has demonstrated that a shift from Th2 helper T cell-mediated inflammatory responses to Th1-stimulated cytotoxic responses is associated with successful immunotherapy. Achieving this shift may be facilitated by co-treatment with cisplatin, which has positive effects on the immune system as well as acting as an antimitotic agent. Such immunochemotherapy may thus combine the exquisite specificity of the immune system with the blunt force cytotoxicity of chemotherapy to dramatically improve patient outcomes.

Specificity is the watchword of the immune system in normal immune responses to foreign entities such as bacteria and viruses and in responses to tumors. These responses are initiated by the recognition of soluble or cell surface immunogenic structures. A surprising finding has been that tumor cells, which are formed initially from normal cells in the body, i.e., from "self," can be recognized as "non-self" and attacked by the immune system. Recognition is mediated through tumor-associated antigens (TAAs), which become targets for cellular and humoral immune responses. In the past, TAAs have been identified through painstaking study of specific tumor cells. However, "data-driven science," science initiated by computerized analysis of massive data sets and hypothesis formulation therefrom, now offers an alternative method for tumor targeting. Algorithms developed to interrogate data produced from DNA sequencing of exomes and transcriptomes of tumor cells now enable bioinformatics specialists to identity candidate neoantigens and to prioritize them for study based on their abundance and predicted immunogenicity. Neoantigens identified in this manner may be used for the development of cancer vaccines, T-cell adoptive cellular therapy, and immunomonitoring. In Chapter 2, Boegel et al. discuss the steps and challenges involved in computerized neoantigen discovery.

TAAs exist and the immune system recognizes and responds to them. Why then do humans still develop and succumb to tumors? One answer is tumor cells have developed an astoundingly large number of immune escape mechanisms. Zahavi and Weiner, in Chapter 3, present a broad review of these mechanisms, with emphasis on the ability of tumor cells to lose antigenicity and immunogenicity, as well as create an immunosuppressive microenvironment. These abilities are a major impediment to the successful use of immunotherapy for cancer therapy. He et al. then discuss one immune escape mechanism, that involving a specific subset of regulatory T cells (Tregs), CD4+Foxp3+ Tregs. These cells can be activated by tumor necrosis factor (TNF) interacting with TNF2 receptors on these cells, suggesting that antagonists of this binary interaction might be effective in facilitating antitumor immune responses.

Immune recognition of TAAs and subsequent activation of a protective immune response are a critical goal in cancer immunotherapy. However, what if the cells recognized are not the actual source of tumor but simply a result of an unrecognized aberrant cellular process? This describes the biology of many tumors, which comprise a diverse array of tumor cells, including a small population of stem cells that give rise to the majority of cells recognized by the immune system but are themselves not, or poorly, recognized. In Chapter 5, Ruiu et al. discuss cancer stem cells, which are resistant to radio- and chemotherapy and serve as reservoirs for continued production of tumor cells that form local and metastatic tumor masses. Targeting these cells would eliminate tumor cells at their source, preventing post-treatment reemergence of disease. This approach has great promise, but many difficulties, both of which are discussed in detail in this chapter.

In Chapter 6, we move from bench to bedside, discussing efforts to treat pancreatic ductal adenocarcinoma, a cancer of particularly poor prognosis. The median survival time for untreated advanced pancreatic cancer is about 3.5 months, but the treatment may only increase this to ~8 months. The American Cancer Society reports a 5-year survival rate of only 7%. Pancreatic cancers often are detected at late stages and show a remarkable resistance to chemotherapy, radiotherapy, and targeted therapy, which in part explains their poor prognosis. However, this does not mean that new therapies that are more effective cannot be developed. Immunotherapy is one. In this chapter, Sunami and Kleef discuss recent preclinical and clinical studies of immunotherapy involving natural killer cells, dendritic cells, and chimeric antigen receptor T cells. They also discuss how "reprogramming the tumor immune microenvironment" may be efficacious. This would involve elimination of circulating and infiltrating monocytes/macrophages, a strategy that has shown some success in tumor model systems. It also involves targeting factors produced by cells that block the immune response. These may include the tumor cells themselves as well as cancer-associated fibroblasts and stellate cells. Sunami and Kleef suggest that treatment of pancreatic cancer may be improved by combinatorial therapies that include antimitotics, focal adhesion kinase inhibitors, or PD1 and CTLA4 antagonists.

In Chapter 7, Ajina and Maher discuss the synergistic effects of combining oncolytic virotherapy with CAR T-cell therapy on treatment outcomes for solid tumors. These authors suggest that engineering oncolytic viruses and CAR T cells with customizable transgenes may, in the future, herald a day when patient-specific, combined immune system-directed therapy may be a routine and effective approach for the treatment of solid tumors. Indeed, recombinant oncolytic virotherapy for patients with malignant melanoma, and chimeric antigen receptor (tisagenlecleucel) therapy for children and young adults with relapsed or refractory B-cell acute lymphoblastic leukemia recently have been approved for clinical use.

Khan et al. continue and broaden the scope of the discussion of immunotherapy of solid tumors in Chapter 8. They begin by summarizing the current state of CAR T-cell therapy in hematologic malignancies, for which at least two FDA-approved therapies now are available, tisagenlecleucel, for treating diffuse large B-cell lymphoma (DLBCL) and acute lymphoblastic leukemia (ALL), and axicabtagene ciloleucel, for treating certain types of B-cell lymphoma. They contrast the successes of these treatments with the challenges faced in the arena of solid tumors, which exist in an immunosuppressive tumor microenvironment that effectively blocks immune attack. This environment is characterized by poor vascularization, hypoxia, immunosuppressive cell types, including tumor-associated macrophages and neutrophils, myeloid-derived suppressor cells, and regulatory T cells, immunosuppressive soluble factors such as transforming growth factor β, adenosine, and interleukin 10, and the upregulation of immunoinhibitory checkpoint ligands such as programmed death ligands 1 and 2. Khan et al. discuss a variety of methods for overcoming these challenges to the development of effective immunotherapy for solid tumors.

The recognition of cancer as a disease 3000 years ago and the progress made in understanding and treating this disease have been remarkable achievements. We have moved from "blunt instrument," unspecific, crude therapeutic approaches with substantial iatrogenic morbidity to fine-scale, cell- and patient-specific treatments that promise to eliminate existing tumor with minimal or no side effects and maintain the tumor-free state through long-lasting, effective immune surveillance. The future is bright indeed.

David B. Teplow, PhD
Department of Neurology
David Geffen School of Medicine at UCLA
Los Angeles, CA, United States

CHAPTER ONE

From the Hellstrom paradox toward cancer cure

Karl Erik Hellstrom*, Ingegerd Hellstrom
Department of Pathology, University of Washington, Harborview Medical Center, Seattle, WA, United States
*Corresponding author: e-mail address: hellsk@uw.edu

Contents

1. Cancer immunology has a long history 2
2. A Th2-type inflammation promotes tumor growth 4
3. Treating tumor-bearing mice with immunomodulatory mAbs 5
4. Mechanisms responsible for the therapeutic efficacy of immunomodulating mAbs 8
5. Synergistic effect combining immunomodulatory mAbs with cisplatin 12
6. In vitro evaluation of immunomodulatory mAbs 15
7. Outlook 15
Acknowledgments 18
References 18

Abstract

Several decades ago we published some of the first papers showing that both murine and human cancers are recognized in vitro as immunologically foreign and that this is the case also in the presence of a growing tumor. The latter situation, sometimes referred to as the Hellstrom paradox, implies that the tumor is protected in vivo by a highly immunosuppressive environment. After many disappointments, the discovery that tumor-related immunosuppression can be counteracted by administrating monoclonal antibodies (mAbs) to checkpoint inhibitors such as CTLA-4, PD-1, and PD-L1 is now revolutionizing cancer therapy. Over the past several years we have applied mouse models in attempts to further improve the ability of such mAbs to cause long-term complete tumor rejection. This review is focused on that work and emphasizes that successful immunotherapy is associated with a shift from a tumor-promoting Th2 inflammation to a tumor-inhibiting Th1 response.

Progress in Molecular Biology and Translational Science, Volume 164
ISSN 1877-1173
https://doi.org/10.1016/bs.pmbts.2018.11.002

1. Cancer immunology has a long history

In 1896, Coley[1] reported that some human cancer patients whose tumors had been injected with bacterial toxins underwent remission including an abscopal effect on tumors that had not been injected. Although remarkable, Coley's findings were ignored for many years in favor of radiotherapy and chemotherapy, probably because the mechanism of action was unknown and the therapeutic procedures were not well standardized.

More than 50 years later, Gross,[2] Foley,[3] Prehn,[4] and others showed that inbred mice can be immunized against a chemically induced syngeneic sarcoma so that they reject transplanted cells from the same tumor but not from a different one, and Klein's group demonstrated that this is true also for primary such sarcomas, proving that the immune response is, indeed, tumor specific.[5] At about the same time, tumors induced by an oncogenic virus, such as polyoma, were found to share rejection antigens[6,7] and Sjogren showed that expression of these antigens is closely involved in the neoplastic transformation.[8] However, similar immunization protocols did not protect against transplanted cells from spontaneously arising tumors, which for many years were considered to be nonimmunogenic.[9,10]

In the mid-1960s our group started a series of experiments in which we examined the ability of lymphoid cells from mice immunized against a syngeneic tumor to prevent plated cells from the same tumor from attaching, dividing, and forming colonies in vitro.[11] We found that lymphocytes from mice whose chemically induced sarcoma had been removed inhibited the plating efficiency of cells from the same sarcoma but not from a different one and that colony formation also was inhibited when the lymphocytes came from a mouse with a growing tumor.[12] We next plated cells from short-term cultures of several different human carcinomas together with lymphoid cells from the autologous patients' peripheral blood and found that their plating efficiency was inhibited and that this was the case also when the patient had a large tumor.[13–15] These findings were repeatedly confirmed[16–21] but remained controversial for many years,[22–24] probably because the discordant experiments measured cytolytic responses against tumor cell lines rather than inhibited adhesion and division of cells from short-term tumor cultures.[11]

The fact that large tumors also can be recognized in vitro as immunologically foreign has sometimes been referred to as the Hellstrom paradox.[25,26] It is related to the in vivo phenomenon concomitant tumor immunity, which is the ability of a subject who has a large tumor to reject a small number of cells from that tumor when transplanted outside the original site.[27,28] The paradox implies that the in vivo tumor microenvironment is highly immunosuppressive and must be overcome for immunotherapy to be effective.[29]

Initial data indicated that tumor antigens and antigen–antibody complexes in tumor-bearers' sera were primarily responsible for inhibiting ("blocking") a tumor-destructive immune response in vivo,[30–36] a situation mimicking the critical role of antigen for generating and maintaining immunological tolerance to allografts.[37] We found that the highest concentration of the "blocking" factors was at the tumor site.[38] Importantly, counteracting the blocking effect by administering antibodies to tumor antigens could induce tumor regression in mice with Moloney virus-induced sarcoma[38] and in rats with primary or transplanted polyoma tumors.[33,39] Further, a tumor-directed delayed-type hypersensitivity reaction and tumor regression could be induced by antiidiotypic antibodies.[40,41]

Subsequent work showed that the situation was more complex than originally anticipated. For example, most neoplastic cells do not express the key costimulatory molecules CD80 and CD86[42] and thus display their antigens in a way that downregulates the immune response. Engaging costimulatory signals by vaccination with tumor cells transfected to express costimulatory CD80 molecules,[43,44] or by administering agonistic anti-CD137 mAb,[45] causes tumor rejection in some models.

The power of a costimulated immune response was also illustrated by experiments in which lymphoid cells from the blood of human patients with advanced cancer could proliferate in vitro, produce high levels of tumor necrosis factor alpha (TNFα) and IFNγ, and generate tumor-destructive cytolytic T cells (CTL) if they were cultured together with autologous tumor cells and stimulated by a combination of mAbs to CD28 and CD3.[46] An analogous approach was not effective in vivo, perhaps because the mAb doses that could be safely given did not overcome the immunosuppressive mechanisms.

Furthermore, neoplastic cells, in addition to shedding or secreting tumor antigens, make a variety of immunosuppressive factors,[47] which include the PDL-1 and -2 ligands to the PD-1 receptor,[48,49] members of the transforming growth factor (TGFβ) family,[50] IDO (indoleamine 2,3-dioxygenase),[51,52] prostaglandin,[53] AhR (blockade of IDO-kynurenine-AhR [aryl hydrocarbon receptor] metabolic circuitry), NO, and others. It will become important to learn which immunosuppressive cells and factors play the key roles, how they interact, and how they are best inhibited.

It is noteworthy that the first identified rejection antigens were in tumors that had been induced by a large dose of chemical carcinogen, that they were unique to each individual tumor, and that tumors induced by a small dose of the same carcinogen were less immunogenic and sometimes even stimulated to grow in immunized mice.[54] These findings are most likely explained by postulating that the mutagenic carcinogen creates more target epitopes when applied at a large dose and thus induces tumors that are more antigenic.[55] These individually unique rejection antigens, in addition to antigens encoded by oncogenic viruses, provide the most tumor-specific targets for a therapeutic immune response. The mutation-induced tumor epitopes display great variability so a vaccine targeting them has to be personalized, but this variability is no problem for treatment with immunomodulatory mAbs.

2. A Th2-type inflammation promotes tumor growth

There is a strong support for Virchow's postulate that inflammation promotes both carcinogenesis and tumor progression.[56–58] Invasive cervical cancer (ICC) provides an ideal system to investigate this. The etiological agent, high-risk human papillomavirus (hr-HPV), is known,[59] its E6 and E7 genes encode tumor-specific epitopes that can be recognized by immune T lymphocytes to cause cell destruction,[60–63] expression of these epitopes is intimately associated with the neoplastic transformation,[64] and the lesions preceding ICC are well defined, including cervical intraepithelial neoplasia (CIN) grades 1, 2, and 3/carcinoma in situ (CIS).[65–69] Chronic Th2-type inflammation is commonly seen during persistent infection with hr-HPV and promotes tumor progression,[70] and immunological markers have been reported to predict regression of CIN2–3 lesions.[71]

Feng et al. applied immunohistochemistry to archived cervical samples to characterize the local immunological environment during the gradual progression from hr-HPV-infected epithelial cells to ICC.[72] Lymphoid cells were examined for the expression of FoxP3, a marker of regulatory T cells,[73,74] CD20, a marker for B lymphocytes,[75] CD138,[76] a marker expressed on plasma cells and some epithelial cells, and CD32B,[77] which is expressed primarily by antigen-presenting cells and can downregulate an immune response after uptake of immune complexes.[78] Epithelial cells were examined for the expression of indoleamine 2,3-dioxygenase 1 (IDO1), an enzyme that induces immunosuppression.[51,79] Feng et al. also studied the expression of two transcription factors, GATA3 and T-bet, which play key roles in determining T-cell differentiation.[80,81] GATA3 promotes Th2 differentiation and induces Th2 cytokine production, while T-bet is essential for Th1 differentiation by upregulating Th1 cytokines, downregulating Th2 cytokines, and it can induce a shift from Th2 to Th1 dominance. The relative expression of GATA3 and T-bet (G:T) has been proposed to determine the balance between Th1 and Th2 immune responses.[80,81]

Already at the stage of CIN1, the mucosa was shown to be infiltrated by CD20+ and CD138+ cells, and infiltration increased in CIN3/CIS and ICC, where it strongly correlated with infiltration by CD32B+ and FoxP3+ lymphocytes. GATA3+ and T-bet+ lymphoid cells were increased in ICC compared to normal, and expression in epithelial cells of the Th2 inflammation-promoting IDO1 was higher in CIN3/CIS and ICC. Thus, hr-HPV initiates a local Th2 inflammation at an early stage, involving antibody forming cells, and fosters an immunosuppressive microenvironment that aids tumor progression.

3. Treating tumor-bearing mice with immunomodulatory mAbs

In view of the role of a Th2-type inflammation in facilitating tumor induction and progression, Dai et al. hypothesized that shifting a Th2 response to the Th1 type would promote tumor destruction. To test this hypothesis, they first worked with the MOSEC clone ID8 mouse ovarian carcinoma which grows intraperitoneally (i.p.) and shares many characteristics with human ovarian cancer.[82,83] Mice implanted i.p. with ID8 cells had small nests of tumor cells within approximately 10–15 days, which was the time when they were first treated. In the controls, these nests developed rapidly into solid tumors and malignant ascites.

The first objective of this study was to compare the therapeutic efficacy of IgG mAbs to CD137,[45] PD-1,[84] and CTLA4[85] with that of mAbs to other immunoregulatory (TIM-3, LAG-3) or costimulatory (OX40, GITR, CD40) molecules, administering them as single agents and in various combinations. IgG mAbs to irrelevant molecules were used at the same protein dose[82,83] as controls. Treatment with a combination of mAbs to CD137 + PD-1 increased survival from 24 days in the control to 55 days, which was similar for mAbs derived from two anti-CD137 clones, 2A and lob 12.3. Inclusion of a mAb to CTLA4 further increased survival to 74 days. None of these 3 mAbs prolonged survival when used alone, and neither did several other combinations (anti-CTLA4/PD-1, anti-CTLA4/PD-1/CD40, anti-CTLA4/PD-1/TIM3, anti-CTLA4PD-1/LAG3, or anti-CTLA4/PD-1/LAG3/CD40). Interestingly, when used as single agents, anti-CD137 and anti-PD-1 mAb prevented outgrowth of 1D8 cells in mice that had been implanted just 3 days before treatment with 60% and 30%, respectively, of the mice surviving tumor free. Combining these two mAbs increased survival to 78%.[82,83]

These studies were expanded to include three additional models, the SW1 clone of the K1735 melanoma, the B16 melanoma, and the TC1 lung carcinoma.[82] SW1 was selected because it expresses a low level of the major histocompatibility class I complex and, perhaps because of that, responds poorly to several forms of immunotherapy.[86] For the majority of experiments, mice were injected 7–10 days after subcutaneous (s.c.) transplantation of 10^6 tumor cells, at which time the tumor nodules had a mean diameter of 4–5 mm.

Subcutaneous transplantation of tumor cells rapidly increased the numbers of CD19 cells in tumor-draining lymph nodes (TLNs).[87] This is in agreement with published evidence for a tumor-promoting role of B lymphocytes,[88] and it is noteworthy that tumor regression after administration of mAbs to CD137/PD-1/CTLA4 (referred to as the 3 mAb combination) was associated with a significantly decreased number of CD19 cells in tumors, TLNs and spleen,[82,87] probably because of the anti-B cell activity of agonistic mAbs to CD137.[89] Dai et al. therefore hypothesized that the therapeutic efficacy could be improved by also including a mAb to CD19. To test this hypothesis, mice were injected with a 4–5 mm SW1 tumor on the right side of the back and a 2–3 mm such tumor on the left with either the 3 mAb combination or anti-CD137/PD-1/CTLA4/CD19 ("the 4 mAb combination") into the right tumor. Mice whose right tumor was injected with the 4 mAb combination had an overall survival of 59 days compared to 35 days with the 3 mAb combination and 18 days in controls. Importantly, 60% of mice receiving the 4 mAb combination were alive, tumor free, 100 days after the first treatment when the mice getting the 3 mAb combination were all dead, as were mice given mAbs to CD19, CTLA-4, or PD-1. Subsequent studies showed that the 4 mAb combination also was more effective in the B16 and TC1 models.[87,90]

Dai et al. also compared the therapeutic efficacy of injecting mAbs intraperitoneally (i.p.) or intratumorally (i.t.) to mice that had s.c. SW1 melanoma. Given i.t., the 3 mAb combination induced CR in 65% of mice as compared to 20% CR in mice when given i.p., and in the B16 model, the 4 mAb combination produced 70% CR when given i.t. as compared to 20% when given i.p.[87] In another set of experiments, SW1 cells were transplanted on both sides of the back of mice, after which the 4 mAb combination was injected either i.t. or i.p. The i.t. injection had the best therapeutic activity both against the injected and the untreated tumor in the same mouse, i.e., it produced a more efficacious local and systemic response than the i.p. administration. The reasons for this need to be explored.[82,87]

To evaluate the therapeutic efficacy against larger tumors, experiments were performed with mice whose s.c. melanoma had a mean diameter around 8–9 mm at the onset of therapy.[87] The 3 mAb combination prolonged survival of the SW1-bearing mice from 16 days in the control to 50 days with 10% CR, while the 4 mAb combination induced 60% CR. In mice with B16 tumors, the 3 mAb combination prolonged survival from 10 to 30 days and the 4-mAb combination induced long-lasting CR in 50% of mice. However, it was much easier to induce regression of small SW1, B16, and TC1 tumors than of large ones and this was reflected by a greater difficulty in inducing a shift from a Th2 to a Th1 immunological landscape as tumors were larger.[91]

Mice whose SW1 tumors had undergone CR were transplanted s.c. with SW1 cells 140 days after the first treatment, as were age-matched naive controls. The SW1 cells were rejected by 80% of regressor mice, which remained tumor free when the experiment was terminated after more than 200 days,[90] i.e., treatment had induced a long-term response.

Fig. 1 summarizes two of the key findings: (1) i.t. delivery is more therapeutically efficacious than systemic (i.p.) administration; and (2) the 4 mAb combination is more efficacious than the 3 mAb combination. Both of these findings were statistically significant in several models.

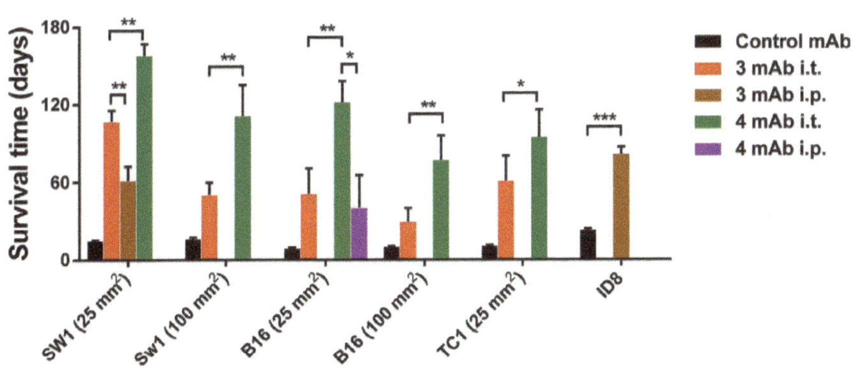

Fig. 1 Significantly prolonged survival of mice with established tumors following i.t. or i.p. administration of mAb combinations. $^{*}P< 0.05$; $^{**}P< 0.01$; $^{***}P < 0.001$.

4. Mechanisms responsible for the therapeutic efficacy of immunomodulating mAbs

Subcutaneous transplantation of syngeneic mice with SW1 or B16 melanoma or TC1 lung carcinoma significantly increased the number of CD19 cells in TLNs as early as 1 day later, reminiscent of the fact that HPV-infected human cervix uterus is already infiltrated by B lymphocytes at the CIN1 state. As in humans, the number of CD3 and CD8 cells concomitantly decreased,[87] and there were a significantly increased number of CD11Gr-1 MDSCs (myeloid-derived suppressor cells) in spleens. RNAs were extracted from TLN 2 days after transplantation of B16 cells and evaluated by qRT-PCR. There was a statistically significant twofold decrease in RNA levels for IFNγ and TNFα and a twofold increase in the RNA level for IL4, while mice transplanted with cultured fibroblasts were not different from naïve mice according to flow cytometry or PCR. Implanted tumor cells thus rapidly induced changes typical of a Th2-type inflammation.

To investigate the role of CD4+, CD8+, and NK cells in the therapy-induced antitumor response, mice injected with the 3 mAb combination were also injected i.p. with mAbs to the respective cell populations. In the ID8 model, depletion of CD4+ or CD8+ cells abrogated the antitumor effect, while depletion of NK cells had a marginal effect. In the SW1 model, CD4+ cells were needed for therapeutic efficacy, while the differences were not statistically significant for CD8 and NK cells. It is noteworthy that CD4 cells were primarily responsible for tumor rejection also in previous studies by our group in which mice with SW1 melanoma were immunized with SW1 cells which had been transfected to engage CD137.[90]

Mice implanted i.p. with ID8 ovarian carcinoma cells as in the therapy experiments were injected i.p. 10 days later with a combination of anti-PD1/CD137 mAbs or either mAb alone. Seven days later, the percentage, absolute numbers, and effector function markers of lymphocytes in the spleens were analyzed by flow cytometry.[83] Administration of the two mAbs resulted in a significantly increased frequency and number of CD3+ and CD8+ T cells and decreased numbers of CD4+ and FoxP3+ Treg cells as well as GR-1+CD11b+ MDSCs in the group receiving the 2 mAbs. The levels of CD44+CD62L− effector/memory cells and of CD44+ CD62L+ central memory cells were also elevated as was the number of IFNγ-producing effector CD8+ T cells, while there were decreased numbers of CD8+ T cells producing IL-10. The data thus indicated that combining the 2 mAbs generated a systemic immune response dominated by significantly increased numbers of CD8+ effector T cells and less immunosuppressive cells.

Administration of anti-CD137 mAb as a single agent also increased the population of CD8+ T cells, but these cells had an "exhausted" phenotype, failing to produce effector cytokines upon polyclonal stimulation.[83] This may be attributed to upregulation of coinhibitory PD-1 and TIM-3 molecules,[92] as in a study in which agonist CD137 mAb induced the expression of PD-1 on CD8 and CD4 tumor-infiltrating cells from mice with B16 melanoma that had been immunized with a cell-based vaccine.[93] Combined administration of anti-PD-1/CD137 mAbs blocked the upregulation of PD-1 and TIM-3 molecules on peritoneal C4 and CD8 T cells.[83]

Tumor rejection after administration of the 3 mAb combination to mice with ID8 tumors correlated with a decreased frequency and number of CD19+ cells and an increased frequency and number of CD3+, CD4+, and CD8+ cells, the last of which made large amounts of IFNγ upon polyclonal stimulation. The frequency of IFNγ+ TNFα+ CD4+ T cells in the peritoneal lavage of treated mice tripled and that of CD8+ cells doubled as did the frequency of IFNγ+ TNFα+ CD4+ and CD8+ tumor-infiltrating lymphoid cells (TILs). The ratio of CD4+Foxp3− to CD4+Foxp3+ Treg cells increased as did that of CD4+Foxp3− cells to either CD8+Foxp3+ Tregs or CD11b+Gr-1+ MDSC. The CD86+CD11c+ mature dendritic cells increased five times relative to the control. Administration of anti PD-1 mAb as a single agent had a little effect on CD4+ and CD8+ T cells or on CD19+ B cells but significantly decreased the populations in spleens of immunosuppressive Treg and MDSC.[83]

Wei et al.[83] as well as Dai et al.[82] also investigated whether the induced immune response had any tumor antigen-specific component by taking advantage of the fact that the ID8 tumor expresses a mouse homologue of human mesothelin, which is one of the biomarkers of human ovarian carcinomas. Splenocytes from treated mice and from controls were cultured in the presence of H-2D^b-restricted mesothelin-specific peptide or control HPV-E7 peptide for 3 days, after which the culture supernatants were assayed for IFNγ by ELISA. Splenocytes from the treated mice, as compared to controls, secreted increased levels of IFNγ when stimulated with the mesothelin peptide as compared to the HPV peptide.[82,94] Whether splenocytes had cytolytic activity was determined by harvesting splenocytes from mice treated with anti-PD-1/CD137 mAbs and stimulating them with UV-irradiated ID8 cells for 4 days before cytolytic T cell

(CTL), after which assays were performed. Splenocytes from anti-PD-1/CD137-treated mice were cytolytic to EL4 cells pulsed with mesothelin but not to cells pulsed with HPV-E7 peptide. Preincubation with anti-CD8 antibody suppressed the killing.[83] Other evidence that the therapy-induced response has a tumor antigen-specific component comes from studies applying the tetramer technology with the TC1 lung carcinoma.[91]

TLNs and spleens were also harvested from mice with B16 melanoma which either had completely regressed ("regressors") or were growing progressively "progressors" after three i.t. administrations of the 3 mAb combination. Regressors had significantly more CD44CD62L− effector memory cells than progressors, which was consistent with shifting of the tumor microenvironment from an immunosuppressive Th2 to an immunostimulatory Th1 type.[82] Further evidence that a therapeutic response was caused by such a shift came from experiments in which the 3 mAb combination was injected once i.t. into SW1 tumors of 4–6 mm diameter and the mice euthanized 7 days later. TLNs and spleens were enlarged as compared to controls and there was a dramatic increase in the number of CD3, CD8, and CD4 cells and of CD11CD86 mature DC in both TLN and spleen. The number of cells expressing CD137 was also increased, while the populations of cells expressing PD-1 or CTLA-4 were unaffected. The treated mice had significantly increased numbers of CD4 and CD8 cells producing IFNγ and TNFα and significantly increased numbers of CD11CD86+ mature DCs. Infiltration by CD19 cells was decreased compared to controls.

Dai et al. also compared B16 melanomas which had either started to regress ("responders") or were growing progressively ("nonresponders") when the mice were euthanized 7 days after the second i.t. injection of the 3 mAb combination. Responding tumors contained much fewer CD19 cells and more CD3, CD4, and CD8 cells than nonresponding ones. Approximately 90% of the infiltrating CD8 cells in responders were CD44CD62L− effector memory cells and were present at higher numbers than in nonresponders or controls. Moreover, 6-fold higher mRNA levels of IFNγ and TNFα, 2-fold higher mRNA levels of perforin, and a 15-fold increase of granzyme B mRNA were found in the responding tumors. Responders had significantly decreased numbers of CD19+ cells at tumor sites, increased IFNγ and TNFα-producing CD4+ and CD8+ T cells and mature CD86+ DC, and increased ratios of effector CD4+ and CD8+ T cells to CD4+Foxp3+ Treg and to CD11b+Gr-1+ MDSC.

According to PCR array analysis of TLNs from mice with SW1 tumors, the 3 mAb combination, as compared to controls, upregulated transcription of the Th1-related molecules IFNγ, Stat4, and TBX21 and downregulated transcription of the Th2 molecules IL-4, Stat6, and GATA3.[82] These findings are consistent with the increased percentage of IFNγ-producing T cells according to intracellular cytokine staining. There was also an increase of the Th1-associated chemokine/receptor CXCL9–11/CXCR3 and a decrease of the B-cell chemotactic chemokine/receptor CXCL13/CXCR5, which was concordant with the increase of Th1-type CD4+ and CD8+ cells and the decrease of B cells in TLNs. Upregulation of CXCL9 in tumor and of CXCL10 and CXCL11 was confirmed by quantitative real-time PCR in both tumors and TLNs from treated mice in a repeat experiment demonstrating a shift of the immune response toward Th1 type by the 3 mAb treatment. These findings are consistent with shifting the tumor microenvironment from a Th2 to a Th1 type.[82]

Dai et al. also investigated the immunological profiles of mice with B16 tumors 7 days after an i.t. injection of either the 3 or 4 mAb combinations or of individual mAbs. Anti-CD19 mAb depleted more than 95% of CD19- and CD20-positive cells as did the 4 mAb combination. There were many more CD3, CD4, CD8, and CD80CD11c cells, and much fewer CD19 cells, in TLNs from mice whose tumors were injected with the 4 rather than the 3 mAb combination, and the 4 mAb combination induced more CD8 cells in the spleen and a higher frequency of CD44CD62L− ratio of T effector (CD4Foxp3−) to Treg (CD4Foxp3+) cells in tumors. Furthermore, relative to the 3 mAb combination, the 4 mAb treatment more effectively increased the number of TILs that expressed CD137 and CD86, whereas the expression of PD-1, CTLA-4, PD-L1, and PD-L2 did not differ. Frequencies of CD44CD62L− effector memory cells in TLNs were similar in mice whose tumors were injected with either combination, while the 4 mAb combination increased the number of CD44CD62L central memory CD8 T cells compared to the 3 mAb combination. In addition, as measured by qPCR, the 4 mAb treatment more effectively increased the shift from a Th1 to a Th2 response with an increased Tbx21/Gata3 ratio, higher mRNA expression of IFNγ, and lower mRNA expression of IL4.[91]

The therapy-induced Th1 response declined within weeks after mAb administration, which probably explains why approximately 20% of macroscopically regressed tumors relapsed within 2 months following treatment. The relapsed tumor cells retained full sensitivity to the original treatment

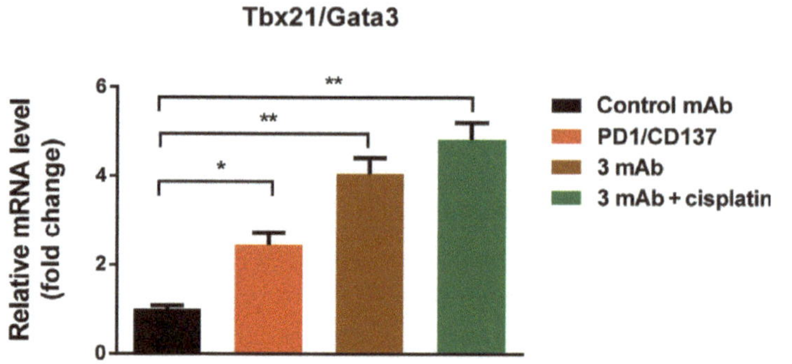

Fig. 2 Increased Tbx21/Gata3 ratio after the mAb treatments. $^{*}P < 0.05$; $^{**}P < 0.01$.

protocol, while TILs, TLNs, and spleens of relapsing mice had the characteristics of a Th2-type inflammation, i.e., the relapses were caused by failure of the immune system to retain a tumor-destructive Th1 tumor environment.[91]

The ability of immunomodulatory mAbs to shift the tumor microenvironment from Th2 to Th1 decreased with tumor size as did the therapeutic efficacy.[91] Most likely, immunosuppressive molecules contributed by both the tumor cells and the tumor-bearing host, such as PDL, IDO, TGFβ, IL10, and tumor antigens, were responsible for this decrease, but the mechanisms involved need to be studied further.

Macrophages also play an important role in the tumor microenvironment where they differentiate into M1 or M2 subtypes, which are characterized by their Th1/Th2 cytokine profiles. In our experiments, tumor regression was consistently associated with a dominance by M1 macrophages.[91]

Fig. 2 illustrates a direct relationship between therapeutic efficacy and higher Tbet/GATA ratios, with the 3 mAb protocol having higher efficacy than the 2 mAbs and addition of cisplatin to the 3 mAbs further improving the Th1-type response.

5. Synergistic effect combining immunomodulatory mAbs with cisplatin

Stimulated by the report that anti-CD137 mAb synergizes with cisplatin to induce regression in a mouse colon cancer model,[95] Wei et al. investigated whether cisplatin, which is commonly given to women with advanced ovarian carcinoma, synergizes with a combination of immunomodulatory mAbs.[83] Their objective was to obtain preclinical data which could be translated to a clinical protocol for such patients. The approach was encouraged by the demonstration that i.p. injection of mAbs to CD137 plus PD-1 doubled survival of mice with the ID8 ovarian carcinoma.[82]

Ten days after implanting C57BL mice with ID8 cells the mice were injected i.p. with 10 mg/kg cisplatin followed by two doses of anti-PD-1/CD137 mAbs at a 4-day interval. Controls received the same protein dose of an irrelevant mAb. Confirming previous data, a doubled survival time was observed with the mAb combination, but there were no long-term survivors, and no survivors among mice only given the drug. However, 8 of 10 mice receiving both cisplatin and the 2 mAb combination survived healthy and tumor free when the experiment was terminated after a 100-day observation period.[83,96] The surviving mice rejected challenge with ID8 cells, while cells from the syngeneic but antigenically different TC1 lung carcinoma grew progressively.

Importantly, treated mice had a significantly increased total number of CD8 + T cells in the peritoneal lavage and spleens and these cells were functional, as demonstrated by antigen-specific cytolytic activity and IFNγ production, and by the demonstration that removal of CD8 + T cells abrogated the antitumor effect.[83,91] Administration of anti-CD137 mAb as a single agent similarly increased CD8 + T cells, but these had no functional activity due to upregulation of coinhibitory PD-1 and TIM-3. Combined CTLA4 blockade and CD137 triggering were not therapeutically effective either, indicating that PD-1-mediated negative signaling plays a more important role than immunosuppression via CTLA-4 in the evasion of ID8 cells from the immunological control.

Subsequently, mice with s.c. TC1 tumors of 4–5 mm diameter were injected i.p. with cisplatin, 10 mg/kg, immediately followed by various mAb combinations given i.t. and repeated two times weekly. While mAbs to CTLA4 plus PD-1 were therapeutically ineffective in this model, their combination with cisplatin induced long-term CR in most of the mice, and therapeutic efficacy could be detected also against large TC1 tumors by giving cisplatin together with the 3 or 4 mAb combinations.[91] The ability of cisplatin to counteract the immunosuppressive tumor microenvironment,[97] including a population of tolerogenic and IDO-positive MDSC,[98] probably contributed to the observed beneficial effects.[91] We hypothesize that the therapeutic efficacy can be further improved by drugs that counteract the induction of IDO by IFNγ, whose level is increased by the immunomodulating mAbs,[99] and that therapeutic vaccination, radiotherapy, or various other drugs can synergize with mAbs to checkpoint inhibitors. For example, a combination of radiotherapy with immunomodulatory mAbs was shown to act synergistically.[100–102]

A long-term tumor dormancy of mouse tumors subjected to immunotherapy has recently been described and involves a minor "stem-like tumor subpopulation" via an IDO-kynurenine-aryl-hydrocarbon-receptor (AhR)-p27-dependent mechanism which protects from IFNγ-induced apoptosis and temporarily slows down proliferation,[99] which is followed by recurrent tumor growth.

Translation of these findings to the clinic should be considered. Although no toxicity beyond occasional hair loss was observed in CD137-treated tumor-bearing mice, caution must be exercised since CD137 agonistic antibodies have been observed to cause toxicity in both preclinical models[103] and clinical trials.[104] Restricting the biodistribution of the mAbs to the peritoneal cavity, e.g., by entrapment in nanoparticles, may provide one targeted approach to elicit effective antitumor immunity while minimizing systemic toxicity.[105]

Fig. 3 summarizes data showing that addition of cisplatin significantly increases the therapeutic response in experiments where treatment with only mAb combinations does not have significant antitumor activity.

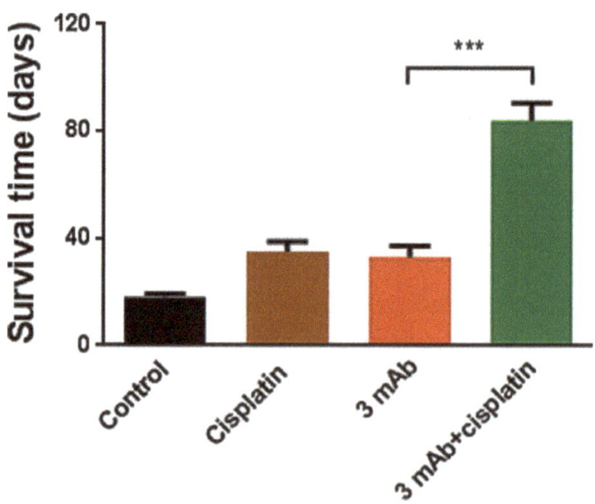

Fig. 3 In a TC1 two-tumor mouse model, cisplatin plus the 3 mAb combination significantly prolonged the survival time.***$P < 0.001$.

6. In vitro evaluation of immunomodulatory mAbs

Predicting those tumor patients who will respond and those who will not can guide personalized therapy. To this end, Dai et al. developed an in vitro model that can reflect the same Th2 to Th1 shifts and predict tumor regression and may further elucidate the responsible mechanisms.[106] Lymphoid cells from tumor-bearing mice were cultured for 3 or 5 days together with cells from the respective tumor in the presence of various combinations of immunomodulatory mAbs whose therapeutic efficacy against the same tumors had been previously established.[87]

Three tumor models were chosen: the TC1 lung carcinoma and the SW1 and B16 melanomas. Lymphoid cells from the spleen or TLNs were harvested when the tumors had an approximate size of 25 mm,[2] and the ability of the mAbs to induce tumor cell destruction in vitro was compared with their therapeutic efficacy in vivo. The data strongly correlated ($r = 0.9$, 0.89, and 0.91, respectively) with the therapeutic efficacy of the respective mAb combinations.

In this model, tumor destruction was associated with a shift from a Th2 to a Th1 response and included a dramatic increase of long-term memory cells and mature dendritic cells. A combination of mAbs to CD137/PD-1/CTLA4/CD19 was most efficacious, again illustrating the improved response by including anti-CD19 mAb in the combination. A similar approach should be translatable to human cancers and it may lend itself to investigation of the mechanisms by which mAbs and other agents, including lymphokines and small molecules, can facilitate the generation of a tumor-destructive immune response. The in vitro sensitization model can be applied to evaluate a variety of mAbs, lymphokines, and small molecules, alone and in combinations, for generation of a tumor-destructive response.

7. Outlook

After many years of disappointment there are several reasons why the value of immunotherapy for cancer is now being widely recognized.[107–110] Immunological mechanisms can destroy tumors by T lymphocytes,[111] NK cells,[112] and macrophages[113] as well as by antibodies,[114] TNFα,[115] and other cytokines. The multitude of possible immunological attacks compensates for the fact that tumor antigens and their presentation via MHC are sometimes lost during tumor progression.[116] While the high mutability of cancer cells[55,117] causes a problem for therapeutic agents with only one target, the mutations create new epitopes that can be recognized by T lymphocytes,[117] and immunomodulatory mAbs make possible a tumor-destructive immune response targeting many different epitopes. Although relatively few patients have been cured, provocative clinical results have been obtained with several forms of immunotherapy and particularly with mAbs to checkpoint inhibitors.[109,118–121] Fig. 4 illustrates our thoughts about mechanisms promoting or inhibiting tumor growth and how they interact.

Carcinogenesis and tumor progression are associated with Th2 mechanisms and regression with a shift to a Th1 response. This shift is more difficult to achieve with larger tumors. One is reminded of the fact that rejection of tumor cells transplanted onto an immunized mouse is often overcome by just increasing the transplanted tumor challenge with a couple of logs,

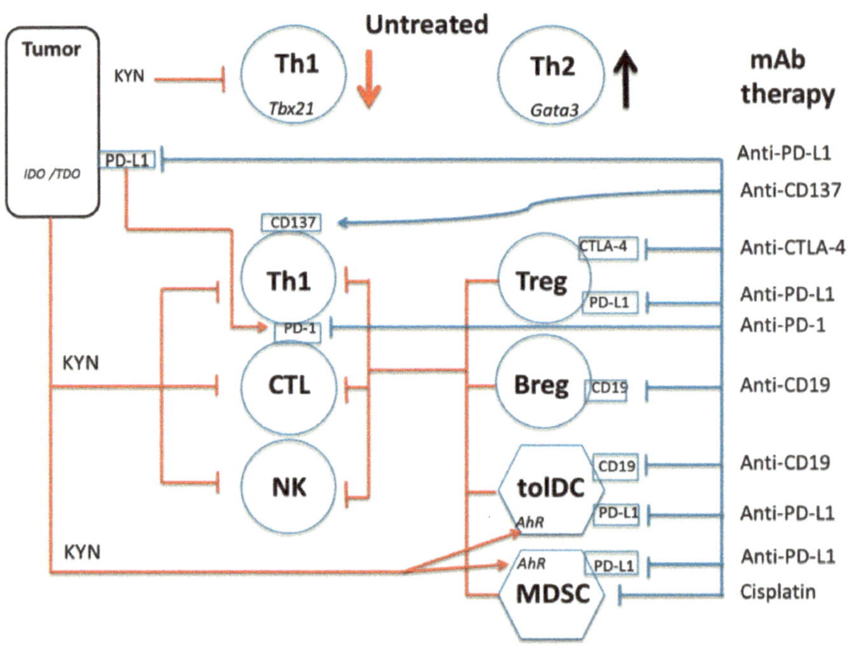

Fig. 4 Illustrates some of the interactions that either facilitate tumor growth via a Th2-type response or inhibit it via a Th1 microenvironment. Tumors induce an immunosuppressive environment through the kynenurine (KYN) and/or PD1/PDLw pathway. The metabolites KYN and KYNA, catalyzed by IDO1 and TDO2, are endogenous agonists of the aryl hydrocarbon receptor (AhR) and block the activation of a Th1 response by inducing a Th2-type environment with Tregs, Bregs, TolDCs, and MDSCs. By combining mAbs, one can target several of these pathways to shift a Th2 to a Th1 response and cause tumor regression.

reflecting the power of tumor-associated immunosuppressive mechanisms. Fortunately, the Th2 to Th1 shift can be facilitated by combining the immunomodulatory mAbs with the anticancer drug cisplatin which, in addition to killing tumor cells, also interacts with the immune system.[29,122] We are encouraged that a combination of three clinically approved agents, anti-PD1 and anti-CTLA4 mAbs plus cisplatin, induced CRs in mice with TC1 tumors, while neither the drug nor the mAbs were effective[91] by themselves. The 3 mAb combination plus cisplatin further increases the efficacy. It will be important to learn whether the combination of immunomodulatory mAbs with inhibitors of IDO, AHR, PDL-1 or -2, TGFβ, and/or other immunosuppressive molecules, or with antitumor mAbs (e.g., Herceptin or Erbitux), will increase efficacy. We expect that information on that will come forth from ongoing preclinical and clinical work.

In our studies, i.t. injection of immunomodulatory mAbs was superior to systemic (i.p.) delivery. We need to learn whether the fact that some of these mAbs were from a foreign species (rat, hamster) contributed to this finding by inducing an immune response which inhibited the injected tumors as "bystanders," as in studies by Edmund Klein decades ago where skin cancers were injected with a strong foreign antigen.[122] If that was the case, one may further improve the therapeutic efficacy by also injecting a more potent antigen at the tumor site to increase efficacy.

For i.t. delivery, enclosing the injected material in nanoparticles[105] may prolong its stay at the tumor site. For targeting to tumors after systemic delivery, the surface of the nanoparticles may be "decorated" by antitumor mAbs or one may use bispecific antibody constructs. If small molecules can be identified with similar functions as the respective immunomodulating mAbs, it should also be possible to make immunoconjugates by coupling them with antitumor mAbs. One also should be aware that while systemic administration is preferable for practical reasons, i.t. delivery has improved the efficacy of some clinically applied therapeutic agents.[123]

So far, the most impressive preclinical and clinical responses to immunomodulating mAbs have been seen with mAbs to PD-1, PD-L1, and CTLA4, i.e., with mAbs that target tumor-related immunosuppression. Furthermore, mAbs to CD137, which are effective in preclinical models, interfere with B-cell activity[89] in addition to activating CD4+, CD8+, and NK cells.[124] In contrast, agonistic mAbs to costimulatory receptors such as CD28 have been less therapeutically beneficial and display much toxicity.

In our recent studies with the TC1 mouse lung carcinoma, tumor relapses were seen in about 20% after macroscopic rejection.[91] It is noteworthy that these relapses were caused by a shift from a tumor-inhibitory Th1-type response to the Th2 type and that the relapsing tumor cells retained their sensitivity to treatment with immunomodulatory mAbs. It will be important to investigate whether one can decrease the frequency of relapses by prolonging mAb and/or drug administration or by combination with therapeutic vaccination or adoptively transferred T cells.

Major efforts should be devoted to investigate side effects, especially focusing on autoimmunity. Although toxicity has not caused a large problem in our studies, it is a primary concerns when treating human patients with immunomodulatory mAbs.[125–127] There is a need for approaches to detect and treat autoimmune side effects early.

Acknowledgments

We much appreciate initial contributions by Dr. Huafeng Wei, several years of stimulating collaboration with Dr. Min Dai, skillful technical contributions by Yuen Yee Yip, and many helpful suggestions during a lifetime of scientific interaction with Dr. Hans Olov Sjogren and more recent frequent discussions with Dr. Paul Abrams. There were no conflicts of interest.

References

1. Coley W. Further observations upon the treatment of malignant tumors with the toxins of erysipelas and Bacillus prodigiosus with a report of 160 cases. *Bull Johns Hopkins Hosp.* 1896;7:157.
2. Gross L. Intradermal immunization of C3H mice against a sarcoma that originated in an animal of the same line. *Cancer Res.* 1943;3:326–333.
3. Foley EJ. Antigenic properties of methylcholanthrene-induced tumor in mice of the strain of origin. *Cancer Res.* 1953;13:835–837.
4. Prehn R, Main D. Immunity to methylcholanthrene-induced sarcomas. *J Natl Cancer Inst.* 1957;18:769–778.
5. Klein G, Sjögren HO, Klein E, Hellström KE. Demonstration of resistance against methylcholanthrene-induced sarcomas in the primary autochthonous host. *Cancer Res.* 1960;20:1561–1572.
6. Sjogren HL, Hellstrom I, Klein G. Transplantation of polyoma virus-induced tumors in mice. *Cancer Res.* 1961;21:329–337.
7. Habel K. Immunological determinants of polyoma virus oncogenesis. *J Exp Med.* 1962;115:181–193.
8. Sjogren HO. Studies on specific transplantation resistance against polyoma virus-induced tumors. IV. Stability of the polyoma cell antigen. *J Natl Cancer Inst.* 1964;32:661–666.
9. Hewitt HB, Blake ER, Walder AS. A critique of the evidence for active host defence against cancer, based on personal studies of 27 murine tumours of spontaneous origin. *Br J Cancer.* 1976;33(3):241–259.
10. Klein G, Klein E. Rejectability of virus-induced tumors and nonrejectability of spontaneous tumors: a lesson in contrasts. *Transplant Proc.* 1977;9(1):1095–1104.
11. Hellstrom I. A colony inhibition (CI) technique for demonstration of tumor cell destruction by lymphoid cells in vitro. *Int J Cancer.* 1967;2:65–68.

12. Hellstrom I, Hellstrom KE. Cell-bound immunity to autologous and syngeneic mouse tumors induced by methylcholanthrene and plastic discs. *Science*. 1967;156(3777): 981–983.
13. Hellstrom I, Hellstrom KE, Pierce GE, Bill AH. Demonstration of cell-bound and humoral immunity against neuroblastoma cells. *Proc Natl Acad Sci USA*. 1968; 60:1231–1238.
14. Hellstrom I, Hellstrom KE, Pierce GE, Yang JP. Cellular and humoral immunity to different types of human neoplasms. *Nature*. 1968;220(5174):1352–1354.
15. Hellström I, Hellström KE, Sjögren HO, Warner GA. Demonstration of cell-mediated immunity to human neoplasms of various histological types. *Int J Cancer*. 1971;7:1–16.
16. Bubenik J, Perlmann P, Helmstein K, Moberger G. Cellular and humoral immune responses to human urinary bladder carcinomas. *Int J Cancer*. 1970;5(3):310–319.
17. Baldwin RW, Embleton MJ, Jones JS, Langman MJS. Cell mediated and humoral immune reactions to human tumors. *Int J Cancer*. 1973;12:73–83.
18. Halliday WJ, Maluish AE, Little JH, Davis NC. Leukocyte adherence inhibition and specific immunoreactivity in malignant melanoma. *Int J Cancer*. 1975;16(4):645–658.
19. McCoy JL, Jerome LF, Dean JH, et al. Inhibition of leukocyte migration by tumor-associated antigens in soluble extracts of human breast carcinoma. *J Natl Cancer Inst*. 1974;53(1):11–17.
20. de Vries JE, Rumke P. Tumour-associated lymphocyte cytotoxicity superimposed on "spontaneous" cytotoxicity in melanoma patients. *Int J Cancer*. 1976;17(2):182–190.
21. Steele G, Sjogren HO, Stadenberg I. In vitro cell-mediated immune reactions of melanoma and colorectal carcinoma patients demonstrated by long-term chromium assays. *Int J Cancer*. 1976;17:27–39.
22. Takasugi M, Mickey MR, Terasaki M. Reactivity of lymphocytes from normal persons on cultured tumor cells. *Cancer Res*. 1973;33:2898–2902.
23. Herberman RB, Oldham RK. Problems associated with study of cell-mediated immunity to human tumors by microcytotoxicity assays. *J Natl Cancer Inst*. 1975;55:749–753.
24. Hellstrom I, Hellstrom KE. Cell-mediated reactivity to human tumor-type associated antigens; does it exist? *J Biol Response Mod*. 1983;2:310–320.
25. Dong H. The basic concepts in cancer immunology and immunotherapy. In: Dong H, Markovic SN, eds. *The Basics of Cancer Immunotherapy*. Springer International Publishing. 2018:1–19.
26. Sitkovsky M, Kjaergaard J, Lukashev D, Ohta A. Hypoxia-adenosinergic immunosuppression: tumor protection by T regulatory cells and cancerous tissue hypoxia. *Clin Cancer Res*. 2008;14:5947–5952.
27. Gershon RK, Carter RL, Kondo K. On concomitant immunity in tumour-bearing hamsters. *Nature*. 1967;213(77):674–676.
28. Southam CM. Evidence for cancer-specific antigens in man. *Prog Exp Tumor Res*. 1967;9:1–39.
29. Hellstrom KE, Hellstrom I. Vaccines to treat cancer—an old approach whose time has arrived. *J Cell Biochem*. 2007;102(2):291–300.
30. Hellstrom KaH I. Immunological enhancement as studied by cell culture techniques. *Annu Rev Microbiol*. 1970;24:373–398.
31. Hellstrom I, Hellstrom KE, Sjogren HO. Serum mediated inhibition of cellular immunity to methylcholanthrene-induced murine sarcomas. *Cell Immunol*. 1970;1:18–30.
32. Sjogren HO, Hellstrom I, Bansal SC, Hellstrom KE. Suggestive evidence that the "blocking antibodies" of tumor-bearing individuals may be antigen—antibody complexes. *Proc Natl Acad Sci USA*. 1971;68(6):1372–1375.
33. Bansal SC, Sjogren HO. Counteractions of the blocking of cell-mediated tumor immunity by inoculation of unblocking sera and splenectomy: immunotherapeutic effects on primary polyoma tumors in rats. *Int J Cancer*. 1972;9(3):490–509.

34. Baldwin RW, Price MR, Robins RA. Blocking of lymphocyte-mediated cytotoxicity for rat hepatoma cells by tumour-specific antigen-antibody complexes. *Nature*. 1972;238:185–187.
35. Vaage J. Influence of tumor antigen on maintenance versus depression of tumor-specific immunity. *Cancer Res*. 1973;33(3):493–503.
36. Alexander P. Escape from immune destruction by the host through shedding of surface antigens: is this a characteristic shared by malignant and embryonic cells? *Cancer Res*. 1974;34:2077–2082.
37. Wright PW, Hargreaves RE, Bansal SC, Bernstein ID, Hellstrom KE. Allograft tolerance: presumptive evidence that serum factors of tolerant animals which block lymphocyte-mediated immunity in vitro are soluble antigen-antibody complexes. *Proc Natl Acad Sci USA*. 1973;70:2539–2543.
38. Hellstrom I, Hellstrom KE, Pierce GE, Fefer A. Studies on immunity to autochthonous mouse tumors. *Transplant Proc*. 1969;1(1):90–94.
39. Bansal SC, Sjogren HO. "Unblocking" serum activity in vitro in the polyoma system may correlate with antitumour effects of antiserum in vivo. *Nat New Biol*. 1971;233(37):76–78.
40. Forstrom JW, Nelson KA, Nepom GT, Hellstrom I, Hellstrom KE. Immunization to a syngeneic sarcoma by a monoclonal auto-anti-idiotypic antibody. *Nature*. 1983;303: 627–629.
41. Nelson KA, George E, Swenson C, Forstrom JW, Hellstrom KE. Immunotherapy of murine sarcomas with auto-anti-idiotypic monoclonal antibodies which bind to tumor-specific T cells. *J Immunol*. 1987;139:2110–2117.
42. Linsley PS, Ledbetter JA. The role of the CD28 receptor during T cell responses to antigen. *Annu Rev Immunol*. 1993;11:191–212.
43. Chen L, Ashe S, Brady WA, et al. Costimulation of antitumor immunity by the B7 counterreceptor for the T lymphocyte molecules CD28 and CTLA-4. *Cell*. 1992; 71(7):1093–1102.
44. Chen L, McGowan P, Ashe S, et al. Tumor immunogenicity determines the effect of B7 costimulation on T cell-mediated tumor immunity. *J Exp Med*. 1994;179:523–532.
45. Melero I, Shuford WW, Newby SA, et al. Monoclonal antibodies against the 4-1BB T-cell activation molecule eradicate established tumors. *Nat Med*. 1997;3(6):682–685.
46. Hellstrom I, Ledbetter JA, Scholler N, et al. CD3-mediated activation of tumor-reactive lymphocytes from patients with advanced cancer. *Proc Natl Acad Sci USA*. 2001;98:6783–6788.
47. Kiessling R, Wasserman K, Horiguchi S, et al. Tumor-induced immune dysfunction. *Cancer Immunol Immunother*. 1999;48:353–362.
48. Dong H, Strome SE, Salomao D, et al. Tumor-associated B7-H1 promotes T-cell apoptosis. A potential mechanism of immune-evasion. *Nat Med*. 2002;8:793–800.
49. Freeman G, Long A, Iwai Y, et al. Engagement of the PD-1 immunoinhibitory receptor by a novel B7 family member leads to negative regulation of lymphocyte activation. *J Exp Med*. 2000;192:1027–1034.
50. Liu P, Jaffar J, Zhou J, Yang Y, Hellstrom I, Hellstrom KE. Inhibition of TGFbeta 1 makes nonimmunogenic tumor cells effective for therapeutic vaccination. *J Immunother*. 2009;32:232–239.
51. Munn DH. Indoleamine 2,3-dioxygenase, Tregs and cancer. *Curr Med Chem*. 2011; 18(15):2240–2246.
52. Mellor AL, Munn DH. IDO expression by dendritic cells: tolerance and tryptophan catabolism. *Nat Rev Immunol*. 2004;4:762–774.
53. Baratelli F, Lin Y, Zhu L, et al. Prostaglandin E2 induces FOXP3 gene expression and T regulatory cell function in human CD4+ T cells. *J Immunol*. 2005;175(3): 1483–1490.

54. Prehn RT. The immune reaction as a stimulator of tumor growth. *Science*. 1972; 176(31):170–171.
55. Bielas JH, Loeb KR, Rubin BP, True LD, Loeb LA. Human cancers express a mutator phenotype. *Proc Natl Acad Sci USA*. 2006;103:18238–18242.
56. Balkwill F, Mantovani A. Inflammation and cancer: back to Virchow? *Lancet*. 2001;357(9255):539–545.
57. Grivennikov SI, Greten FR, Karin M. Immunity, inflammation, and cancer. *Cell*. 2010;140(6):883–899.
58. Coussens LM, Werb Z. Inflammation and cancer. *Nature*. 2002;420(6917):860–867.
59. zur Hausen H. Viruses in human cancers. *Eur J Cancer*. 1999;35(14):1878–1885.
60. Chen L, Mizuno MT, Singhal MC, et al. Induction of cytotoxic T lymphocytes specific for a syngeneic tumor expressing the E6 oncoprotein of human papillomavirus type 16. *J Immunol*. 1992;148(8):2617–2621.
61. Chen LP, Thomas EK, Hu SL, Hellstrom I, Hellstrom KE. Human papillomavirus type 16 nucleoprotein E7 is a tumor rejection antigen. *Proc Natl Acad Sci USA*. 1991; 88(1):110–114.
62. Kenter GG, Welters MJ, Valentijn AR, et al. Vaccination against HPV-16 oncoproteins for vulvar intraepithelial neoplasia. *N Engl J Med*. 2009;361(19): 1838–1847.
63. van den Hende M, Redeker A, Kwappenberg KM, et al. Evaluation of immunological cross-reactivity between clade A9 high-risk human papillomavirus types on the basis of E6-Specific CD4+ memory T cell responses. *J Infect Dis*. 2010;202(8):1200–1211.
64. zur Hausen H. Papillomaviruses in the causation of human cancers—a brief historical account. *Virology*. 2009;384(2):260–265.
65. Cuzick J, Meijer CJ, Walboomers JM. Screening for cervical cancer. *Lancet*. 1998;351(9113):1439–1440.
66. Sedlacek TV. Advances in the diagnosis and treatment of human papillomavirus infections. *Clin Obstet Gynecol*. 1999;42(2):206–220.
67. Stoler MH, Schiffman M. Interobserver reproducibility of cervical cytologic and histologic interpretations: realistic estimates from the ASCUS-LSIL Triage Study. *JAMA*. 2001;285(11):1500–1505.
68. Kiviat NB, Koutsky LA, Paavonen JA, et al. Prevalence of genital papillomavirus infection among women attending a college student health clinic or a sexually transmitted disease clinic. *J Infect Dis*. 1989;159(2):293–302.
69. Sherman ME, Schiffman MH, Lorincz AT, et al. Toward objective quality assurance in cervical cytopathology. Correlation of cytopathologic diagnoses with detection of high-risk human papillomavirus types. *Am J Clin Pathol*. 1994;102(2):182–187.
70. Sheu BC, Lin RH, Lien HC, Ho HN, Hsu SM, Huang SC. Predominant Th2/Tc2 polarity of tumor-infiltrating lymphocytes in human cervical cancer. *J Immunol*. 2001;167(5):2972–2978.
71. Ovestad IT, Gudlaugsson E, Skaland I, et al. Local immune response in the microenvironment of CIN2-3 with and without spontaneous regression. *Mod Pathol*. 2010;23(9):1231–1240.
72. Feng Q, Wei H, Morihara J, et al. Th2 type inflammation promotes the gradual progression of HPV-infected cervical cells to cervical carcinoma. *Gynecol Oncol*. 2012; 127(2):412–419.
73. Zou W. Immunosuppressive networks in the tumour environment and their therapeutic relevance. *Nat Rev Cancer*. 2005;5(4):263–274.
74. Byrne WL, Mills KH, Lederer JA, O'Sullivan GC. Targeting regulatory T cells in cancer. *Cancer Res*. 2011;71(22):6915–6920.
75. Nelson B. CD20+ B cells: the other tumor-infiltrating lymphocytes. *J Immunol*. 2010;185(9):4977–4982.

76. Sanderson RD, Børset M. Syndecan-1 in B lymphoid malignancies. *Ann Hematol.* 2002;81(3):125–135.
77. Selvaraj P, Fifadara N, Nagarajan S, Cimino A, Wang G. Functional regulation of human neutrophil Fc gamma receptors. *Immunol Res.* 2004;29(1–3):219–230.
78. Ravetch J, Bolland S. IgG Fc receptors. *Annu Rev Immunol.* 2001;19:275–290.
79. Balachandran VP, Cavnar MJ, Zeng S, et al. Imatinib potentiates antitumor T cell responses in gastrointestinal stromal tumor through the inhibition of Ido. *Nat Med.* 2011;17(9):1094–1100.
80. De Monte L, Reni M, Tassi E, et al. Intratumor T helper type 2 cell infiltrate correlates with cancer-associated fibroblast thymic stromal lymphopoietin production and reduced survival in pancreatic cancer. *J Exp Med.* 2011;208(3):469–478.
81. Pedroza-Gonzalez A, Xu K, Wu TC, et al. Thymic stromal lymphopoietin fosters human breast tumor growth by promoting type 2 inflammation. *J Exp Med.* 2011;208(3):479–490.
82. Dai M, Wei H, Yip Y, et al. Long-lasting complete regression of established mouse tumors by counteracting Th2 inflammation. *J Immunother.* 2013;36:248–257.
83. Wei H, Zhao L, Li W, et al. Combinatorial PD-1 blockade and CD137 activation has therapeutic efficacy in murine cancer models and synergizes with cisplatin. *PLoS One.* 2013;8(12):e84927. https://doi.org/10.1371/journal.pone.0084927.
84. Chen LP. Lymphocyte co-signal molecules in cancer immunotherapy. *J Immunother.* 2006;29:664.
85. Leach DR, Krummel MF, Allison JP. Enhancement of antitumor immunity by CTLA-4 blockade. *Science.* 1996;271(5256):1734–1736.
86. Yang SC, Yang Y, Raycraft J, et al. Melanoma cells transfected to express CD83 induce anti-tumor immunity that can be increased by also engaging CD137. *Proc Natl Acad Sci USA.* 2004;101:4990–4995.
87. Dai M, Yip YY, Hellstrom I, Hellstrom KE. Curing mice with large tumors by locally delivering combinations of immunomodulatory antibodies. *Clin Cancer Res.* 2015;21(5):1127–1138.
88. de Visser K, Eichten A, Coussens LM. Paradoxical roles of the immune system during cancer development. *Nat Rev Cancer.* 2006;6:24–37.
89. Mittler RS, Bailey TS, Klussman K, Trailsmith MD, Hoffmann MK. Anti-4-1BB monoclonal antibodies abrogate T cell-dependent humoral immune responses in vivo through the induction of helper T cell anergy. *J Exp Med.* 1999;190(10): 1535–1540.
90. Ye Z, Hellstrom I, Hayden-Ledbetter M, Dahlin A, Ledbetter JA, Hellstrom KE. Gene therapy for cancer using single-chain Fv fragments specific for 4-1BB. *Nat Med.* 2002;8(4):343–348.
91. Dai M, Yip Y, Sjogren H, Hellstrom K. Tumor regression and cure depends on sustained Th1 responses. *J Immunother.* 2018;41(8):369–378. https://doi.org/10.1097/CJI.0000000000000231.
92. Fourcade J, Sun Z, Benallaoua M, et al. Upregulation of TIM-3 and PD-1 expression is associated with tumor antigen-specific CD8+ cell dysfunction in melanoma patients. *J Exp Med.* 2010;207:2175–2186.
93. Curran M, Kim MR, Montalvo W, Al-Shamkhani A, Allison AC. Combination CTLA-4 blockade and 4-1BB activation enhances tumor rejection by increasing T-cell infiltration, proliferation and cytokine production. *PLoS One.* 2011;6(4), e19499.
94. Wei H, Zhao L, Li W, et al. Combinatorial PC-1 blockade and CD137 activation has therapeutic efficacy in murine cancer models and synergizes with cisplatin. *PLoS One.* 2013;8(12):e84927.

95. Kim Y, Choi B, Kim K, Kang S, Kwon B. Combination therapy with cisplatin and anti-4-1-BB: synergistic anticancer effects and amelioration of cisplatin-induced nephrotoxicity. *Cancer Res.* 2008;68:7264–7269.
96. Wei H, Zhao L, Hellstrom I, Hellstrom K, Guo Y. Dual targeting of CD137 co-stimulatory and PD-1 co-inhibitory molecules for ovarian cancer immunotherapy. *Oncoimmunology.* 2014;3, e28248. https://doi.org/10.4161/onci.28248.
97. de Biasi AR, Villena-Vargas J, Adusumilli PS. Cisplatin-induced antitumor immunomodulation: a review of preclinical and clinical evidence. *Clin Cancer Res.* 2014; 20(21):5384–5391.
98. Baban B, Hansen AM, Chandler PR, et al. A minor population of splenic dendritic cells expressing CD19 mediates IDO-dependent T cell suppression via type I IFN signaling following B7 ligation. *Int Immunol.* 2005;17:909–919.
99. Liu Y, Liang X, Yin X, et al. Blockade of IDO-kynurenine-AhR metabolic circuitry abrogates IFN-gamma-induced immunologic dormancy of tumor-repopulating cells. *Nat Commun.* 2017;8, 15207.
100. Hwang WL, Niemierko A, Hwang KL, et al. Clinical outcomes in patients with metastatic lung cancer treated with PD-1/PD-L1 inhibitors and thoracic radiotherapy. *JAMA Oncol.* 2018;4(2):253–255. https://doi.org/10.1001/jamaoncol.2017.3808.
101. Oweida A, Lennon S, Calame D, et al. Ionizing radiation sensitizes tumors to PD-L1 immune checkpoint blockade in orthotopic murine head and neck squamous cell carcinoma. *Oncoimmunology.* 2017;6(10), e1356153.
102. Sundahl N, De Wolf K, Rottey S, et al. A phase I/II trial of fixed-dose stereotactic body radiotherapy with sequential or concurrent pembrolizumab in metastatic urothelial carcinoma: evaluation of safety and clinical and immunologic response. *J Transl Med.* 2017;15(1):150.
103. Zhang H, Knutson KL, Hellstrom KE, Disis ML, Hellstrom I. Antitumor efficacy of CD137 ligation is maximized by the use of a CD137 single-chain Fv-expressing whole-cell tumor vaccine compared with CD137-specific monoclonal antibody infusion. *Mol Cancer Ther.* 2006;5(1):149–155.
104. Ascierto PA, Simeone E, Sznol M, Fu YX, Melero I. Clinical experiences with anti-CD137 and anti-PD1 therapeutic antibodies. *Semin Oncol.* 2010;37(5):508–516.
105. Lei C, Liu P, Chen B, et al. Local release of highly loaded antibodies from functionalized nanoporous support for cancer immunotherapy. *J Am Chem Soc.* 2010; 132:6906–6907.
106. Dai M, Yip Y, Hellstrom I, Hellstrom K. An in vitro model that predicts the therapeutic efficacy of immunomodulatory antibodies. *J Immunother.* 2016;39:298–305.
107. Prendergast GC, Jaffee EM. Cancer immunologists and cancer biologists: why we didn't talk then but need to now. *Cancer Res.* 2007;67(8):3500–3504.
108. Chen D, Mellman I. Oncology meets immunology: the cancer immunity cycle. *Immunity.* 2013;39:1–10.
109. Pardoll D, Drake C. Immunotherapy earns its spot in the ranks of cancer therapy. *J Exp Med.* 2012;209(2):201–209.
110. Rosenberg SA, Yang JC, Restifo NP. Cancer immunotherapy: moving beyond current vaccines. *Nat Med.* 2004;10:909–915.
111. Yee C, Greenberg PD. Monitoring T-cell immunity to tumours: new strategies for monitoring T-cell responses. *Nat Rev.* 2002;2:409–419.
112. Karre K. Express yourself or die: peptides, MHC molecules, and NK cells. *Science.* 1995;267(5200):978–979.
113. Fidler IJ, Poste G. Macrophage-mediated destruction of malignant tumor cells and new strategies for the therapy of metastatic disease. *Springer Semin Immunopathol.* 1982; 5:161–174.

114. Fagerberg J, Hjelm AL, Ragnhammar P, Frodin JE, Wigzell H, Mellstedt H. Tumor regression in monoclonal antibody-treated patients correlates with the presence of anti-idiotype-reactive T lymphocytes. *Cancer Res.* 1995;55(9):1824–1827.
115. Kolb W, Granger G. Lymphocyte in vitro toxicity: characterization of human lymphotoxin. *Proc Natl Acad Sci USA.* 1968;61:1250–1255.
116. Ferrone S, Marincola FM. Loss of HLA class I antigens by melanoma cells: molecular mechanisms, functional significance and clinical relevance. *Immunol Today.* 1995; 16(10):487–494.
117. Schumacher TN, Schreiber RD. Neoantigens in cancer immunotherapy. *Science.* 2015;348(6230):69–74.
118. Brahmer JR, Tykodi SS, Chow LQ, et al. Safety and activity of anti-PD-L1 antibody in patients with advanced cancer. *N Engl J Med.* 2012;366(26):2455–2465.
119. Topalian SL, Hodi FS, Brahmer JR, et al. Safety, activity, and immune correlates of anti-PD-1 antibody in cancer. *N Engl J Med.* 2012;366(26):2443–2454.
120. Lipson EJ, Sharfman WH, Drake CG, et al. Durable cancer regression off-treatment and effective reinduction therapy with an anti-PD-1 antibody. *Clin Cancer Res.* 2013; 19(2):462–468.
121. Sznol M, Chen L. Antagonist antibodies to PD-1 and B7-H1 (PD-L1) in the treatment of advanced human cancer. *Clin Cancer Res.* 2013;19(5):1021–1034.
122. Zitvogel L, Apetoh I, Ghiringhelli F, Kroemer G. Immunological aspects of cancer chemotherapy. *Nat Rev Immunol.* 2008;8:59–73.
123. Sagiv-Barfi I, Czerwinski DK, Levy S, et al. Eradication of spontaneous malignancy by local immunotherapy. *Sci Transl Med.* 2018;10(426).
124. Melero I, Johnston JV, Shufford WW, Mittler RS, Chen L. NK1.1 cells express 4-1BB (CDw137) costimulatory molecule and are required for tumor immunity elicited by anti-4-1BB monoclonal antibodies. *Cell Immunol.* 1998;190(2):167–172.
125. Brennan FR, Morton LD, Spindeldreher S, et al. Safety and immunotoxicity assessment of immunomodulatory monoclonal antibodies. *MAbs.* 2010;2(3):233–255.
126. Naidoo J, Page DB, Li BT, et al. Toxicities of the anti-PD-1 and anti-PD-L1 immune checkpoint antibodies. *Ann Oncol.* 2015;26(12):2375–2391.
127. Boutros C, Tarhini A, Routier E, et al. Safety profiles of anti-CTLA-4 and anti-PD-1 antibodies alone and in combination. *Nat Rev Clin Oncol.* 2016;13(8):473–486.

CHAPTER TWO

Bioinformatic methods for cancer neoantigen prediction

Sebastian Boegel[a,*], John C. Castle[c], Julia Kodysh[b], Timothy O'Donnell[b], Alex Rubinsteyn[b,*]

[a]Johannes Gutenberg University, Mainz, Germany
[b]Icahn School of Medicine at Mount Sinai, New York, NY, United States
[c]Agenus Inc., Lexington, MA, United States
*Corresponding authors: e-mail address: seb.boegel@gmail.com; alex@openvax.org

Contents

1. Introduction 26
2. What do we mean by "immunogenicity"? 28
3. Neoantigen sources 28
4. Identifying neoantigens from high-throughput sequencing data 29
   4.1 Sequencing platform 29
   4.2 Personalized reference genomes 29
   4.3 SNV calling 30
   4.4 Indel variant calling 30
   4.5 Structural variants and chromosomal fusions 33
   4.6 Abnormal splicing 33
   4.7 Circular RNA 34
   4.8 RNA editing 34
   4.9 Human endogenous retroviruses 34
   4.10 When variants collide: Co-occurrence of germline and somatic variants (phasing proximal variants) 34
5. Which protein changes result in neoantigens? 35
   5.1 MHC I binding prediction 35
   5.2 MHC I antigen processing 37
   5.3 MHCII binding prediction 38
   5.4 TCR recognition and similarity to self 40
   5.5 MHC allele considerations 41
6. Multi-step neoantigen prediction workflows 41
   6.1 pVACseq 42
   6.2 MuPeXI 43
   6.3 TIminer 44
   6.4 OpenVax 44
   6.5 NeoEpiScope 45
   6.6 Epi-Seq 45

Progress in Molecular Biology and Translational Science, Volume 164
ISSN 1877-1173
https://doi.org/10.1016/bs.pmbts.2019.06.016

7. Identifying neoantigen responsive T-cell receptors 46
8. Conclusion 47
Acknowledgment 48
References 48

Abstract

Tumor cells accumulate aberrations not present in normal cells, leading to presentation of neoantigens on MHC molecules on their surface. These non-self neoantigens distinguish tumor cells from normal cells to the immune system and are thus targets for cancer immunotherapy. The rapid development of molecular profiling platforms, such as next-generation sequencing, has enabled the generation of large datasets characterizing tumor cells. The simultaneous development of algorithms has enabled rapid and accurate processing of these data. Bioinformatic software tools encoding the algorithms can be strung together in a workflow to identify neoantigens. Here, with a focus on high-throughput sequencing, we review state-of-the art bioinformatic tools along with the steps and challenges involved in neoantigen identification and recognition.

1. Introduction

Tumor cells accumulate aberrations not present in normal cells.[1] Altered protein sequences that can arise from DNA mutations include single nucleotide variations (SNVs), insertions and deletions (indels), structural variations and gene fusions. Additional non-self proteins can arise from tumor-specific mRNA splicing, RNA editing, post-translational modifications and human endogenous retrovirus expression (Fig. 1). The human leukocyte antigen protein (HLA; also referred to as MHC) presents peptides on the cell surface to T cells recognizing a peptide-HLA complex and that can initiate an immune response. While few aberrations directly promote a tumor phenotype, an aberration that changes a protein sequence can lead to peptide presentation that distinguishes tumor cells from normal cells to the immune system—a neoantigen.[2]

Neoantigens can be used therapeutically for the development of cancer vaccines, T cell adoptive cellular therapy, and immunomonitoring. Aberration-targeting immunotherapies exploit the landscape of tumor-specific aberrations displayed by a tumor (i.e., the mutanome).[3–8] Immune checkpoint inhibition has been tremendously beneficial for some patients[9] and the effect of immune checkpoint modulation is mediated by neoantigen T-cell responses.[10] Tumor mutational burden (TMB), neoantigen burden,[11–13] HLA type[14] and HLA expression[15] have been associated with response to checkpoint modulation. Lastly, neoantigen recognizing T cells can be tracked as biomarkers of post-treatment checkpoint blockade efficacy.[16,17]

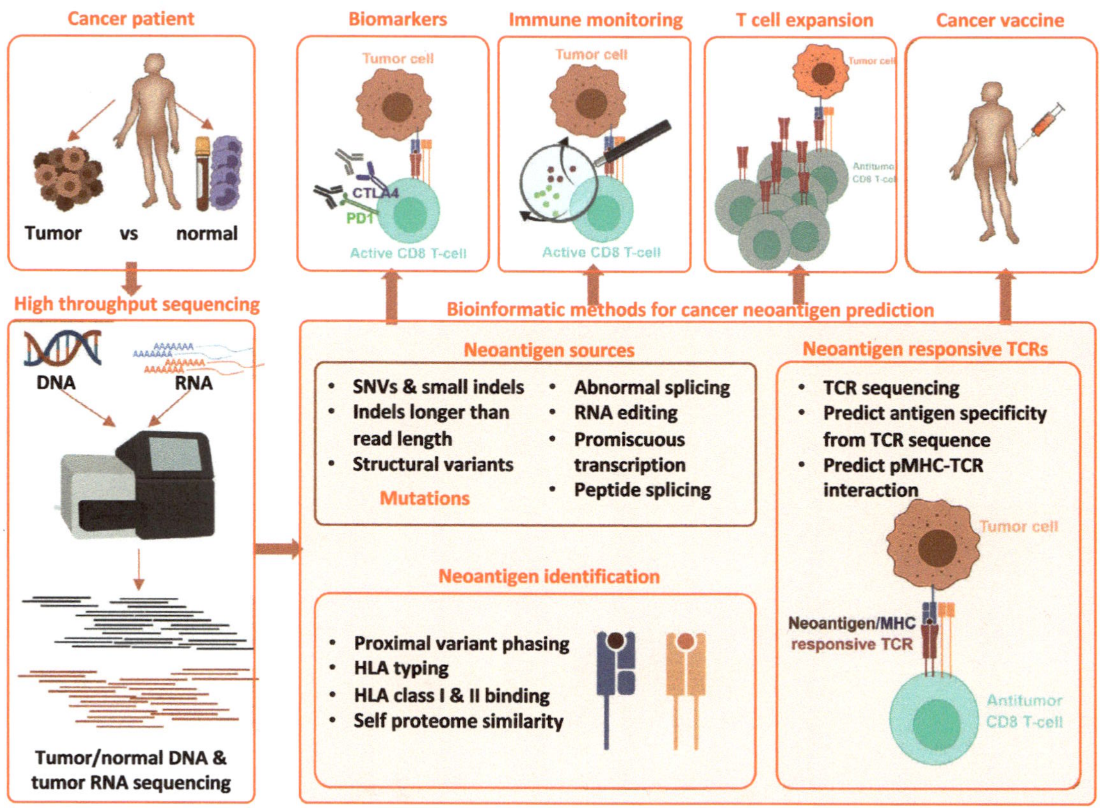

Fig. 1 Process overview of discovering neoantigen candidates for cancer immunotherapy. *Figure made with biorender (https://biorender.com/).*

Technology platforms and algorithms enable identification of tumor neoantigens, such as the use of next-generation sequencing to screen the exome and transcriptome of a patient tumor. The process of neoantigen identification typically entails (i) identification of tumor-specific protein sequences and (ii) prioritization of candidate neoantigens according to their abundance, presence on patient HLA molecules, and ability to elicit a T-cell response. Here, we discuss the challenges involved in each of the steps and review the state-of-the art bioinformatic tools and computational pipelines developed for neoantigen identification and TCR sequencing with a focus on high-throughput sequencing data.

2. What do we mean by "immunogenicity"?

Neoantigen prediction tools often focus on the identification of peptides presented on tumor MHC molecules. To understand the results and utility of such tools, it is useful to define the term "immunogenicity"[18] and to distinguish it from the sometimes conflated term "antigenicity."[19] Immunogenicity is defined as the "the ability of a molecule or substance to provoke an immune response"[20] and is "mostly mediated by T cells."[21] The immunogenicity of a tumor depends—among other thing—on its antigenicity, which is defined as the ability to be recognized by T or B cells without necessarily leading to an ouvert immune responses.[21]

Furthermore, it is important to distinguish between several, often conflated meanings of immunogenicity by asking "How did the antigen-specific immune response arise?" In spontaneous anti-tumor T-cell responses, which occur frequently, the tumor is the source of the antigen. In contrast, in vaccine-induced T-cell responses, the vaccine encodes the antigen which imparts an immune response. A vaccine can, for example, elicit an immune response that does not recognize tumor cells. The ground truth for neoantigen prediction is typically an T-cell immune monitoring assay measuring the immunogenicity of a particular peptide-MHC complex, as reviewed by Britten et al.[22] The identification of immunogenic neoantigens is a surrogate endpoint in the neoantigen prioritization as immunogenicity does not necessarily lead to tumor control. For example, a neoantigen-recognizing CD8+ T cell has been identified that showed immune response in vivo but was functionally irrelevant.[23]

3. Neoantigen sources

Tumor neoantigens are non-self peptides presented on the surface of tumor cells by MHC molecules (Fig. 1). Common genomic alterations in cancer include single nucleotide variants (SNVs) and small insertions and deletions (indels). Small mutations are relatively easy to detect with the current short-read sequencing platforms, and thus likely over-represented in genomic analyses. Beyond small mutations, there is a vast landscape of genomic events which can also potentially change the translated amino acid sequence of a protein. These include "medium" sized indels (>10 bp), larger deletions which can exceed the sequencing read length, expansion of repetitive genomic elements, and structural variants such as chromosomal fusions and inversions. Here, we highlight existing tools and approaches which have been used to identify neoantigens from these sources.

Other neoantigen classes include abnormal splicing, circular RNA, RNA editing and human endogenous retroviruses. Neoantigens such as non-canonical reading frames,[24,25] post-translational amino acid modifications[26,27] and proteasomal peptide splicing[28,29] are identified with mass spectrometry[30] and not a subject of this review.

4. Identifying neoantigens from high-throughput sequencing data

4.1 Sequencing platform

The sequencing platform used for tumor and normal samples constrains the identified somatic alterations. Short read whole exome sequencing (WES) on tumor and normal tissue is common, as well as short read tumor RNA sequencing. Short read sequencing is preferred because of the current cost, infrastructure, and error rates of long read platforms such as SMRT[31] and Nanopore.[32] Exome sequencing is typically used because it allows for detection of small coding variants at a fraction of the cost of whole genome sequencing. On the other hand, whole-genome sequencing (WGS) enables more sensitive and accurate variant calling.[33,34] Furthermore, copy number estimates from whole exome sequencing can be unreliable,[35] which in turns affects the quality of clonality estimates for individual mutations. Lastly, only a small fraction of structural variants can be identified by exome sequencing, typically cases where a breakpoint occurs between two exons.[36,37] The integrated use of whole genome and exome capture sequencing can generate better results.[38]

4.2 Personalized reference genomes

The choice of genomic reference has an impact on mapping and variant calling.[39,40] The use of a recently defined reference assembly, such as GRCh38, yields consistent improvements in quality of genomic alignments.[41]

Beyond choosing a high quality reference genome, the use of a "personalized" reference genome that includes hereditary variants can improve the alignment of tumor DNA and RNA sequence reads,[42] improving both tumor mutation identification and transcript quantification.[43] To construct a personalized reference genome, germline variants are first identified from the normal DNA and then patched into the reference genome sequences, enabling the creation of both "reference" and "personalized" genomes. Tumor reads are aligned against both genomes and variants called against the personalized genome are shifted back to the reference coordinate system.

4.3 SNV calling

Single nucleotide variations (SNVs) are the most common mutation for cancer vaccines[4,5] and biomarker identification.[11–13,44] SNVs are usually identified by comparing the tumor to the normal DNA reads (usually WES) after alignment to the reference genome. A plethora of SNV calling tools have been developed, with common tools listed in Table 1. Of note, including tumor RNA-Seq data (i.e., "which mutation is expressed?") for variant detection has been shown to result in additional supportive mutation information.[65,66]

4.4 Indel variant calling

Indels are also commonly identified in most cancers, although they occur less frequently than SNVs and are more difficult to unambiguously detect. Indel detection rate is affected by the mapping parameters of a short read aligner, normalization scheme for representing indel alignments, as well biases resulting from the use of targeted capture exome sequencing.[33] As a result, the concordance rate for indel detection tools from short read exome sequencing can be low.[67]

For example, sequencing reads containing indels are prone to ambiguous alignment, where the same underlying genomic event can give rise to many different pattern of alignment against the reference. If a hypothetical reference chromosome consisted of the nucleotide sequence "CCCTAATAGGG" and, in a hypothetical cancer mutation, the middle three nucleotides "AAT" are deleted, the result sequence is "CCCTAGGG." Reads supporting this mutation might be aligned as a deletion of "AAT" (the correct deletion) but might also be aligned as a deletion of "TAA" or "ATA," which would give equally good mapping scores. Multiple reads supporting the same mutation might all be aligned differently, making confident identification of a single variant difficult.

Table 1 Bioinformatic tools for identification of different categories of neoantigen sources.

Pileup variant callers		
MuTect[45]	SNV calling using tumor and normal sequencing data	https://github.com/broadinstitute/mutect
Strelka[46]	SNV and small indel calling using matched tumor-normal samples	ftp://strelka@ftp.illumina.com/
Microassembly variant callers		
MuTect2	Combination of MuTect and HaplotypeCaller, part of the GATK framework, including microassembly variant calling	https://software.broadinstitute.org/gatk/documentation/tooldocs/3.6-0/org_broadinstitute_gatk_tools_walkers_cancer_m2_MuTect2.php
Strelka2[47]	Including normal sample contamination, germline mutations and microassembly variant calling	https://github.com/Illumina/strelka
scalpel[48]	Performs localized microassembly for detecting indels	scalpel.sourceforge.net
lancet[49]	SNV and indel calling from tumor/normal pairs using localized microassembly	github.com/nygenome/lancet
Indel realigners		
ABRA2[50]	Supports joint realignment of both DNA and RNA	github.com/mozack/abra2
GATK 3.x IndelRealigner[51]	DNA only realigner, no longer available in newer versions of GATK	software.broadinstitute.org/gatk/download/archive
Gene fusion callers		
deFuse[52]	Gene fusion discovery using RNA-Seq data	https://bitbucket.org/dranew/defuse
INTEGRATE[53]	Fusion discovery using WGS and RNA-Seq	https://sourceforge.net/projects/integrate-fusion/

Continued

Table 1 Bioinformatic tools for identification of different categories of neoantigen sources.—cont'd

Tools for detecting alternative splicing		
neoantigenR[54]	Identifies isoforms from RNA-Seq not in reference proteome, uses these sequences for neoantigen prediction	github.com/tangshao2016/neoantigenR
SplAdder[55]	Identifies broad category of alternative splicing events from RNA-Seq	github.com/ratschlab/spladder
MiSplice[56]	Identifies mutation associated splicing alterations	github.com/ding-lab/misplice
KeepMeAround (KMA)[57]	Detects intron retention from RNA-Seq	github.com/pachterlab/kma
iRead[58]	Detect intron retention from RNA-Seq	http://www.genemine.org/iread.php
MMsplice[59]	Predicts whether a genomic variant affects splicing	github.com/gagneurlab/MMSplice
Tools for RNA editing		
DeepRed[60]	Identification of RNA editing events from standard RNA-Seq data without prior-knowledge-based filtering steps or genomic annotations	https://github.com/wenjiegroup/DeepRed
REDItools[61]	RNA editing detection using (i) RNA-Seq and DNA-Seq data from the same sample, or (ii) RNA-Seq data alone	https://sourceforge.net/projects/reditools/
RNAEditor[62]	Analyzes RNA editing events from standard RNA-Seq data	http://rnaeditor.uni-frankfurt.de/
RES-Scanner[63]	Detects and annotates RNA-editing sites using matching RNA-Seq and DNA-Seq data	https://github.com/ZhangLabSZ/RES-Scanner
Identification of human endogenous retroviruses (hERVs)		
hervQuant[64]	Workflow to identify active hERVs in RNA-Seq data	https://unclineberger.org/vincent/resources/

To overcome this difficulty, several computational tools do "indel realignment" by performing a local assembly of reads around a candidate indel and then identifying the mutation by aligning the assembled sequence against the reference. Several variant callers (Table 1), including Mutect2 and Strelka2, use microassembly, an alternative algorithm, which integrates examination of candidate variant loci as part of the variant calling process, improving both SNV and indel identification. Indeed, with the advent of microassembly variant callers, indel realignment began to be phased out of recommendations for genomics pipelines. However, a recently published indel realignment program, ABRA2,[50] showed significant improvements in accuracy, even for variant callers which implement their own internal microassembly. Furthermore, the ability of the ABRA2 to jointly realign both DNA and RNA samples improves downstream analysis of allele-specific expression of neoantigens in RNA.

4.5 Structural variants and chromosomal fusions

Gene fusions are a promising source of highly potent neoantigens.[68] Nine years before the publication of the first draft of the human genome, tumor-specific fusion proteins in leukemias were identified that generated T-cell responses.[69] In 2001, the antigenicity of fusion proteins from sarcoma-associated chromosomal translocations was examined, finding evidence for MHC class I binding sarcoma-associated chimeras that might serve as neoantigens.[70] Analysis of TCGA data found that 16% of cancers have a gene fusion as one of their drivers.[71] Candidate neoantigens associated with fusions have been identified in osteosarcomas[72] using a combination of deFuse and trinity[73] for gene fusion detection, seq2HLA and Optitype for HLA typing, and NetMHC3.0 for HLA binding. Of note, INTEGRATE[53] detects fusion events from WGS and RNA-Seq data, while INTEGRATE-neo[74] is a gene fusion neoantigen discovery pipeline.

4.6 Abnormal splicing

In addition to the genomic mutations, splicing events can produce novel protein sequences. Alternative splicing is a common mechanisms in cells for maintaining diversity of the produced proteins[75] and tumor cells show enormous alterations in their transcriptome, caused in part by mutations that produce tumor-specific isoforms.[76] Indeed, >20 years ago, a cancer-specific melanoma antigen arising from intron retention in TRP2 was originally identified.[77] Intron retention is common in many cancers[78] and can lead to neoantigens.[79] Further, applying splAdder[55] to TCGA data to analyze splicing events in 8705 patients across 32 cancer types, Kahles et al.[80] found up to 30% more alternative splicing events in tumors compared to normal samples. Using misplice, Jayasinghe et al.[56] identified 1964 splice-site-creating mutations creating alternative splice junctions across 8656 TCGA tumors, predicted neoantigens and suggested that splice-site-creating mutations may be more immunogenic than missense mutations. Common tools for the detection of alternative splicing are listed in Table 1.

4.7 Circular RNA

Circular RNA may form when a transcript is back-spliced on itself[81] and in limited cases these can be translated into novel peptides using internal ribosomal entry sites.[82] However, though circRNAs may be identified using software such as CIRCExplorer2,[83] translation appears to be rare and difficult to predict from sequencing data.

4.8 RNA editing

RNA editing is a post-transcriptional process in which nucleotide changes are introduced into a RNA sequence, many of which can thus contribute to proteomic sequence variation.[84] Dysregulated RNA editing has been found in different types of cancers,[85] the associated epitopes have been identified on MHC molecules, and several elicit immune responses.[86] Several bioinformatic tools identifying RNA editing events in RNA-Seq data (Table 1). REDItools[61] was used with RNA-Seq data from blood samples from systemic lupus erythematosus patients to identify abnormally high levels of RNA editing, some may generate novel neoantigens.[87]

4.9 Human endogenous retroviruses

Human endogenous retroviruses (hERVs) are elements of retroviral DNA sequences that integrated into the human genome throughout evolution. Using hervQuant (Table 1) and TCGA RNA-Seq tumor data, transcriptionally active hERVs have been identified in tumors, including several potential tumor-specific hERV epitopes.[64]

4.10 When variants collide: Co-occurrence of germline and somatic variants (phasing proximal variants)

A neoantigen-encoding peptide sequence cannot be determined solely from a single mutation relative to a reference, as one must account for the co-occurrence of other proximal somatic and germline variants. There are several approaches for phasing variants in cancer samples: explicit phasing from DNA using a tool such as HapCUT2,[88] explicit phasing from RNA using a tool such as phASER,[89] or implicit phasing by assembly of RNA reads to determine a coding sequence. Recently published neoantigen prediction pipelines phasing variants using DNA include.[90,91]

5. Which protein changes result in neoantigens?

In this section, we describe criteria that have been proposed for computationally identifying the protein-changing aberrations that elicit a T-cell response. We focus first on the standard practice of MHC I affinity prediction, then address a number of emerging criteria: antigen processing, ability to bind MHC II, ability to be recognized by a TCR, and considerations based on MHC allele.

5.1 MHC I binding prediction

There is near-universal agreement that a tumor neoantigen should bind at least one of a patient's MHC I alleles for recognition by cytotoxic CD8+ T cells. The three classical human MHC I genes HLA-A, HLA-B, and HLA-C are extremely polymorphic, with thousands of alleles identified in the human population.[92] Each allele exhibits a bias for presenting peptides with a particular sequence motif, which varies from allele to allele. Since the pioneering work by Sette and others nearly 30 years ago,[93,94] computational prediction of peptide/MHC binding has received significant attention and today is remarkably accurate, with typical area under the curve (AUC) scores of >0.90 for most alleles. These predictors are a standard tool in the fields of infectious disease, allergy, autoimmunity, immuno-oncology, and vaccine development.[95,96]

To apply computational MHC I ligand prediction, the patient's MHC genotype must be determined. This is standard practice in the organ transplant setting and can be performed in the laboratory using PCR or sequencing-based analysis. Multiple computational tools can process WES or RNA-Seq to determine a patient's HLA type. Commonly used tools (Table 2) include Optitype,[97] which calls MHC I alleles using DNA or RNA sequence reads, and seq2hla,[98] which calls MHC I and II alleles using RNA. Other tools have been developed,[104–106] but Optitype or seq2hla are nearly optimal in most settings.

Table 2 Neoantigen prediction or prioritization tools.

Neoantigen prediction or prioritization tool	Description	URL
Optitype[97]	Genotypes HLA class I using RNA or DNA	https://github.com/FRED-2/OptiType
seq2hla[98]	Genotypes HLA class I and II alleles using RNA	https://github.com/TRON-Bioinformatics/seq2HLA
NetMHCpan 4.0[99]	Predict peptide/HLA class I binding	http://www.cbs.dtu.dk/services/NetMHCpan/
NetChop[100]	Proteasomal cleavage prediction	http://www.cbs.dtu.dk/services/NetChop/
NetMHCIIpan[101]	Predict peptide/HLA class II binding	http://www.cbs.dtu.dk/services/NetMHCIIpan/
MixMHC2pred[102]	Predict peptide/HLA class II binding	https://github.com/GfellerLab/MixMHC2pred
antigen.garnish[103]	Neoantigen prioritization (multiple approaches)	https://github.com/immune-health/antigen.garnish

The NetMHC suite of computational tools developed by Morten Nielsen's group at the Technical University of Denmark is the most popular package for peptide/MHC binding affinity prediction.[99,107–111] The NetMHC tools are based on machine learning approaches (ensembles of simple neural networks) and are fit to affinity measurements largely derived from the Immune Epitope Database (IEDB).[112] The most recent version of NetMHCpan (version 4.0) additionally includes MS-identified MHC ligands in its training set. Other tools using similar approaches have recently been developed[113] as well as tools whose training data relies entirely on (MS)-identified peptide ligands.[114,115] However, as yet no tool has consistently shown significantly improved accuracy over NetMHCpan 4.0 in independent benchmarks. An advantage of alternate tools is that they are open source and in many cases are easier to integrate into bioinformatic pipelines. Approaches applying physics-based modeling of the peptide/MHC structural interaction have also been investigated, but show lower accuracy than machine-learning approaches.[116,117]

Tools such as NetMHC (Table 2) return a predicted nanomolar binding affinity for each queried peptide/MHC pair. There is no obvious threshold for what it means to "bind," and thus there is substantial room for variation, even across groups using the same prediction software. A common approach is to require that peptide/MHC complexes bind with a predicted affinity of 500 nM or lower.[118] As different MHC alleles show different distributions of binding affinity and promiscuity, it is also common to select peptides in terms of their percentile rank (e.g., <2.0%) from a sample of random peptides. The Sette group, in contrast, has recommended specific nanomolar cutoffs for each allele.[119] An additional point of variation is that the NetMHCpan 4.0 models produce two outputs, a predicted nM binding affinity and a probability of identification in mass spectrometry (MS) HLA ligand-elution experiments. While these outputs are highly correlated, the question of which to use for neoantigen prediction is still open.

Finally, we note that peptide/MHC *affinity*—the ratio of the off-rate to on-rate for the binding reaction in solution—may not be the most biologically relevant parameter describing peptide/MHC interactions. A more useful property may be the off-rate, which dictates the duration the pMHC remains intact on the surface of a cell. Indeed, several studies have suggested that the half life of the pMHC complex is a better predictor of immunogenicity than pMHC affinity.[120,121] While the netMHCstab tool has been developed to predict pMHC stability,[122] most users continue to apply affinity prediction as affinity predictors have the advantage of much larger training sets. Recently introduced high-throughput assays for pMHC stability, however, may create an opportunity to revisit stability prediction.[123]

5.2 MHC I antigen processing

Our understanding of the steps required before a peptide is loaded onto an MHC class I molecule have improved dramatically over the past two decades, enabling efforts to characterize the impact of these steps on the epitope repertoire. Most neoantigen prediction pipelines do not attempt to account for antigen processing as there is surprisingly little evidence that these tools improve the accuracy of epitope predictions. However, this view may need revision in the near future as new tools based on MS training datasets are developed.

Intracellular proteins are degraded by the proteasome, a multi-enzyme complex with multiple cleavage specificities. The resulting fragments are transported into the endoplasmic reticulum (ER) via transporter associated with antigen presentation (TAP), where they can be further trimmed on the N-terminus by one of two endoplasmic reticulum aminopeptidase (ERAP) enzymes and loaded onto MHC I molecules. These steps all show some degree of sequence specificity, making them candidates for prediction. Tools have been developed for this purpose (Table 2), such as NetChop[100] for C-terminal proteasomal cleavage prediction as well as lesser-used tools for TAP transport.[124,125] However, most benchmarks have found these tools do not improve accuracy when used in conjunction with MHC I binding predictors. One possible reason for this is that each step in peptide processing may have evolved to preferentially produce peptides that match the specificity of the subsequent step, resulting in redundancy in the predictors.[126] For NetChop, released in 2005, an important limitation may also be its small training set.

Recent large datasets of MHC ligands identified by MS provide an opportunity to more closely evaluate the predictability of antigen processing. In a 2017 study, over 24,000 MHC ligands were identified across 16 alleles in a B cell line.[127] The authors found evidence for processing signals, both within the MHC ligand and in its flanking sequence in the source protein, that differ substantially from those predicted by NetChop. While analyses such as these must carefully account for biases in the MS-identifiability of peptides (such as a depletion of cysteines), we expect that MS datasets may soon enable the development of a next generation of improved antigen processing predictors. In some sense, this is already happening, as NetMHCpan 4.0 may partially account for certain aspects of antigen processing as its training set includes MS-identified peptides, thus integrating multiple steps in the antigen processing machinery. Software that further blurs the lines by integrating processing prediction, MHC binding prediction, and T cell recognition is likely to emerge.

5.3 MHCII binding prediction

CD4 T cells respond to epitopes presented by specialized antigen presenting cells on MHC II. While helper CD4 T cells are likely not strictly required for cytotoxic (MHC I-restricted) antitumor T cell responses, they may support the long term survival of cytotoxic T cells[128,129] and evidence is accumulating that they are involved in potentiating effective antitumor responses. Helper T cells recognizing neoantigens can be found infiltrating melanoma lesions,[130] and neoantigen vaccination on the basis of MHC II-predicted binding has been shown to be effective in a mouse model of a personalized mRNA vaccine.[131] Most provocatively, in the case of a B cell lymphoma (which expresses MHC II), one study has found MHC II-restricted epitopes that are directly targeted by a *cytotoxic* CD4+ anti-tumor response.[132]

Despite persuasive evidence that class II-restricted neoantigen responses are crucial at least for some antitumor responses, most neoantigen discovery pipelines do not prioritize neoantigens on the basis of predicted MHC II binding due to the lower accuracy of currently available predictors. This poor performance is a consequence of both a paucity of training data and an intrinsically more difficult modeling problem. As of early 2019, there are approximately 80,000 MHC II ligand entries in IEDB, compared to 350,000 for MHC I. Class II prediction is a more difficult problem because the peptide-binding groove of MHC II is open at both ends, allowing long peptides of variable length to bind.[133] Models must identify the part of an MHC II-binding peptide that lies in the binding groove not only to generate predictions, but—in a severe complication to model fitting procedures—also during training, as only rarely is the binding core experimentally determined.

The most commonly used MHC II ligand predictor is NetMHCIIpan (Table 2) from the NetMHC suite.[101] NetMHCIIpan uses similar neural network architectures as NetMHCpan, but differs in training procedure, using an iterative approach to attempt to predict the binding core for each peptide during training. NetMHCIIpan does not make use of MS currently because insufficient monoallelic (i.e., associated with a single known allele) MHC II MS data have been published, although this is likely to change in the near future as data becomes available.[134] An analysis using NetMHCIIpan is very similar to one using NetMHCpan, with results reported as nanomolar affinity values for each peptide and allele. A less restrictive affinity cutoff (e.g., 1000 nM or top 10% rank) is typically applied.[135]

Several more recently introduced tools also show promise for class II prediction, although in each case further validation will be required. The MixMHC2pred tool uses multiallelic MS datasets to train class II predictors by clustering MS hits by motif, then associating clusters with alleles based on the pattern of shared clusters among individuals with some alleles in common.[102] This approach has also been pursued by the developers of the NetMHC suite, who have recently extended the idea to include a mix of monoallelic and multiallelic MS training data and to use neural networks for motif identification in a tool called NNAlign_MA.[136] The BOTA package is notable for attempting to predict immunodominant bacterial MHC II epitopes by combining MHC binding prediction with a model of class II antigen processing and immunogenicity.[137] As BOTA focuses on epitopes of bacterial origin, however, it would require adaptation for use in neoantigen prediction.

5.4 TCR recognition and similarity to self

Not all mutated peptides displayed on an individual's MHC molecules will have a corresponding T-cell clone capable of recognition. While some pMHC may be non-immunogenic due to fundamental biochemical properties of the TCR/pMHC interaction, an important factor is immunological tolerance. T cells recognizing self-peptides are eliminated or rendered unresponsive during T cell development, creating "holes" in the TCR repertoire around self-peptides.[138] Since each TCR likely cross reacts with many pMHC, tolerance may render large parts of the antigenic landscape unrecognizable.[139]

A range of criteria have been proposed for prioritizing non-self MHC I-presented peptides for likelihood of T cell recognition. The simplest approach is to eliminate mutated peptides that are found elsewhere in the human proteome using an exact search. Although this removes few candidate peptides from consideration, this step is easy and should probably always be applied. The search may be made more sensitive by using sequence distance metrics (e.g., weighted "edit distance") that attempt to account for T-cell cross-reactivity profiles, for example by weighting changes in TCR-contacting residues in the peptide more than others.[139] Another approach, termed the differential agretopicity index (DAI) uses thresholds on the ratio of the MHC affinities of the mutated and wild type peptides.[103,140] Studies have come to widely varying conclusions on the predictive power of DAI.[141–143] To eliminate pMHC that are unrecognizable due to intrinsic, rather than tolerance-induced holes in the repertoire, it has been suggested that peptides with hydrophobic TCR-contacting residues (typically the central residues) are more immunogenic.[144]

An approach that considers the similarity between peptides and known-immunogenic epitopes from infectious diseases deposited in IEDB has shown remarkable predictive power.[145,146] Rather than identifying neoantigens that are immunogenic due to genuine cross reactivity with infectious disease epitopes to which the patient was exposed, this approach may implicitly combine an avoidance of self-peptides with biochemical properties that tend to make peptides immunogenic.

Unfortunately, few robust tools currently exist to perform the filtering approaches described here, although some approaches are straightforward to implement manually as needed. The tool *antigen.garnish*[103] is notable for its ability to annotate predicted neoantigens for DAI and similarity to T cell epitopes from IEDB.

5.5 MHC allele considerations

We highlight two considerations relating to the MHC genotype of a tumor that may inform neoantigen predictions. These factors are not considered by existing neoantigen pipelines to our knowledge, but we suggest that their inclusion may warrant investigation.

First, MHC alleles vary considerably in expression levels. Overall, HLA-C is expressed at less than one tenth the level of HLA-A or HLA-B.[147] There is also substantial variation at the allele level, attributed to differences in methylation across HLA-A alleles and differences in regulation by a micro-RNA for HLA-C alleles.[148,149] For example, HLA-A*24 alleles are expressed at nearly four times the level as HLA-A*03 alleles.[150] Neoantigen prediction pipelines may benefit from prioritizing putative neoantigens predicted to bind highly expressed MHC alleles. For this, HLA expression analysis in combination with HLA typing is desirable.[151]

Second, the MHC genotype may undergo somatic mutation in the tumor leading to deletion or loss of function. Alleles undergoing these changes in an individual may be best excluded or down-weighted when predicting neoantigens. Loss of all or part of an MHC haplotype is a relatively frequent occurrence; a tool called LOHHLA is capable of identifying these events.[152] Point mutations in MHC can be identified by the POLYSOLVER tool, whose authors also found evidence for loss of function mutations.[153]

6. Multi-step neoantigen prediction workflows

Many multi-step workflows have been developed integrating methods for performing various steps of the identification of neoantigen vaccine candidates as mentioned above (Table 3).

These tools are optimized for different use cases and offer varying features that reflect their divergent priorities for neoantigen selection. Many of these workflows require the user to perform a large amount of data preprocessing. For example, most do not support automated somatic or germline variant calling on raw sequencing data, expecting as input an existing variant call format (VCF) file that the user must generate. Many tools require the user to perform reference genome alignment and preprocessing of tumor RNA-Seq data.

There are several common features where existing neoantigen prediction workflows differ. Most pipelines handle SNV variants, but only some are able to predict neoantigens that arise from small indel variants and still

Table 3 Multi-step neoantigen prediction workflows.

Neoantigen pipeline	Tool used for phasing	URL
pVACtools[90]	Not implemented, pre-processing required with GATK ReadBackedPhasing	https://github.com/griffithlab/pVACtools
neoepiscope[91]	HapCUT2	https://github.com/pdxgx/neoepiscope
Epi-Seq[140]	RefHap[162]	https://dna.engr.uconn.edu/?page_id=470
Mupexi[156]	Not implemented	https://github.com/ambj/MuPeXI
TIminer[157]	Not implemented	https://icbi.i-med.ac.at/software/timiner/timiner.shtml
Vaxrank[163] as part of OpenVax	Isovar	https://github.com/openvax/vaxrank https://github.com/openvax

fewer support structural variants such as gene fusions. There is also variable support for variant phasing, filtering vs. ranking candidate neoepitopes, and self-similarity calculations. Some of these workflows focus mainly and more narrowly on neoantigen prediction, whereas others are broader tool aggregators for common cancer immunology data mining tasks. Below we describe some frequently used neoantigen prediction workflows.

6.1 pVACseq

pVACseq[154] is a well-documented neoantigen prediction tool which has recently been augmented and expanded into the pVACtools cancer immunotherapy tool suite. The program applies a set of user-specified filters sequentially, paring down the input variants to a set of passing predicted epitopes. Filter types and thresholds are configurable, which makes this a useful pipeline for experimenting with varying ways of predicting neoantigens.

The pVACseq tool, as well as the rest of the tools made available as part of the pVACtools suite, supports many useful features for neoantigen prediction. Supported filters include variant expression (using both bulk and allele-specific expression), variant allelic fraction (VAF), and peptide binding affinity using MHC class I and class II predictions. DNA-based proximal variant phasing is supported. Optional extensions to the main pVACseq prediction algorithm include stability predictions for predicted epitopes using NetMHCStabPan. To create vaccine contents from neoantigen predictions, pVACtools includes a tool called pVACvector which computes DNA-based vaccine sequences using the pVACseq epitope prediction. Lastly, while pVACseq can predict neoantigens resulting from SNVs and small indels, the PVACtools suite also includes a wrapper around a gene fusion variant caller (INTEGRATE-Neo) that predicts neoepitopes arising from fusion variants. This is a unique feature of this pipeline; most neoantigen prediction workflows have no support for structural variants.

A drawback to the pVACseq workflow is the relatively high amount of data preprocessing that must be performed before using this tool. While pVACseq supports many filters and types of processing, the user must provide the information to enable the tool to use it in neoantigen prediction. Like most pipelines described here, pVACseq requires the user to call somatic variants and annotate them with predicted protein effects. In order to use the filtering functionality related to variant expression and coverage, the user must provide the necessary data by first running other tools such as Kallisto[155] and bam-readcount (https://github.com/genome/bam-readcount) to estimate those quantities. Phased variants are handled correctly only if the user provides phasing information from GATK's ReadBacked-Phasing tool.[51] Additionally, the user must provide HLA types to pVACseq as in silico prediction using sequencing reads is not supported.

6.2 MuPeXI

The MuPeXI workflow[156] has a similar set of requirements to pVACseq: the user must provide somatic variants, gene expression estimates, and HLA types to the software. Conceptually, the MuPeXI epitope prediction approach is similar to that used by pVACseq, using similar features for filtering, and adds self-dissimilarity filtering criterion. One important difference is that instead of simply applying sequential filtering to a candidate neoepitope list, MuPeXI uses a multiplicative ranking function to create a ranked list of potential neoepitopes arising from the input variants. This built-in neoepitope ranking is useful particularly in cases where the user needs to prioritize possible neoepitopes for a vaccine. In addition to built-in neoepitope ranking, another advantage of the MuPeXI tool is its use of NetMHCpan 4.0[99] for MHC class I binding predictions, allowing the user to benefit from a predictor trained on both binding affinity and eluted ligand data.

There are several caveats to consider in using the MuPeXI workflow. Phasing proximal variants is not supported by the software at this point. Additionally, since ranking is applied on the level of minimal epitopes rather than the variants that give rise to those epitopes, the user must decide on an approach to aggregate the results in cases where multiple epitopes resulting from one variant are interleaved with epitopes from a different variant in the ranked list.

6.3 TIminer

The TIminer[157] tool, like most others, requires the user to input a preexisting set of somatic variants as input. One advantage over other workflows is that TIminer is able to process raw RNA-Seq data and extract several types of information that are relevant to neoantigen prediction. To that end, the TIminer software contains and runs many other tools, including Kallisto for estimating transcript expression, GSEA[158] and Immunophenogram[159] for measuring immune infiltrate, OptiType[97] for in silico HLA typing, and NetMHCpan 3.0 to determine MHC binding predictions for candidate neoantigens.

The final neoantigen prediction output of TIminer is almost identical to the info provided by NetMHCpan 3.0, one line for each predicted MHC ligand corresponding to an input variant. These results are limited to variants occurring in expressed genes, and augmented with an extra column for that gene expression information. Optionally, the tool also supports filtering by allele-specific expression, keeping only variants that are supported by least five overlapping RNA-Seq reads. The command-line version of the tool uses non-configurable thresholds; however, the authors provide a programmatic API that allows for more configuration. Finally, TIminer only supports neoantigen prediction for SNVs.

6.4 OpenVax

The OpenVax neoantigen prediction pipeline evolved from an earlier workflow called Epidisco,[160] which was more focused on computer cluster parallelism. The OpenVax pipeline automates read processing and somatic variant calling, supporting SNVs and indel variants. This is an end-to-end workflow that starts with raw DNA and RNA FASTQ data and generates mutation-containing peptides ranked for vaccine inclusion as the final output. Currently, HLA types must be provided to the software.

Similar to MuPeXI, the neoantigen prediction component of OpenVax uses a multiplicative ranking function. However, the function itself is simpler and contains only two features: variant-specific expression based on the supporting RNA-Seq read count, and an MHC class I affinity score.

This ranking is applied to variants as a whole, rather than treating the neoepitopes arising from those variants as separate entities. The OpenVax pipeline is configurable, allowing the user to specify which MHC class I binding predictor to use, as well as thresholds for MHC binding affinity and variant allele-specific expression.

The OpenVax workflow relies heavily on the RNA-Seq data for its neoantigen prediction. While somatic variants are called from DNA data, the mutant peptide sequence for each variant is assembled based on RNA reads. Additionally, because of this RNA sequence assembly method, variant phasing information is preserved.

The final result of the OpenVax pipeline is a set of ranked synthetic long peptides (SLPs) of a user-specified length, optimized for use in an SLP vaccine. We note that the OpenVax workflow can also be used simply for read processing and variant calling.

6.5 NeoEpiScope

The NeoEpiScope[91] workflow prioritizes handling phased variants correctly, both somatic/germline and somatic/somatic combinations. As for several tools, this workflow expects a set of somatic variants as input. In order to use the phasing functionality, the user must first run another external tool to assemble haplotypes from the DNA sequencing data, after performing germline variant calling. Currently, NeoEpiScope supports input from the results of GATK's ReadBackedPhasing and HapCUT2.[88]

Like most common tools, NeoEpiScope supports neoantigen prediction for SNVs and indels. It then uses phased variant information and one of several MHC binding predictors to compute candidate neoantigens.

6.6 Epi-Seq

Unlike the other workflows described here so far, Epi-Seq[140] works entirely with tumor RNA-Seq data, supporting variant calling from RNA and neoantigen prediction from those variants. This pipeline can be particularly useful in cases where only RNA-Seq data is available, and somatic variant calls from the DNA are unknown.

Epi-Seq is essentially a wrapper around two main tools: SNVQ[161] for RNA variant calling, and RefHap[162] for single individual haplotyping from which variant phasing is inferred. Only SNV variants are supported for neoantigen prediction. Epi-Seq processes one MHC allele at a time, and uses NetMHC 3.0 for binding affinity prediction. The tool also adds related variant information such as coverage of reference/alternate alleles, nearby mutations, and genotype. It does not appear to perform any neoantigen filtering, therefore acting as a neoantigen prediction information aggregator and leaving filtering decisions to the user.

7. Identifying neoantigen responsive T-cell receptors

The T-cell receptor (TCR) of CD8 and CD4 T cells recognizes peptides presented on HLA class I and class II molecules, respectively. The TCR alpha and beta sequences determine the TCR peptide-HLA recognition. The diversity of distinct TCR beta sequences in an individual, referred to as the TCR repertoire, is estimated in the range of 1×10^6 (Ref. 164) to 3×10^6 (Ref. 165).

Advancements in high throughput sequencing protocols and bioinformatic tools allows us to retrieve these repertoires, retrieve alpha-beta pairings, and deorphanize TCR pMHC recognition[166,167] (Table 4). For example, IMSEQ[168] was used to analyze melanoma antigen-specific TCR repertoires[179] and to define an immune DNA signature of T-cell infiltration in breast tumor exomes.[180] MiXCR[169] enabled the comparative analysis of murine TCR repertoires[181] and—together with tcr[170]—was used to examine TCR clonality of TILs in non-small cell lung cancer treated with PD-1 blockade.[182] Mass-spectrometry can directly detect HLA-presented peptides (i.e., the ligandome); the growing number of identified ligands enables us to understand the rules of peptide-MHC binding and presentation. Similarly, deciphering the TCR repertoire is enabling us to identify the rules of TCR-pMHC recognition, understand the tolerance-induced hole, and better understand what is immunogenic.[183]

Studies have identified tumor-recognizing T cells, particularly from TILs, recovered the associated TCR alpha-beta sequences, and used these sequences to identify the recognized tumor neoantigens. Curated catalogs of TCR alpha-beta sequences and the pMHC they recognize include VDJdb,[184] McPAS-TCR[185] and ATLAS.[186] Works leveraging these catalogs to find patterns mapping TCR sequences to pMHC complexes include[173–176,187] (Table 4). The integration of rules modeling pMHC-TCR interactions to prioritize neoantigens for the MHC alleles and the TCR repertoire in a patient has been done for HLA-A*02:01[177] and MHC class II epitopes[178] (Table 4).

Table 4 Tools for analysis of TCR sequences, antigen specificity from TCR sequences and pMHC-TCR interactions.

(a) TCR sequence identification	
IMSEQ[168]	http://www.imtools.org/
MiXCR[169]	https://github.com/milaboratory/mixcr
tcR[170]	http://imminfo.github.io/tcr/
LymAnalyzer[171]	https://sourceforge.net/projects/lymanalyzer/
RTCR[172]	http://uubram.github.io/RTCR/along
(b) Antigen specificity from TCR sequences	
tcr-dist[173]	github.com/phbradley/tcr-dist
GLIPH[174]	github.com/immunoengineer/gliph
TCRex[175]	https://tcrex.biodatamining.be/
TCRclusteringPaper[176]	github.com/pmeysman/TCRclusteringPaper
(c) pMHC-TCR interaction	
NetTCR[177]	http://www.cbs.dtu.dk/services/NetTCR/ https://github.com/mnielLab/netTCR
ITcell[178]	http://salilab.org/itcell

8. Conclusion

Genomic variations and inflammation are a hallmarks of cancer.[188] The advent of high throughput sequencing platforms generating big data in combination with the growing number of bioinformatic algorithms gives us the unique opportunity to detect somatic genetic alterations, phenotype the infiltrating lymphocytes, and to utilize the results for new generations of cancer immunotherapies.[6,189–191] Although studies in mouse models and in human clinical trials have shown that a substantial fraction of neoantigens identified by described workflows elicit therapeutic activity,[192–194] there are many opportunities to improve the accuracy and process.

Studies have highlighted the role of HLA class II neoantigens for therapeutic vaccination.[131,193] However, while prediction of HLA class I binding peptides yields high accuracies,[195,196] HLA class II binding prediction is still challenging. New MS datasets from naturally processed HLA class II ligands[197] will guide our understanding of HLA class II ligand processing[198] and presentation.[114,199] MS-derived MHC ligandome data are spread across publications: combining these data in a centralized location would be enabling. The IEDB[112] and the recently established project SysteMHC[200] are integrating and making MS MHC ligandome data available. While mutations are expressed on mRNA level according to their DNA frequency,[201] mRNA expression is not the same as protein expression[202,203] and thus proteogenomics provides valuable insights for neoantigen identification.[196] Neoantigen identification leverages peptide-HLA affinities to identify which peptides are presented. We expect benefit using a patient's TCR repertoire, including alpha-beta pairs, to identify which peptide-MHC complexes are recognized. Finally, as the platforms continue to improve, we may be able to directly identify neoantigens on each patient's tumor, including presented neoantigens[204,205] on class I[196,206] and class II molecules.[207]

In summary, the last decade has seen an explosion of innovative technological platforms (HTS and MS) generating massive datasets that, together with advanced analysis algorithms, have enabled us to better understand how the immune system recognizes tumor cells. We are now witnessing the development of platforms to profile single cells that will unleash novel treatment and diagnosis revolutions.

Acknowledgment

We thank John Castle for reviewing the manuscript and many discussions.

References

1. González S, Volkova N, Beer P, Gerstung M. Immuno-oncology from the perspective of somatic evolution. *Semin Cancer Biol.* 2018;52:75–85. https://doi.org/10.1016/j.semcancer.2017.12.001.
2. Schumacher TN, Scheper W, Kvistborg P. Cancer neoantigens. *Annu Rev Immunol.* 2018. https://doi.org/10.1146/annurev-immunol-042617-053402.
3. Castle JC, Kreiter S, Diekmann J, et al. Exploiting the mutanome for tumor vaccination. *Cancer Res.* 2012;72:1081–1091. https://doi.org/10.1158/0008-5472.CAN-11-3722.
4. Schumacher TN, Schreiber RD. Neoantigens in cancer immunotherapy. *Science.* 2015;348:69–74. https://doi.org/10.1126/science.aaa4971.
5. Sahin U, Türeci Ö. Personalized vaccines for cancer immunotherapy. *Science.* 2018;359:1355–1360. https://doi.org/10.1126/science.aar7112.
6. Vormehr M, Diken M, Boegel S, Kreiter S, Türeci Ö, Sahin U. Mutanome directed cancer immunotherapy. *Curr Opin Immunol.* 2016;39:14–22. https://doi.org/10.1016/j.coi.2015.12.001.
7. Gattinoni L. Adoptive T cell transfer: imagining the next generation of cancer immunotherapies. *Semin Immunol.* 2016;28:1–2. https://doi.org/10.1016/j.smim.2016.03.019.

8. Cohen CJ, Gartner JJ, Horovitz-Fried M, et al. Isolation of neoantigen-specific T cells from tumor and peripheral lymphocytes. *J Clin Invest.* 2015;125:3981–3991. https://doi.org/10.1172/JCI82416.
9. Ribas A, Wolchok JD. Cancer immunotherapy using checkpoint blockade. *Science.* 2018;359:1350–1355. https://doi.org/10.1126/science.aar4060.
10. Gubin MM, Zhang X, Schuster H, et al. Checkpoint blockade cancer immunotherapy targets tumour-specific mutant antigens. *Nature.* 2014;515:577–581. https://doi.org/10.1038/nature13988.
11. Rizvi NA, Hellmann MD, Snyder A, et al. Cancer immunology. Mutational landscape determines sensitivity to PD-1 blockade in non-small cell lung cancer. *Science.* 2015;348:124–128. https://doi.org/10.1126/science.aaa1348.
12. Brown SD, Warren RL, Gibb EA, et al. Neo-antigens predicted by tumor genome meta-analysis correlate with increased patient survival. *Genome Res.* 2014;24:743–750. https://doi.org/10.1101/gr.165985.113.
13. Yarchoan M, Hopkins A, Jaffee EM. Tumor mutational burden and response rate to PD-1 inhibition. *N Engl J Med.* 2017;377:2500–2501. https://doi.org/10.1056/NEJMc1713444.
14. Chowell D, Morris LGT, Grigg CM, et al. Patient HLA class I genotype influences cancer response to checkpoint blockade immunotherapy. *Science.* 2018;359:582–587. https://doi.org/10.1126/science.aao4572.
15. Roemer MGM, Redd RA, Cader FZ, et al. Major histocompatibility complex class II and programmed death ligand 1 expression predict outcome after programmed death 1 blockade in classic Hodgkin lymphoma. *J Clin Oncol.* 2018;36:942–950. https://doi.org/10.1200/JCO.2017.77.3994.
16. Danilova L, Anagnostou V, Caushi JX, et al. The mutation-associated neoantigen functional expansion of specific T cells (MANAFEST) assay: a sensitive platform for monitoring antitumor immunity. *Cancer Immunol Res.* 2018;6:888–899. https://doi.org/10.1158/2326-6066.CIR-18-0129.
17. Peng S, Zaretsky JM, Bethune MT, et al. Sensitive, non-destructive detection and analysis of neoantigen-specific T cell populations from tumors and blood. *Cell.* 2018; https://doi.org/10.2139/ssrn.3155791.
18. Mahanty S, Prigent A, Garraud O. Immunogenicity of infectious pathogens and vaccine antigens. *BMC Immunol.* 2015;16:31. https://doi.org/10.1186/s12865-015-0095-y.
19. Srivastava PK, Duan F. Harnessing the antigenic fingerprint of each individual cancer for immunotherapy of human cancer: genomics shows a new way and its challenges. *Cancer Immunol Immunother.* 2013;62:967–974. https://doi.org/10.1007/s00262-013-1422-x, Springer-Verlag.
20. Murphy KM. *Janeway's Immunobiology.* Internet, Garland Science; 2011. Available: https://books.google.com/books/about/Janeway_s_Immunobiology.html?hl=&id=WDMmAgAAQBAJ.
21. Blankenstein T, Coulie PG, Gilboa E, Jaffee EM. The determinants of tumour immunogenicity. *Nat Rev Cancer.* 2012;12:307–313. https://doi.org/10.1038/nrc3246.
22. Britten CM, van der Burg SH, Gouttefangeas C. A framework for T cell assays. *Oncotarget.* 2015;6:35143–35144. https://doi.org/10.18632/oncotarget.6181.
23. Vormehr M, Reinhard K, Blatnik R, et al. A non-functional neoepitope specific CD8 T-cell response induced by tumor derived antigen exposure. *Oncoimmunology.* 2019;8:1553478. https://doi.org/10.1080/2162402X.2018.1553478.
24. Klatt MG, Mun SS, Socci ND, Korontsvit T, Dao T, Scheinberg DA. Epigenetic drug treatment induces presentation of new class of non-exonic, cryptic neoantigens in acute myeloid leukemia cells. *Blood.* 2018;132:2717. https://doi.org/10.1182/blood-2018-99-113691, American Society of Hematology.

25. Laumont CM, Vincent K, Hesnard L, et al. Noncoding regions are the main source of targetable tumor-specific antigens. *Sci Transl Med*. 2018;10. https://doi.org/10.1126/scitranslmed.aau5516, eaau5516.
26. Cobbold M, De La Peña H, Norris A, et al. MHC class I-associated phosphopeptides are the targets of memory-like immunity in leukemia. *Sci Transl Med*. 2013;5:203ra125. https://doi.org/10.1126/scitranslmed.3006061.
27. Penny S, Malaker S, Steadman L, et al. Glycosylated and methylated peptides as neoantigens in leukaemia. *Eur J Cancer*. 2016;61:S217. https://doi.org/10.1016/S0959-8049(16)61765-3.
28. Mylonas R, Beer I, Iseli C, et al. Estimating the contribution of proteasomal spliced peptides to the HLA-I ligandome. *Mol Cell Proteomics*. 2018;17:2347–2357. https://doi.org/10.1074/mcp.RA118.000877.
29. Liepe J, Marino F, Sidney J, et al. A large fraction of HLA class I ligands are proteasome-generated spliced peptides. *Science*. 2016;354:354–358. https://doi.org/10.1126/science.aaf4384.
30. Creech AL, Ting YS, Goulding SP, et al. The role of mass spectrometry and proteogenomics in the advancement of HLA epitope prediction. *Proteomics*. 2018;1700259:1–10. https://doi.org/10.1002/pmic.201700259.
31. Rhoads A, Au KF. PacBio sequencing and its applications. *Genomics Proteomics Bioinformatics*. 2015;13:278–289. https://doi.org/10.1016/j.gpb.2015.08.002.
32. Brown CG, Clarke J. Nanopore development at Oxford nanopore. *Nat Biotechnol*. 2016;34:810–811. https://doi.org/10.1038/nbt.3622.
33. Fang H, Wu Y, Narzisi G, et al. Reducing INDEL calling errors in whole genome and exome sequencing data. *Genome Med*. 2014;6:89https://doi.org/10.1186/s13073-014-0089-z.
34. Belkadi A, Bolze A, Itan Y, et al. Whole-genome sequencing is more powerful than whole-exome sequencing for detecting exome variants. *Proc Natl Acad Sci USA*. 2015;112:5473–5478. https://doi.org/10.1073/pnas.1418631112.
35. Nam J-Y, Kim NKD, Kim SC, et al. Evaluation of somatic copy number estimation tools for whole-exome sequencing data. *Brief Bioinform*. 2016;17:185–192. https://doi.org/10.1093/bib/bbv055.
36. Yang L, Lee M-S, Lu H, et al. Analyzing somatic genome rearrangements in human cancers by using whole-exome sequencing. *Am J Hum Genet*. 2016;98:843–856. https://doi.org/10.1016/j.ajhg.2016.03.017.
37. Tattini L, D'Aurizio R, Magi A. Detection of genomic structural variants from next-generation sequencing data. *Front Bioeng Biotechnol*. 2015;3:92. https://doi.org/10.3389/fbioe.2015.00092.
38. Griffith M, Miller CA, Griffith OL, et al. Optimizing cancer genome sequencing and analysis. *Cell Syst*. 2015;1:210–223. https://doi.org/10.1016/j.cels.2015.08.015.
39. Li H. Toward better understanding of artifacts in variant calling from high-coverage samples. *Bioinformatics*. 2014;30:2843–2851. https://doi.org/10.1093/bioinformatics/btu356.
40. Guo Y, Dai Y, Yu H, Zhao S, Samuels DC, Shyr Y. Improvements and impacts of GRCh38 human reference on high throughput sequencing data analysis. *Genomics*. 2017;109:83–90. https://doi.org/10.1016/j.ygeno.2017.01.005.
41. Schneider VA, Graves-Lindsay T, Howe K, et al. Evaluation of GRCh38 and de novo haploid genome assemblies demonstrates the enduring quality of the reference assembly. *Genome Res*. 2017;27:849–864. https://doi.org/10.1101/gr.213611.116.
42. Cao C, Mak L, Jin G, Gordon P, Ye K, Long Q. PRESM: personalized reference editor for somatic mutation discovery in cancer genomics. *Bioinformatics*. 2018;35:1445–1452.
43. Munger SC, Raghupathy N, Choi K, et al. RNA-Seq alignment to individualized genomes improves transcript abundance estimates in multiparent populations. *Genetics*. 2014;198:59–73. https://doi.org/10.1534/genetics.114.165886.

44. Hellmann MD, Nathanson T, Rizvi H, et al. Genomic features of response to combination immunotherapy in patients with advanced non-small-cell lung cancer. *Cancer Cell*. 2018;33:843–852.e4. https://doi.org/10.1016/j.ccell.2018.03.018.
45. Cibulskis K, Lawrence MS, Carter SL, et al. Sensitive detection of somatic point mutations in impure and heterogeneous cancer samples. *Nat Biotechnol*. 2013;31:213–219. https://doi.org/10.1038/nbt.2514.
46. Saunders CT, Wong WSW, Swamy S, Becq J, Murray LJ, Cheetham RK. Strelka: accurate somatic small-variant calling from sequenced tumor-normal sample pairs. *Bioinformatics*. 2012;28:1811–1817. https://doi.org/10.1093/bioinformatics/bts271.
47. Kim S, Scheffler K, Halpern AL, et al. Strelka2: fast and accurate calling of germline and somatic variants. *Nat Methods*. 2018;15:591–594. https://doi.org/10.1038/s41592-018-0051-x.
48. Fang H, Bergmann EA, Arora K, et al. Indel variant analysis of short-read sequencing data with scalpel. *Nat Protoc*. 2016;11:2529–2548. https://doi.org/10.1038/nprot.2016.150.
49. Narzisi G, Corvelo A, Arora K, et al. Genome-wide somatic variant calling using localized colored de Bruijn graphs. *Commun Biol*. 2018;1:20. https://doi.org/10.1038/s42003-018-0023-9.
50. Mose LE, Perou CM, Parker JS. Improved Indel detection in DNA and RNA via realignment with ABRA2. *Bioinformatics*. 2019. https://doi.org/10.1093/bioinformatics/btz033.
51. Van der Auwera GA, Carneiro MO, Hartl C, et al. From FastQ data to high confidence variant calls: the genome analysis toolkit best practices pipeline. *Curr Protoc Bioinformatics*. 2013;43:11.10.1–11.10.33. https://doi.org/10.1002/0471250953.bi1110s43.
52. McPherson A, Hormozdiari F, Zayed A, et al. deFuse: an algorithm for gene fusion discovery in tumor RNA-Seq data. *PLoS Comput Biol*. 2011;7:e1001138. https://doi.org/10.1371/journal.pcbi.1001138.
53. Zhang J, White NM, Schmidt HK, et al. INTEGRATE: gene fusion discovery using whole genome and transcriptome data. *Genome Res*. 2016;26:108–118. https://doi.org/10.1101/gr.186114.114.
54. Tang S, Madhavan S. neoantigenR: an annotation based pipeline for tumor neoantigen identification from sequencing data. *bioRxiv*. 2017;171843. https://doi.org/10.1101/171843.
55. Kahles A, Ong CS, Zhong Y, Rätsch G. SplAdder: identification, quantification and testing of alternative splicing events from RNA-Seq data. *Bioinformatics*. 2016;32:1840–1847. https://doi.org/10.1093/bioinformatics/btw076.
56. Jayasinghe RG, Cao S, Gao Q, et al. Systematic analysis of splice-site-creating mutations in cancer. *Cell Rep*. 2018;23:270–281.e3. https://doi.org/10.1016/j.celrep.2018.03.052.
57. Pimentel H, Conboy JG, Pachter L. Keep me around: intron retention detection and analysis. *arXiv [q-bioGN]*. 2015. Available, http://arxiv.org/abs/1510.00696.
58. Li H-D, Funk CC, Price ND. iREAD: a tool for intron retention detection from RNA-seq data. *bioRxiv*. 2017;135624. https://doi.org/10.1101/135624.
59. Cheng J, Nguyen TYD, Cygan KJ, et al. MMSplice: modular modeling improves the predictions of genetic variant effects on splicing. *Genome Biol*. 2019;20:48. https://doi.org/10.1186/s13059-019-1653-z.
60. Ouyang Z, Liu F, Zhao C, et al. Accurate identification of RNA editing sites from primitive sequence with deep neural networks. *Sci Rep*. 2018;8:6005. https://doi.org/10.1038/s41598-018-24298-y.
61. Picardi E, Pesole G. REDItools: high-throughput RNA editing detection made easy. *Bioinformatics*. 2013;29:1813–1814. https://doi.org/10.1093/bioinformatics/btt287.
62. John D, Weirick T, Dimmeler S, Uchida S. RNAEditor: easy detection of RNA editing events and the introduction of editing islands. *Brief Bioinform*. 2017;18:993–1001. https://doi.org/10.1093/bib/bbw087.

63. Wang Z, Lian J, Li Q, et al. RES-scanner: a software package for genome-wide identification of RNA-editing sites. *Gigascience*. 2016;5:37. https://doi.org/10.1186/s13742-016-0143-4.
64. Smith CC, Beckermann KE, Bortone DS, et al. Endogenous retroviral signatures predict immunotherapy response in clear cell renal cell carcinoma. *J Clin Invest*. 2018;128:4804–4820. https://doi.org/10.1172/JCI121476.
65. Yizhak K, Aguet F, Kim J, et al. A comprehensive analysis of RNA sequences reveals macroscopic somatic clonal expansion across normal tissues. *bioRxiv*. 2018;416339. https://doi.org/10.1101/416339.
66. Coudray A, Battenhouse AM, Bucher P, Iyer VR. Detection and benchmarking of somatic mutations in cancer genomes using RNA-seq data. *PeerJ*. 2018;6:e5362. https://doi.org/10.7717/peerj.5362.
67. O'Rawe J, Jiang T, Sun G, et al. Low concordance of multiple variant-calling pipelines: practical implications for exome and genome sequencing. *Genome Med*. 2013;5:28. https://doi.org/10.1186/gm432.
68. Dixon JR, Xu J, Dileep V, et al. Integrative detection and analysis of structural variation in cancer genomes. *Nat Genet*. 2018;50:1388–1398. https://doi.org/10.1038/s41588-018-0195-8.
69. Pawelec G, Kalbacher H, Bruserud O. Tumor-specific antigens revisited: presentation to the immune system of fusion peptides resulting solely from tumor-specific chromosomal translocations. *Oncol Res*. 1992;4:315–320. Available, https://www.ncbi.nlm.nih.gov/pubmed/1486216.
70. Worley BS, van den Broeke LT, Goletz TJ, et al. Antigenicity of fusion proteins from sarcoma-associated chromosomal translocations. *Cancer Res*. 2001;61:6868–6875. Available, https://www.ncbi.nlm.nih.gov/pubmed/11559563.
71. Gao Q, Liang W-W, Foltz SM, et al. Driver fusions and their implications in the development and treatment of human cancers. *Cell Rep*. 2018;23:227–238.e3. https://doi.org/10.1016/j.celrep.2018.03.050.
72. Rathe SK, Popescu FE, Johnson JE, et al. Identification of candidate neoantigens produced by fusion transcripts in human osteosarcomas. *Sci Rep*. 2019;9:358. https://doi.org/10.1038/s41598-018-36840-z.
73. Haas BJ, Papanicolaou A, Yassour M, et al. De novo transcript sequence reconstruction from RNA-seq using the Trinity platform for reference generation and analysis. *Nat Protoc*. 2013;8:1494–1512. https://doi.org/10.1038/nprot.2013.084.
74. Zhang J, Mardis ER, Maher CA. INTEGRATE-neo: a pipeline for personalized gene fusion neoantigen discovery. *Bioinformatics*. 2017;33:555–557. https://doi.org/10.1093/bioinformatics/btw674.
75. Castle JC, Zhang C, Shah JK, et al. Expression of 24,426 human alternative splicing events and predicted cis regulation in 48 tissues and cell lines. *Nat Genet*. 2008;40:1416–1425. https://doi.org/10.1038/ng.264.
76. El Marabti E, Younis I. The cancer spliceome: reprograming of alternative splicing in cancer. *Front Mol Biosci*. 2018;5:80. https://doi.org/10.3389/fmolb.2018.00080.
77. Lupetti R, Pisarra P, Verrecchia A, et al. Translation of a retained intron in tyrosinase-related protein (TRP) 2 mRNA generates a new cytotoxic T lymphocyte (CTL)-defined and shared human melanoma antigen not expressed in normal cells of the melanocytic lineage. *J Exp Med*. 1998;188:1005–1016. Available, https://www.ncbi.nlm.nih.gov/pubmed/9743519.
78. Dvinge H, Bradley RK. Widespread intron retention diversifies most cancer transcriptomes. *Genome Med*. 2015;7:45. https://doi.org/10.1186/s13073-015-0168-9.
79. Smart AC, Margolis CA, Pimentel H, et al. Intron retention is a source of neoepitopes in cancer. *Nat Biotechnol*. 2018;36:1056–1058. https://doi.org/10.1038/nbt.4239.

80. Kahles A, Lehmann K-V, Toussaint NC, et al. Comprehensive analysis of alternative splicing across tumors from 8,705 patients. *Cancer Cell*. 2018;34:211–224.e6. https://doi.org/10.1016/j.ccell.2018.07.001.
81. Chen S, Huang V, Xu X, et al. Widespread and functional RNA circularization in localized prostate cancer. *Cell*. 2019;176:831–843.e22. https://doi.org/10.1016/j.cell.2019.01.025.
82. Koch L. RNA: translated circular RNAs. Internet, *Nat Rev Genet*. 2017;272–273. https://doi.org/10.1038/nrg.2017.27.
83. Dong R, Ma X-K, Chen L-L, Yang L. Genome-wide annotation of circRNAs and their alternative back-splicing/splicing with CIRCexplorer pipeline. *Methods Mol Biol* 2019;1870:137–149. https://doi.org/10.1007/978-1-4939-8808-2_10.
84. Eisenberg E, Levanon EY. A-to-I RNA editing—immune protector and transcriptome diversifier. *Nat Rev Genet*. 2018;19:473–490. https://doi.org/10.1038/s41576-018-0006-1.
85. Paz-Yaacov N, Bazak L, Buchumenski I, et al. Elevated RNA editing activity is a major contributor to transcriptomic diversity in tumors. *Cell Rep*. 2015;13:267–276. https://doi.org/10.1016/j.celrep.2015.08.080.
86. Zhang M, Fritsche J, Roszik J, et al. RNA editing derived epitopes function as cancer antigens to elicit immune responses. *Nat Commun*. 2018;9:3919. https://doi.org/10.1038/s41467-018-06405-9.
87. Roth SH, Danan-Gotthold M, Ben-Izhak M, et al. Increased RNA editing may provide a source for autoantigens in systemic lupus erythematosus. *Cell Rep*. 2018;23:50–57. https://doi.org/10.1016/j.celrep.2018.03.036.
88. Edge P, Bafna V, Bansal V. HapCUT2: robust and accurate haplotype assembly for diverse sequencing technologies. *Genome Res*. 2017;27:801–812. https://doi.org/10.1101/gr.213462.116.
89. Castel SE, Mohammadi P, Chung WK, Shen Y, Lappalainen T. Rare variant phasing and haplotypic expression from RNA sequencing with phASER. *Nat Commun*. 2016;7:12817. https://doi.org/10.1038/ncomms12817.
90. Hundal J, Kiwala S, Feng Y-Y, et al. Accounting for proximal variants improves neoantigen prediction. *Nat Genet*. 2019;51:175–179. https://doi.org/10.1038/s41588-018-0283-9.
91. Wood MA, Nguyen A, Struck AJ, Ellrott K, Nellore A, Thompson RF. Neoepiscope improves neoepitope prediction with multi-variant phasing. *bioRxiv*. 2018;418129. https://doi.org/10.1101/418129.
92. Robinson J, Halliwell JA, Hayhurst JD, Flicek P, Parham P, Marsh SGE. The IPD and IMGT/HLA database: allele variant databases. *Nucleic Acids Res*. 2015;43:D423–D431. https://doi.org/10.1093/nar/gku1161.
93. Sette A, Buus S, Appella E, et al. Prediction of major histocompatibility complex binding regions of protein antigens by sequence pattern analysis. *Proc Natl Acad Sci USA*. 1989;86:3296–3300. https://doi.org/10.1073/pnas.86.9.3296.
94. Falk K, Rötzschke O, Stevanović S, Jung G, Rammensee HG. Allele-specific motifs revealed by sequencing of self-peptides eluted from MHC molecules. *Nature*. 1991;351:290–296. https://doi.org/10.1038/351290a0.
95. Konstantinou GN. T-cell epitope prediction. In: Lin J, Alcocer M, eds. *Food Allergens: Methods and Protocols*. New York, NY: Springer New York; 2017:211–222. https://doi.org/10.1007/978-1-4939-6925-8_17.
96. Soria-Guerra RE, Nieto-Gomez R, Govea-Alonso DO, Rosales-Mendoza S. An overview of bioinformatics tools for epitope prediction: implications on vaccine development. *J Biomed Inform*. 2015;53:405–414. https://doi.org/10.1016/j.jbi.2014.11.003.

97. Szolek A, Schubert B, Mohr C, Sturm M, Feldhahn M, Kohlbacher O. OptiType: precision HLA typing from next-generation sequencing data. *Bioinformatics*. 2014; 30:3310–3316. https://doi.org/10.1093/bioinformatics/btu548.
98. Boegel S, Löwer M, Schäfer M, et al. HLA typing from RNA-Seq sequence reads. *Genome Med*. 2012;4:102. https://doi.org/10.1186/gm403.
99. Jurtz V, Paul S, Andreatta M, Marcatili P, Peters B, Nielsen M. NetMHCpan-4.0: improved peptide–MHC class I interaction predictions integrating eluted ligand and peptide binding affinity data. *J Immunol*. 2017;199:3360–3368, ji1700893. https://doi.org/10.4049/jimmunol.1700893.
100. Nielsen M, Lundegaard C, Lund O, Keşmir C. The role of the proteasome in generating cytotoxic T-cell epitopes: insights obtained from improved predictions of proteasomal cleavage. *Immunogenetics*. 2005;57:33–41. https://doi.org/10.1007/s00251-005-0781-7.
101. Jensen KK, Andreatta M, Marcatili P, et al. Improved methods for predicting peptide binding affinity to MHC class II molecules. *Immunology*. 2018;154:394–406. https://doi.org/10.1111/imm.12889.
102. Racle J, Michaux J, Rockinger GA, et al. Deep motif deconvolution of HLA-II peptidomes for robust class II epitope predictions. *bioRxiv*. 2019;539338. https://doi.org/10.1101/539338.
103. Rech AJ, Balli D, Mantero A, et al. Tumor immunity and survival as a function of alternative neopeptides in human cancer. *Cancer Immunol Res*. 2018;6(3):276–287. https://doi.org/10.1158/2326-6066.CIR-17-0559.
104. Xie C, Yeo ZX, Wong M, et al. Fast and accurate HLA typing from short-read next-generation sequence data with xHLA. *Proc Natl Acad Sci USA*. 2017;114:8059–8064. https://doi.org/10.1073/pnas.1707945114.
105. Kawaguchi S, Higasa K, Shimizu M, Yamada R, Matsuda F. HLA-HD: an accurate HLA typing algorithm for next-generation sequencing data. *Hum Mutat*. 2017;38:788–797, Available: https://onlinelibrary.wiley.com/doi/full/10.1002/humu.23230.
106. Bai Y, Wang D, Fury WPHLAT. Inference of high-resolution HLA types from RNA and whole exome sequencing. In: Boegel S, ed. *HLA Typing: Methods and Protocols*. New York, NY: Springer New York; 2018:193–201. https://doi.org/10.1007/978-1-4939-8546-3_13.
107. Nielsen M, Lundegaard C, Worning P, et al. Reliable prediction of T-cell epitopes using neural networks with novel sequence representations. *Protein Sci*. 2003;12: 1007–1017. https://doi.org/10.1110/ps.0239403.
108. Nielsen M, Lundegaard C, Blicher T, et al. Quantitative predictions of peptide binding to any HLA-DR molecule of known sequence: NetMHCIIpan. *PLoS Comput Biol*. 2008;4:e1000107. https://doi.org/10.1371/journal.pcbi.1000107.
109. Hoof I, Peters B, Sidney J, et al. NetMHCpan, a method for MHC class I binding prediction beyond humans. *Immunogenetics*. 2009;61:1–13. https://doi.org/10.1007/s00251-008-0341-z.
110. Andreatta M, Karosiene E, Rasmussen M, Stryhn A, Buus S, Nielsen M. Accurate pan-specific prediction of peptide-MHC class II binding affinity with improved binding core identification. *Immunogenetics*. 2015;67:641–650. https://doi.org/10.1007/s00251-015-0873-y.
111. Nielsen M, Andreatta M. NetMHCpan-3.0; improved prediction of binding to MHC class I molecules integrating information from multiple receptor and peptide length datasets. *Genome Med*. 2016;8https://doi.org/10.1186/s13073-016-0288-x.
112. Vita R, Mahajan S, Overton JA, et al. The immune epitope database (IEDB): 2018 update. *Nucleic Acids Res*. 2019;47:D339–D343. https://doi.org/10.1093/nar/gky1006.

113. O'Donnell TJ, Rubinsteyn A, Bonsack M, Riemer AB, Laserson U, Hammerbacher J. MHCflurry: open-source class I MHC binding affinity prediction. *Cell Syst.* 2018;7:1–4. https://doi.org/10.1016/j.cels.2018.05.014. Elsevier.
114. Bassani-Sternberg M, Chong C, Guillaume P, et al. Deciphering HLA-I motifs across HLA peptidomes improves neo-antigen predictions and identifies allostery regulating HLA specificity. *PLoS Comput Biol.* 2017;13:e1005725. https://doi.org/10.1371/journal.pcbi.1005725.
115. Boehm KM, Bhinder B, Raja VJ, Dephoure N, Elemento O. Predicting peptide presentation by major histocompatibility complex class I: an improved machine learning approach to the immunopeptidome. *BMC Bioinf.* 2019;20:7. https://doi.org/10.1186/s12859-018-2561-z.
116. Antes I, Siu SWI, Lengauer T. DynaPred: a structure and sequence based method for the prediction of MHC class I binding peptide sequences and conformations. *Bioinformatics.* 2006;22:e16–e24. https://doi.org/10.1093/bioinformatics/btl216.
117. Fagerberg T, Cerottini J-C, Michielin O. Structural prediction of peptides bound to MHC class I. *J Mol Biol.* 2006;356:521–546. https://doi.org/10.1016/j.jmb.2005.11.059.
118. Sette A, Vitiello A, Reherman B, et al. The relationship between class I binding affinity and immunogenicity of potential cytotoxic T cell epitopes. *J Immunol.* 1994;153:5586–5592, Available, https://www.ncbi.nlm.nih.gov/pubmed/7527444.
119. Paul S, Weiskopf D, Angelo MA, Sidney J, Peters B, Sette A. HLA class I alleles are associated with peptide-binding repertoires of different size, affinity, and immunogenicity. *J Immunol.* 2013;191:5831–5839. https://doi.org/10.4049/jimmunol.1302101.
120. Harndahl M, Rasmussen M, Roder G, et al. Peptide-MHC class I stability is a better predictor than peptide affinity of CTL immunogenicity. *Eur J Immunol.* 2012;42: 1405–1416. https://doi.org/10.1002/eji.201141774.
121. Strønen E, Toebes M, Kelderman S, et al. Targeting of cancer neoantigens with donor-derived T cell receptor repertoires. *Science.* 2016;352:1337–1341. https://doi.org/10.1126/science.aaf2288.
122. Jørgensen KW, Rasmussen M, Buus S, Nielsen M. NetMHCstab—predicting stability of peptide-MHC-I complexes; impacts for cytotoxic T lymphocyte epitope discovery. *Immunology.* 2014;141:18–26. https://doi.org/10.1111/imm.12160.
123. Blaha DT, Anderson SD, Yoakum DM, et al. High-throughput stability screening of neoantigen/HLA complexes improves immunogenicity predictions. *Cancer Immunol Res.* 2019;7:50–61. https://doi.org/10.1158/2326-6066.CIR-18-0395.
124. Diez-Rivero CM, Chenlo B, Zuluaga P, Reche PA. Quantitative modeling of peptide binding to TAP using support vector machine. *Proteins.* 2010;78:63–72, Available: https://onlinelibrary.wiley.com/doi/abs/10.1002/prot.22535.
125. Zhang GL, Petrovsky N, Kwoh CK, August JT, Brusic V. PRED TAP: a system for prediction of peptide binding to the human transporter associated with antigen processing. *Immunome Res.* 2006;2:3. Available, https://immunome-research.biomedcentral.com/articles/10.1186/1745-7580-2-3.
126. Peters B, Bulik S, Tampe R, Van Endert PM, Holzhütter H-G. Identifying MHC class I epitopes by predicting the TAP transport efficiency of epitope precursors. *J Immunol.* 2003;171:1741–1749. Available, https://www.ncbi.nlm.nih.gov/pubmed/12902473.
127. Abelin JG, Keskin DB, Sarkizova S, et al. Mass spectrometry profiling of HLA-associated peptidomes in mono-allelic cells enables more accurate epitope prediction. *Immunity.* 2017;46:315–326. https://doi.org/10.1016/j.immuni.2017.02.007.
128. Novy P, Quigley M, Huang X, Yang Y. CD4 T cells are required for CD8 T cell survival during both primary and memory recall responses. *J Immunol.* 2007;179:8243–8251. Available, https://www.ncbi.nlm.nih.gov/pubmed/18056368.

129. Hu Z, Molloy MJ, Usherwood EJ. CD4(+) T-cell dependence of primary CD8(+) T-cell response against vaccinia virus depends upon route of infection and viral dose. *Cell Mol Immunol.* 2016;13:82–93. https://doi.org/10.1038/cmi.2014.128.
130. Linnemann C, van Buuren MM, Bies L, et al. High-throughput epitope discovery reveals frequent recognition of neo-antigens by CD4+ T cells in human melanoma. *Nat Med.* 2014;21:1–7. https://doi.org/10.1038/nm.3773.
131. Kreiter S, Vormehr M, van de Roemer N, et al. Mutant MHC class II epitopes drive therapeutic immune responses to cancer. *Nature.* 2015;520:692–696. https://doi.org/10.1038/nature14426.
132. Khodadoust MS, Olsson N, Wagar LE, et al. Antigen presentation profiling reveals recognition of lymphoma immunoglobulin neoantigens. *Nature.* 2017;543:723–727. https://doi.org/10.1038/nature21433.
133. Brown JH, Jardetzky TS, Gorga JC, et al. Three-dimensional structure of the human class II histocompatibility antigen HLA-DR1. *Nature.* 1993;364:33–39. https://doi.org/10.1038/364033a0.
134. Barra C, Alvarez B, Paul S, et al. Footprints of antigen processing boost MHC class II natural ligand predictions. *Genome Med.* 2018;10:84. https://doi.org/10.1186/s13073-018-0594-6.
135. Southwood S, Sidney J, Kondo A, et al. Several common HLA-DR types share largely overlapping peptide binding repertoires. *J Immunol.* 1998;160:3363–3373. Available, https://www.ncbi.nlm.nih.gov/pubmed/9531296.
136. Alvarez B, Reynisson B, Barra C, et al. NNAlign_MA; semi-supervised MHC peptidome deconvolution for accurate characterization of MHC binding motifs and improved T cell epitope prediction. *bioRxiv.* 2019;550673, Cold Spring Harbor Laboratory. Available, https://www.researchgate.net/profile/Bruno_Alvarez/publication/331151814_NNAlign_MA_semi-supervised_MHC_peptidome_deconvolution_for_accurate_characterization_of_MHC_binding_motifs_and_improved_T_cell_epitope_prediction/links/5c6ade8c4585156b570695f1/NNAlign-MA-semi-supervised-MHC-peptidome-deconvolution-for-accurate-characterization-of-MHC-binding-motifs-and-improved-T-cell-epitope-prediction.pdf.
137. Graham DB, Luo C, O'Connell DJ, et al. Antigen discovery and specification of immunodominance hierarchies for MHCII-restricted epitopes. *Nat Med.* 2018;24:1762–1772. https://doi.org/10.1038/s41591-018-0203-7.
138. Xing Y, Hogquist KA. T-cell tolerance: central and peripheral. *Cold Spring Harb Perspect Biol.* 2012;4:a006957. https://doi.org/10.1101/cshperspect.a006957.
139. Calis JJA, de Boer RJ, Keşmir C. Degenerate T-cell recognition of peptides on MHC molecules creates large holes in the T-cell repertoire. *PLoS Comput Biol.* 2012;8:e1002412. https://doi.org/10.1371/journal.pcbi.1002412.
140. Duan F, Duitama J, Al Seesi S, et al. Genomic and bioinformatic profiling of mutational neoepitopes reveals new rules to predict anticancer immunogenicity. *J Exp Med.* 2014;211:2231–2248. https://doi.org/10.1084/jem.20141308.
141. Bjerregaard A-M, Nielsen M, Jurtz V, et al. An analysis of natural T cell responses to predicted tumor neoepitopes. *Front Immunol.* 2017;8:1566. https://doi.org/10.3389/fimmu.2017.01566.
142. Ghorani E, Rosenthal R, McGranahan N, et al. Differential binding affinity of mutated peptides for MHC class I is a predictor of survival in advanced lung cancer and melanoma. *Ann Oncol.* 2018;29:271–279. https://doi.org/10.1093/annonc/mdx687.
143. Koşaloğlu-Yalçın Z, Lanka M, Frentzen A, et al. Predicting T cell recognition of MHC class I restricted neoepitopes. *Oncoimmunology.* 2018;7:e1492508. https://doi.org/10.1080/2162402X.2018.1492508.
144. Chowell D, Krishna S, Becker PD, et al. TCR contact residue hydrophobicity is a hallmark of immunogenic CD8+ T cell epitopes. *Proc Natl Acad Sci USA.* 2015;112:E1754–E1762. https://doi.org/10.1073/pnas.1500973112.

145. Balachandran VP, Łuksza M, Zhao JN, et al. Identification of unique neoantigen qualities in long-term survivors of pancreatic cancer. *Nature*. 2017;551:512–516. https://doi.org/10.1038/nature24462.
146. Łuksza M, Riaz N, Makarov V, et al. A neoantigen fitness model predicts tumour response to checkpoint blockade immunotherapy. *Nature*. 2017;551:517–520. https://doi.org/10.1038/nature24473.
147. Apps R, Meng Z, Del Prete GQ, Lifson JD, Zhou M, Carrington M. Relative expression levels of the HLA class-I proteins in normal and HIV-infected cells. *J Immunol*. 2015;194:3594–3600. https://doi.org/10.4049/jimmunol.1403234.
148. O'huigin C, Kulkarni S, Xu Y, et al. The molecular origin and consequences of escape from miRNA regulation by HLA-C alleles. *Am J Hum Genet*. 2011;89:424–431. https://doi.org/10.1016/j.ajhg.2011.07.024.
149. Ramsuran V, Kulkarni S, O'huigin C, et al. Epigenetic regulation of differential HLA-A allelic expression levels. *Hum Mol Genet*. 2015;24:4268–4275. https://doi.org/10.1093/hmg/ddv158.
150. Ramsuran V, Naranbhai V, Horowitz A, et al. Elevated HLA-A expression impairs HIV control through inhibition of NKG2A-expressing cells. *Science*. 2018;359:86–90. https://doi.org/10.1126/science.aam8825.
151. Boegel S, Löwer M, Bukur T, Sorn P, Castle JC, Sahin U. HLA and proteasome expression body map. *BMC Med Genomics*. 2018;11:36. https://doi.org/10.1186/s12920-018-0354-x.
152. McGranahan N, Rosenthal R, Hiley CT, et al. Allele-specific HLA loss and immune escape in lung cancer evolution. *Cell*. 2017;171:1259–1271.e11. https://doi.org/10.1016/j.cell.2017.10.001.
153. Shukla SA, Rooney MS, Rajasagi M, et al. Comprehensive analysis of cancer-associated somatic mutations in class I HLA genes. *Nat Biotechnol*. 2015;33:1152–1158. https://doi.org/10.1038/nbt.3344.
154. Hundal J, Carreno BM, Petti AA, et al. pVAC-Seq: a genome-guided in silico approach to identifying tumor neoantigens. *Genome Med*. 2016;8:11. https://doi.org/10.1186/s13073-016-0264-5.
155. Bray NL, Pimentel H, Melsted P, Pachter L. Near-optimal probabilistic RNA-seq quantification. *Nat Biotechnol*. 2016;34:525–527. https://doi.org/10.1038/nbt.3519.
156. Bjerregaard A-M, Nielsen M, Hadrup SR, Szallasi Z, Eklund AC. MuPeXI: prediction of neo-epitopes from tumor sequencing data. *Cancer Immunol Immunother*. 2017;66:1123–1130. https://doi.org/10.1007/s00262-017-2001-3.
157. Tappeiner E, Finotello F, Charoentong P, Mayer C, Rieder D, Trajanoski Z. TIminer: NGS data mining pipeline for cancer immunology and immunotherapy. *Bioinformatics*. 2017;33:3140–3141. https://doi.org/10.1093/bioinformatics/btx377.
158. Subramanian A, Tamayo P, Mootha VK, et al. Gene set enrichment analysis: a knowledge-based approach for interpreting genome-wide expression profiles. *Proc Natl Acad Sci USA*. 2005;102:15545–15550. https://doi.org/10.1073/pnas.0506580102.
159. Charoentong P, Finotello F, Angelova M, et al. Pan-cancer immunogenomic analyses reveal genotype-immunophenotype relationships and predictors of response to checkpoint blockade. *Cell Rep*. 2017;18:248–262. https://doi.org/10.1016/j.celrep.2016.12.019.
160. Mondet S, Aksoy BA, Rozenberg L, Hodes I, Hammerbacher J. Bioinformatics workflow management With the Wobidisco ecosystem, Internet*bioRxiv*. 2017;213884. https://doi.org/10.1101/213884.
161. Duitama J, Srivastava PK, Măndoiu II. Towards accurate detection and genotyping of expressed variants from whole transcriptome sequencing data. *BMC Genomics*. 2012; 13(Suppl. 2):S6. https://doi.org/10.1186/1471-2164-13-S2-S6.
162. Duitama J, Huebsch T, McEwen G, Suk E-K, Hoehe MR. ReFHap: a reliable and fast algorithm for single individual haplotyping. In: *Proceedings of the First ACM International*

Conference on Bioinformatics and Computational Biology New York, NY, USA, ACM; 2010:160–169. https://doi.org/10.1145/1854776.1854802.
163. Rubinsteyn A, Hodes I, Kodysh J, Hammerbacher J. Vaxrank: a computational tool for designing personalized cancer vaccines. *bioRxiv*. 2018, biorxiv.org. Available, https://www.biorxiv.org/content/early/2018/10/18/142919.abstract.
164. Warren RL, Freeman JD, Zeng T, et al. Exhaustive T-cell repertoire sequencing of human peripheral blood samples reveals signatures of antigen selection and a directly measured repertoire size of at least 1 million clonotypes. *Genome Res*. 2011;21: 790–797. https://doi.org/10.1101/gr.115428.110.
165. Robins HS, Campregher PV, Srivastava SK, et al. Comprehensive assessment of T-cell receptor beta-chain diversity in alphabeta T cells. *Blood*. 2009;114:4099–4107. https://doi.org/10.1182/blood-2009-04-217604.
166. Rosati E, Dowds CM, Liaskou E, Henriksen EKK, Karlsen TH, Franke A. Overview of methodologies for T-cell receptor repertoire analysis. *BMC Biotechnol*. 2017;17:61. https://doi.org/10.1186/s12896-017-0379-9.
167. Heather JM, Ismail M, Oakes T, Chain B. High-throughput sequencing of the T-cell receptor repertoire: pitfalls and opportunities. *Brief Bioinform*. 2018;19:554–565. https://doi.org/10.1093/bib/bbw138.
168. Kuchenbecker L, Nienen M, Hecht J, et al. IMSEQ—a fast and error aware approach to immunogenetic sequence analysis. *Bioinformatics*. 2015;31:2963–2971. https://doi.org/10.1093/bioinformatics/btv309.
169. Bolotin DA, Poslavsky S, Davydov AN, et al. Antigen receptor repertoire profiling from RNA-seq data. *Nat Biotechnol*. 2017;35:908–911. https://doi.org/10.1038/nbt.3979.
170. Nazarov VI, Pogorelyy MV, Komech EA, et al. tcR: an R package for T cell receptor repertoire advanced data analysis. *BMC Bioinformatics*. 2015;16:175. https://doi.org/10.1186/s12859-015-0613-1.
171. Yu Y, Ceredig R, Seoighe C. LymAnalyzer: a tool for comprehensive analysis of next generation sequencing data of T cell receptors and immunoglobulins. *Nucleic Acids Res*. 2016;44:e31. https://doi.org/10.1093/nar/gkv1016.
172. Gerritsen B, Pandit A, Andeweg AC, de Boer RJ. RTCR: a pipeline for complete and accurate recovery of T cell repertoires from high throughput sequencing data. *Bioinformatics*. 2016;32:3098–3106. https://doi.org/10.1093/bioinformatics/btw339.
173. Dash P, Fiore-Gartland AJ, Hertz T, et al. Quantifiable predictive features define epitope-specific T cell receptor repertoires. *Nature*. 2017;547:89–93. https://doi.org/10.1038/nature22383.
174. Glanville J, Huang H, Nau A, et al. Identifying specificity groups in the T cell receptor repertoire. *Nature*. 2017;547:94–98. https://doi.org/10.1038/nature22976.
175. Gielis S, Moris P, De Neuter N, et al. TCRex: a webtool for the prediction of T-cell receptor sequence epitope specificity, Internet*bioRxiv*. 2018;373472. https://doi.org/10.1101/373472.
176. Meysman P, De Neuter N, Gielis S, Bui Thi D, Ogunjimi B, Laukens K. On the viability of unsupervised T-cell receptor sequence clustering for epitope preference. *Bioinformatics*. 2018;35:1461–1468. https://doi.org/10.1093/bioinformatics/bty821.
177. Jurtz VI, Jessen LE, Bentzen AK, et al. NetTCR: sequence-based prediction of TCR binding to peptide-MHC complexes using convolutional neural networks, Internet-*bioRxiv*. 2018;433706. https://doi.org/10.1101/433706.
178. Schneidman-Duhovny D, Khuri N, Dong GQ, et al. Predicting CD4 T-cell epitopes based on antigen cleavage, MHCII presentation, and TCR recognition. *PLoS One*. 2018;13:e0206654. https://doi.org/10.1371/journal.pone.0206654.
179. Simon S, Wu Z, Cruard J, et al. TCR analyses of two vast and shared melanoma antigen-specific T cell repertoires: common and specific features. *Front Immunol*. 2018;9:1962. https://doi.org/10.3389/fimmu.2018.01962.

180. Levy E, Marty R, Gárate Calderón V, et al. Immune DNA signature of T-cell infiltration in breast tumor exomes. *Sci Rep.* 2016;6:30064. https://doi.org/10.1038/srep30064.
181. Izraelson M, Nakonechnaya TO, Moltedo B, et al. Comparative analysis of murine T-cell receptor repertoires. *Immunology.* 2018;153:133–144. https://doi.org/10.1111/imm.12857.
182. Thommen DS, Koelzer VH, Herzig P, et al. A transcriptionally and functionally distinct PD-1 CD8 T cell pool with predictive potential in non-small-cell lung cancer treated with PD-1 blockade. *Nat Med.* 2018;24:994–1004. https://doi.org/10.1038/s41591-018-0057-z.
183. Fink K. Can we improve vaccine efficacy by targeting T and B cell repertoire convergence? *Front Immunol.* 2019;10:110. https://doi.org/10.3389/fimmu.2019.00110.
184. Shugay M, Bagaev DV, Zvyagin IV, et al. VDJdb: a curated database of T-cell receptor sequences with known antigen specificity. *Nucleic Acids Res.* 2018;46:D419–D427. https://doi.org/10.1093/nar/gkx760.
185. Tickotsky N, Sagiv T, Prilusky J, Shifrut E, Friedman N. McPAS-TCR: a manually curated catalogue of pathology-associated T cell receptor sequences. *Bioinformatics.* 2017;33:2924–2929. https://doi.org/10.1093/bioinformatics/btx286.
186. Borrman T, Cimons J, Cosiano M, et al. ATLAS: a database linking binding affinities with structures for wild-type and mutant TCR-pMHC complexes. *Proteins.* 2017;85:908–916. https://doi.org/10.1002/prot.25260.
187. Emerson RO, DeWitt WS, Vignali M, et al. Immunosequencing identifies signatures of cytomegalovirus exposure history and HLA-mediated effects on the T cell repertoire. *Nat Genet.* 2017;49:659–665. https://doi.org/10.1038/ng.3822.
188. Hanahan D, Weinberg RA. Hallmarks of cancer: the next generation. *Cell.* 2011;144:646–674. https://doi.org/10.1016/j.cell.2011.02.013.
189. Coulie PG, Van den Eynde BJ, van der Bruggen P, Boon T. Tumour antigens recognized by T lymphocytes: at the core of cancer immunotherapy. *Nat Rev Cancer.* 2014;14:135–146. https://doi.org/10.1038/nrc3670.
190. Yarchoan M, Johnson 3rd BA, Lutz ER, Laheru DA, Jaffee EM. Targeting neoantigens to augment antitumour immunity. *Nat Rev Cancer.* 2017;17:209–222. https://doi.org/10.1038/nrc.2016.154.
191. Gubin MM, Artyomov MN, Mardis ER, Schreiber RD. Tumor neoantigens: building a framework for personalized cancer immunotherapy. *J Clin Invest.* 2015;125:3413–3421. https://doi.org/10.1172/JCI80008.
192. Sahin U, Derhovanessian E, Miller M, et al. Personalized RNA mutanome vaccines mobilize poly-specific therapeutic immunity against cancer. *Nature.* 2017;547:222–226. https://doi.org/10.1038/nature23003.
193. Sun Z, Chen F, Meng F, Wei J, Liu B. MHC class II restricted neoantigen: a promising target in tumor immunotherapy. *Cancer Lett.* 2017;392:17–25. https://doi.org/10.1016/j.canlet.2016.12.039.
194. Hilf N, Kuttruff-Coqui S, Frenzel K, et al. Actively personalized vaccination trial for newly diagnosed glioblastoma. *Nature.* 2019;565:240–245. https://doi.org/10.1038/s41586-018-0810-y.
195. Marty Pyke R, Thompson WK, Salem RM, Font-Burgada J, Zanetti M, Carter H. Evolutionary pressure against MHC class II binding cancer mutations. *Cell.* 2018;175:1991. https://doi.org/10.1016/j.cell.2018.11.050.
196. Pearson H, Daouda T, Granados DP, et al. MHC class I-associated peptides derive from selective regions of the human genome. *J Clin Invest.* 2016;126:4690–4701. https://doi.org/10.1172/JCI88590.
197. Ritz D, Sani E, Debiec H, Ronco P, Neri D, Fugmann T. Membranal and blood-soluble HLA class II peptidome analyses using data-dependent and independent acquisition. *Proteomics.* 2018;18:e1700246. https://doi.org/10.1002/pmic.201700246.

198. Paul S, Karosiene E, Dhanda SK, et al. Determination of a predictive cleavage motif for eluted major histocompatibility complex class II ligands. *Front Immunol.* 2018;9:1795. https://doi.org/10.3389/fimmu.2018.01795.
199. Barra CM, Alvarez B, Paul S, et al. Footprints of antigen processing boost MHC class II natural ligand binding predictions, Internet, *bioRxiv.* 2018;285767. https://doi.org/10.1101/285767.
200. Shao W, Pedrioli PGA, Wolski W, et al. The SysteMHC Atlas project. *Nucleic Acids Res.* 2018;46:D1237–D1247. https://doi.org/10.1093/nar/gkx664.
201. Castle JC, Loewer M, Boegel S, et al. Mutated tumor alleles are expressed according to their DNA frequency. *Sci Rep.* 2014;4:4743. https://doi.org/10.1038/srep04743.
202. Weinzierl AO, Lemmel C, Schoor O, et al. Distorted relation between mRNA copy number and corresponding major histocompatibility complex ligand density on the cell surface. *Mol Cell Proteomics.* 2007;6:102–113. https://doi.org/10.1074/mcp.M600310-MCP200.
203. Wang D, Eraslan B, Wieland T, et al. A deep proteome and transcriptome abundance atlas of 29 healthy human tissues, Internet, *bioRxiv.* 2018;357137. https://doi.org/10.1101/357137.
204. Laumont CM, Daouda T, Laverdure J-P, et al. Global proteogenomic analysis of human MHC class I-associated peptides derived from non-canonical reading frames. *Nat Commun.* 2016;7:10238. https://doi.org/10.1038/ncomms10238.
205. Fritsche J, Rakitsch B, Hoffgaard F, et al. Translating immunopeptidomics to immunotherapy-decision-making for patient and personalized target selection. *Proteomics.* 2018;18:e1700284. https://doi.org/10.1002/pmic.201700284.
206. Murphy JP, Konda P, Kowalewski DJ, et al. MHC-I ligand discovery using targeted database searches of mass spectrometry data: implications for T-cell immunotherapies. *J Proteome Res.* 2017;16:1806–1816. https://doi.org/10.1021/acs.jproteome.6b00971.
207. Bassani-Sternberg M, Bräunlein E, Klar R, et al. Direct identification of clinically relevant neoepitopes presented on native human melanoma tissue by mass spectrometry. *Nat Commun.* 2016;7:13404. https://doi.org/10.1038/ncomms13404.

CHAPTER THREE

Tumor mechanisms of resistance to immune attack

David J. Zahavi[a,b], Louis M. Weiner[b,*]
[a]Tumor Biology Training Program, Department of Oncology, Georgetown University Medical Center, Washington, DC, United States
[b]Georgetown Lombardi Comprehensive Cancer Center, Department of Oncology, Georgetown University Medical Center, Washington, DC, United States
*Corresponding author: e-mail address: weinerl@georgetown.edu

Contents

1. Introduction 62
2. Loss of antigenicity 66
   2.1 Immune system recognition of transformed cells 66
   2.2 MHC class I loss or downregulation 68
   2.3 Defects in antigen processing 70
   2.4 Tumor-specific antigen loss or downregulation 71
   2.5 Clinical relevance 72
3. Loss of immunogenicity 73
   3.1 Regulation of immune cell activity 73
   3.2 Upregulation of inhibitory immune checkpoints 75
   3.3 Reduced costimulation and other inhibitory signals 76
   3.4 Death receptor signaling and loss of sensitivity 78
   3.5 Clinical relevance 80
4. Immunosuppressive microenvironment 81
   4.1 The tumor microenvironment (TME) 81
   4.2 Immunosuppressive cytokines 83
   4.3 Recruitment of inhibitory immune cells 85
   4.4 Clinical relevance 86
5. Conclusion 87
References 90

Abstract

The immune system plays a key role in the interactions between host and tumor. Immune selection pressure is a driving force behind the sculpting and evolution of malignant cancer cells to escape this immune attack. Several common tumor cell-based mechanisms of resistance to immune attack have been identified and can be broadly categorized into three main classes: loss of antigenicity, loss of immunogenicity, and creation of an immunosuppressive microenvironment. In this review, we will discuss in detail the relevant literature associated with each class of resistance and will describe

Progress in Molecular Biology and Translational Science, Volume 164
ISSN 1877-1173
https://doi.org/10.1016/bs.pmbts.2019.03.009

the relevance of these mechanisms to human cancer patients. To conclude, we will outline the implications these mechanisms have for the treatment of cancer using currently available therapeutic approaches. Immunotherapy has been a successful addition to current treatment approaches, but many patients either do not respond or quickly become resistant. This reflects the ability of tumors to continue to adapt to immune selection pressure at all stages of development. Additional study of immune escape mechanisms and immunotherapy resistance mechanisms will be needed to inform future treatment approaches.

1. Introduction

Cancer is a diverse set of diseases in which accumulation of mutations affecting cell growth in normal cells leads to malignant transformation. Hanahan and Weinberg originally proposed that most, if not all, cancers displayed six hallmark characteristics: self-sufficient growth, insensitivity to anti-growth signaling, unlimited replication, resistance to apoptosis, angiogenesis, and invasion.[1] The role of the immune system in cancer development and progression was not initially included in their list. This may be due in part to the historically controversial nature of the field of tumor immunology. As cancer ultimately arises from "self" there was at first some debate about whether or not the immune system could recognize and control neoplastic growth.

Paul Ehrlich was the first to conceptualize the idea of immunosurveillance, and in 1909 he suggested that the immune system played a protective role against carcinogenesis.[2] He reasoned that a high frequency of tumors would naturally develop in long-lived organisms, but few would become clinically detectable because immune cells would be capable of finding and destroying malignantly transformed cells. Ehrlich was truly ahead of his time, as it would take 50 years for scientific understanding of immunology to progress to a point where evidence in support of his theory could be generated. Throughout the 1950s, inbred mouse strains were used in tumor transplantation models to demonstrate that mice could be immunized against tumor challenge by prior exposure.[3] This crucial discovery of the existence of tumor specific antigens that could be recognized by the immune system provided the foundation for the field of tumor immunology. These advances led to the formation of the immunosurveillance hypothesis, jointly credited to Burnet and Thomas, which formally postulated that the adaptive immune system is responsible for preventing tumor formation in immunocompetent organisms.[4,5] Unfortunately, early studies meant to evaluate the hypothesis led to its abandonment. Researchers first explored whether immunosuppressed mice would have a higher incidence of induced or spontaneous tumors as the hypothesis would predict. However, little consensus was reached on the method of immunosuppression or if immunocompromised hosts were simply more susceptible to tumor forming agents and not the tumors themselves. A large and comprehensive study by Stutman that compared carcinogen induced and spontaneous tumor formation rates between athymic nude and wild-type mice concluded that there were no significant differences.[6] Based on these findings, tumor cells were believed to be too similar to "self" to activate the immune system,[7] and the cancer surveillance hypothesis was rejected. As the field of immunology progressed it was eventually appreciated that nude mice are not completely immunodeficient and that the CBA/H strain used in the Stutman experiments are very sensitive to the MCA carcinogen. Although these new discoveries supported revisiting the idea of immunosurveillance, it would take another 30 years before the scientific community would embrace the concept.

By the 1990s, the characterization of new subsets of immune cells such as natural killer (NK) and $\gamma\delta$ T cells began to renew interest in cancer immunosurveillance. Researchers once again examined whether components of the immune system were capable of abrogating tumor development using immunosuppressed animal models. Mice were found to be more susceptible to tumor transplantation if IFN-γ was inhibited[8] and were also more prone to spontaneously develop tumors when deficient in perforin.[9] Furthermore, RAG-2 mutant mice, which lack NK, T, and B cells, more frequently grew tumors both spontaneously and when challenged with MCA when compared to their wild-type counterparts.[10] Additional experiments conducted using mice with deficiencies in other components of both the innate and adaptive immune systems also confirmed that members of the immune system played a role in preventing cancer formation. Altogether, there was now overwhelming data to support the basic principles of cancer immunosurveillance.

Since all of the studies that confirmed the existence of immunosurveillance to that point had been done in animal models, researchers next asked whether the concept held true in humans. A review of available clinical data reveals that patients who are immunosuppressed either due to AIDS or immunosuppressive drug regimens following organ transplant have an increased incidence of both virally-associated and non-virally associated malignancies.[11] This link between immunodeficiency and cancer susceptibility in humans mimics what is seen in the animal models and suggests that cancer immunosurveillance is a vital function of the immune system.

Analysis of patient sera has demonstrated that humans have the capacity for spontaneous antibody responses to many cancer types due to the recognition of tumor-associated antigens, a finding that also supports the occurrence of immunosurveillance.[12] Perhaps the most striking evidence for cancer immunosurveillance is the existence of tumor infiltrating lymphocytes (TILs). It is apparent by histology that immune cells do infiltrate most tumors in human patients, suggesting that the immune system does recognize tumors as foreign entities. It was first demonstrated in melanoma patients that the number of TILs could be correlated to prognosis, where patients with higher levels of CD8+ T cell infiltration had more favorable outcomes.[13,14] This observation was subsequently confirmed for a variety of different malignancies including ovarian and colon cancer.[15,16] The landmark study by Galon et al. went a step further and revealed that the type, density, and location of the TILs were strong predictors of survival. A review of the emerging patient data by Fridman and colleagues concluded that the immune contexture encountered by a tumor has a significant impact on its clinical course.[17] In fact, an extensive analysis of TILs discovered that the immune system response to a tumor was dynamic and evolved with tumor progression.[18] Therefore, it became widely recognized that cancer immunosurveillance was physiologically relevant and could potentially yield new therapeutic strategies. Interest in the field grew and the cancer immunology renaissance began.

Although the cancer immunosurveillance hypothesis had been confirmed, important questions remained. Chief among them was how cancer was able to progress in spite of the immune response. In the early 2000s, several studies found that tumors grown in immunodeficient mice were more immunogenic and could regress when transplanted into immunocompetent hosts as compared to tumors originally grown in wild-type mice.[19] These results indicated that the immune system was capable of influencing tumor immunogenicity during its development. It was theorized that an immune response against tumor-associated antigens could eventually sculpt the tumor into a less immunogenic variant that is able to progress into clinically relevant disease. In order to address the seemingly conflicting duality of the immune system's role as both tumor-preventive and tumor-promoting, the cancer immunosurveillance hypothesis was revised into the concept of

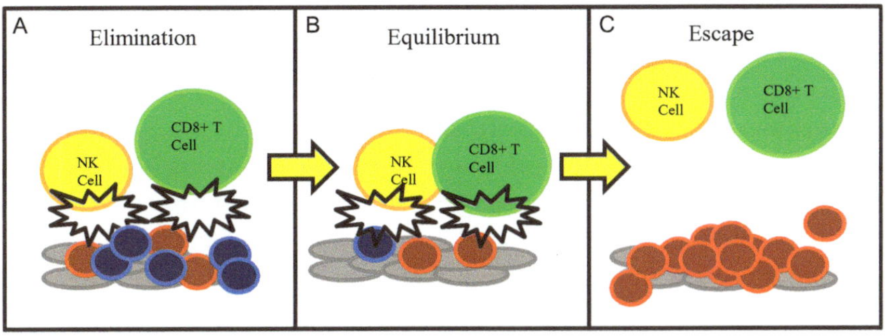

Fig. 1 The three stages of immunoediting. Gray = normal cells, blue = developing tumor cells, red = immunoresistant tumor cell variant, stars = cytotoxic activity. (A) Elimination phase where immune cells carry out immunosurveillance. (B) Equilibrium phase where immune selection pressure promotes variants with escape mechanisms. (C) Escape phase where selected tumor cell variants proliferate in an uncontrolled manner. *Adapted from Dunn G, Bruce A, Ikeda H, Old L, Schreiber R. Cancer immunoediting: from immunosurveillance to tumor escape. Nat Immunol 2002;3(11):991–998.*

immunoediting. Schreiber proposed that cancer immunoediting would consist of three distinct phases: elimination, equilibrium, and escape.[20] These stages of immunoediting (Fig. 1) could account for the varying roles of the immune system as some tumors may be completely eliminated by the first host-protective phase with only a select few malignant cells progressing to equilibrium and tumor-promoting escape. In the equilibrium phase the dynamic interactions between host immune cells and tumor cells could occur over an extended period of time, which accounts for the long estimated lag time between carcinogenic exposure and cancer. It was envisaged that components of both the innate and adaptive immune system would exert powerful Darwinian selection pressure on developing tumors during this equilibrium phase which would ultimately lead to the escape and propagation of tumor cell variants that had developed some mechanism of resistance to immune attack.

Immunoediting has since been confirmed in several experimental models and the expression, recognition, and modification of tumor-specific antigens was shown to be critical to this process.[21] The concept of cancer immunoediting integrated an understanding of immune system function with known observations of tumor development and was quickly adopted by many scientists.

Immune selection pressure is now recognized as a driving force in the sculpting of malignant cancer cells that can lead to host death. Given these well-established facts, Hanahan and Weinberg added escape from immune control as a hallmark of cancer in an updated version of their original landmark paper.[22] Investigators then turned their attention to elucidating the cellular processes that underlie the distinct phases of immunoediting. A more complete understanding of the interplay between the immune system and a developing tumor could potentially lead to new and exciting immunotherapy strategies. While there had been many studies focused on the cancer elimination phase, the cellular effectors and processes involved in equilibrium and escape were not as well characterized. How tumors eventually devise means of evading or blocking destruction by the immune system has become a major research topic. Since it was now accepted that the immune system is capable of eliminating tumor cells, targeting of immune escape mechanisms employed by tumors is an appealing therapeutic strategy. These mechanisms can be broadly divided into three main categories: loss of antigenicity, loss of immunogenicity, and creation of an immunosuppressive microenvironment. In this review, we discuss the relevant literature associated with each category and its relevance to human cancer patients. To conclude, we outline the implications of these mechanisms have for the treatment of cancer and the currently available therapeutic approaches.

2. Loss of antigenicity

2.1 Immune system recognition of transformed cells

The immune system is an extremely important host defense mechanism that has evolved to combat infectious organisms and viruses, and to protect against the growth of malignantly transformed cells. NK cells and CD8+ cytotoxic T cells are the major types of immune cells with the potential to kill host cells they identify as either infected or cancerous. The major histocompatibility complex (MHC) class 1 molecules present on the surface of nearly all nucleated cells in the mammalian body are the fundamental regulators of these immune cells. MHC class 1 molecules form ternary complexes comprised of an alpha heavy chain, β_2-microglobulin light chain, and peptide antigen of eight to nine amino acids in length.[23] Their function is to present antigens representing endogenous cytosolic proteins for recognition by immune cells. While the β_2-microglobulin light chain is not polymorphic, the alpha heavy chain is highly polymorphic and encoded in humans by the human leukocyte antigen (HLA) gene family. The highly genetically variable HLA coding loci allow for a diverse range of antigens to be presented that are mostly unique to an individual. Mutations that alter the structures of cellular proteins lead to the presentation of antigens seen as foreign to the immune system.

The peptides presented by MHC class 1 are generated from the antigen processing of endogenous cellular proteins. Cytosolic proteins are degraded by the proteasome complex, which contains many subunits including the low-molecular-mass polypeptides (LMPs) that are involved in producing the proper sized antigens for loading into MHC class I. The proteasome is the sole producer of antigens for MHC class I display[24] and is stimulated by IFNγ released during immune responses to become what is often termed the "immunoproteasome" and to be more efficient at generating antigens that bind to MHC class I.[25] Peptides created by the proteasome are then transported to the endoplasmic reticulum (ER) by the aptly named transporter associated with antigen processing-1 and -2 (TAP1 and TAP2). In the ER the peptide antigens are loaded onto MHC class I with the help of accessory proteins such as calnexin, calreticulin, and tapasin. The completed trimolecular complex then proceeds through the Golgi secretory pathway for display on the cell surface. Defects in any one of the components in this process will lead to loss of MHC class I and significant changes in cellular interaction with the immune system.

MHC class I molecules are required for the presentation of antigens to CD8+ cytotoxic T cells[26] and also for the immune regulatory activity of NK cells.[27] T cells interact with MHC class I *via* their T cell receptors (TCRs), members of the immunoglobulin superfamily of proteins. The TCR contains a variable region that undergoes genetic recombination in T cells to generate a diverse pool of TCRs each with a unique antigen specificity. After removal of self-reactive T cells during their maturation process, the remaining T cells survey host cells for their recognized antigen. If a TCR engages its cognate antigen in the context of MHC class I, signal transduction occurs to activate the T cell to exert its effects and activate other members of the adaptive immune system. CD8+ cytotoxic T cells are therefore able to find and destroy host cells that contain aberrant proteins resulting from viral infection or mutation. Instead of recognizing antigens presented by MHC class I, NK cells target host cells that are lacking proper MHC class I expression. NK cells have both activating and inhibiting receptors on their surface that control their function. NK cells express killer inhibitory receptors (KIRs) that recognize MHC class I and prevent NK cell activation. Therefore, NK cells search for and destroy host cells that are attempting to hide from T cells. Importantly, alterations in MHC class 1 molecules will affect both T and NK cell immune functions.

It has been known for some time that tumors express their own specific antigens.[28] More recent evidence has shown that tumors can express immunogenic antigens that result from the oncogenic driver mutations that initiate tumor development, while many can also arise from non-driver mutations.[29] Mouse models have demonstrated that not only are CD8+ T cells capable of recognizing these tumor associated antigens, but also that these cytotoxic T lymphocytes (CTLs) can mediate tumor destruction.[30] Thus, one of the ways tumor cells can escape CTL recognition is by altering their antigen presentation.

2.2 MHC class I loss or downregulation

Alterations in MHC class I expression in tumors were first discovered in the 1960s.[31,32] It was later appreciated that decreased MHC class I surface levels could be observed in many experimental and spontaneous tumor models in mice and in human cancer cell lines.[33] In the 1980s it became appreciated that downregulation of MHC class I by tumor cells may represent a major mechanism for tumor evasion of immune attack. Supporting evidence came from a study showing that inoculation of mice with tumor cells possessing experimentally downregulated MHC class I had enhanced growth properties.[34] Immunohistochemical and flow cytometric techniques were used to analyze MHC class I expression in patient tumor samples for many different types of cancer and consistently demonstrated significant rates of MHC class I loss.[35,36] Further analysis revealed that tumor cell populations were very heterogeneous, with coexistence of MHC class I positive and negative cells in most samples. However, it was still estimated that between 40% and 90% of tumors displayed either total or partial loss of MHC class I molecules. Based on the results from different tumor tissues, Garrido et al. proposed seven major MHC class I altered phenotypes for tumors.[37] These phenotypes could all exist within a given tumor, further contributing to complexity. It remains to be determined if any particular cancer type has preference for certain phenotypes. In order to examine the molecular mechanisms underlying these phenotypes, cell populations representative of each phenotype from patient tumor samples were isolated and used to create cell lines.

Total loss of MHC class I is considered to be a frequent phenotype in human cancers and has been reported for many tumor types. Complete loss of MHC class I on the cell surface is most often associated with mutations in the gene encoding the β_2-microglobulin subunit. β_2-microglobulin is an integral part of the ternary complex and transcriptional or translational failure to produce functional β_2-microglobulin leads to a total inability of cells to construct MHC class I in the ER. In fact, in β_2-microglobulin deficient cell lines, the transfection of a wild-type β_2-microglobulin gene resulted in the full restoration of MHC class I expression demonstrating that mutations in β_2-microglobulin are the sole mechanism behind MHC class I loss in these cell lines.[38] In most cases there is still functional β_2-microglobulin detected in these cell lines, indicating a majority of the mutations in the coding gene affect the protein at the post-transcriptional level. Mutational "hot-spots" have been reported for the β_2-microglobulin gene, which potentially reflects the genetic instability inherent in tumor cells.[39] Additionally, typically only one mutant allele for β_2-microglobulin is found, demonstrating that the lack of functional β_2-microglobulin is due to the loss of heterozygosity common in tumor cells. β_2-microglobulin mutations were also discovered in patient tumor tissues that had a complete loss of MHC class I, providing evidence that β_2-microglobulin mutations are driver mechanisms *in vivo* as well.[40] In addition to alterations in β_2-microglobulin, a total loss of MHC class I could be explained by defects in antigen processing machinery or mutations in members of the HLA gene family. As MHC class I can only be assembled if all three components are present, interference in antigen transport to the ER *via* changes to TAP1 or TAP2 could result in the complete loss phenotype. A more detailed discussion of defects in antigen processing will be presented later in this review. Mutations in the HLA gene family would induce a total loss of MHC class I, however these defects have yet to be fully examined in tumor cells.

In addition to the complete loss of MHC class I, several other resistant phenotypes commonly seen in tumor cells involve changes in HLA haplotype, gene product, and allelic loss. Loss of an HLA haplotype is a common mechanism for the reduction of MHC class I observed in many tumor types. Microsatellite markers on chromosomes 6 and 15 were used to identify altered HLA phenotypes in a variety of cancer cell lines.[41] Tumor cells may also decrease MHC class I by downregulating the expression of HLA genes in the absence of mutations that would cause a complete loss. Complex regulatory mechanisms control the levels of MHC class I on the cell surface and altered binding of associated transcriptional regulators has been found in tumor cells. For example, the transcription factors NF-κB and KBF1 bind to an enhancer sequence important for the transcription of the HLA gene family, and their activity was reduced in a cancer cell line.[42] In neuroblastoma cell lines, the oncogene N-myc was implicated as mediating the reduction in NF-κB binding capacity.[43] HLA allelic loss is believed to be due to point mutations or recombination but has yet to be fully confirmed in most cell lines. Yet another phenotype has been described as insufficient response to IFNs. Defects in the IFNγ signaling pathway lead to an inadequate upregulation of TAP1 and LMP2 needed for induction of the immunoproteasome.[44] Ultimately, in many tumors, a compound phenotype is observed that is the product of several of these mechanisms occurring simultaneously.

Total loss or downregulation of MHC class I has been established as a widespread mechanism by which tumor cells can evade immune recognition. Many of the alterations in MHC class I are due to epigenetic changes in the tumor cells, and MHC class 1 phenotypic changes can be reversed by inhibitors targeting histone modifiers.[45] T cell selection pressure *in vivo* drives the appearance of tumor cell variants deficient in MHC class I.[46] It may therefore be important to characterize the MHC class I expression of tumors in patients in order to assess the viability of T cell-based immunotherapy. One group demonstrated that peptide immunization strategies failed in some patients because they lacked MHC class I molecules at the tumor cell surface capable of presenting those antigens.[47] Although removal of MHC class I is a method by which tumor cells can evade T cells, the total loss phenotype should induce a response by NK cells. Tumor cells may aberrantly express non-classical MHC class I such as HLA G that are capable of inhibiting NK cells while remaining invisible to T cells. Unfortunately, studies have been unable to confirm that hypothesis.[48] Clearly, in tumors that have adapted to immune attack by eliminating their MHC class I, additional immune escape mechanisms exist that prevent a response by NK cells. These other escape mechanisms are likely a combination of the other strategies discussed elsewhere in this review.

2.3 Defects in antigen processing

One of the other common mechanisms by which tumor cells alter their antigen presentation is manipulation of antigen processing. In contrast to mutations in β_2-microglobulin or HLA genes, defective antigen processing may not cause changes in surface levels of MHC class I.[49] LMP2 and LMP7 are important subunits of the proteasome for generating peptide antigens destined for loading on MHC class I and the transporters TAP1 and TAP2 play critical roles in ferrying those antigens to the ER for complexing with MHC class I. Unsurprisingly, abnormalities in any of these proteins, sometimes occurring simultaneously, have been identified in many cancer cell lines.[50,51] Interestingly, in those studies, many of the cell line deficiencies in these proteins could be reversed upon treatment with IFNγ, suggesting altered regulatory mechanisms rather than mutations in the genes themselves. Although defects in LMP2, LMP7, TAP1, and TAP2 have been widely reported, very little is known about the molecular basis for them in tumor cells. Mutations in their genes are exceedingly rare and only a few examples of regulatory changes such as methylation of the TAP1 promoter have been reported.[52] As seen with MHC class I, epigenetic regulation of antigen processing genes may underlie immune escape phenotypes.[53] Oncogenic viruses have also been demonstrated to be capable of inducing a downregulation in both LMP2/7 and TAP1/2, again through an unknown mechanism.[54] Defects in antigen transport were confirmed to affect tumor immunogenicity and growth in an *in vivo* model.[55] Mutations in the accessory and chaperone proteins responsible for aiding MHC class I loading have also been found in many cancer types. In particular, frameshift mutations in calnexin, calreticulin, tapasin, and ERP57 were common.[56] These mutant chaperone variants led to an impairment of MHC class I loading and reduction in cell surface expression.[57] Although alternative proteasome subunits and TAP independent transport can maintain some surface levels of MHC class I, mutations in proteins involved with the antigen processing machinery in tumor cells can result in suboptimal peptide presentation in the remaining MHC class I.[57]

2.4 Tumor-specific antigen loss or downregulation

CTLs are only able to clear tumor cells if they recognize tumor antigens *via* their TCR interactions with MHC-mediated presentation of immunogenic peptides derived from those antigens. One way tumor cells may evade CTLs is by altering their expression of targetable antigens by downregulating them, modifying their epitopes, or masking them using immunodominant antigens. The first evidence for antigen loss came from a mouse model of mastocytoma. Tumor cells that escaped immune destruction *in vivo* had mutations in the gene encoding the mastocytoma-specific tumor rejection antigen, resulting in variants that were no longer recognizable by CTLs.[58] Melanoma cell lines have been widely used for studies regarding alterations in tumor-associated antigen expression due to their well characterized expression of MART-1/MelanA. MART-1/MelanA is highly expressed by most melanomas when measured by mRNA, however melanoma lesions from patients may lack such antigens on the cell surface. As MART-1/MelanA is a major target for CTLs, tumor cell manipulation of its post-transcriptional processing is a likely mechanism for evasion.[59] Loss of tumor antigens as an evasion mechanism was also confirmed to occur *in vivo* in melanoma patients.[60] Tumors that result from oncogenic viruses provide another good model for examining tumor-associated antigen expression changes. Viral transformation leads to the expression of protein products not normally seen in the host, and therefore may stimulate effective CTL responses. Mutations in the transforming gene products during tumor progression caused disruption of the epitopes that were recognized by CTLs, leading to immune evasion.[61] It has further been theorized that heterogeneous populations of tumor cells exhibit a diversity of tumor-associated antigens. Even though CTLs exist to a number of tumor-associated antigens, immune response *in vivo* to many of these antigens is prevented by the so-called "immunodominant" antigen to which the majority of the immune response is directed.[62] Tumor cell populations can escape detection by continuously altering the antigens that can be seen by the immune system. Tumor cells can respond to monoclonal antibody therapy in a similar manner by downregulating expression of the targeted antigen.[63]

2.5 Clinical relevance

Loss of antigenicity through loss of MHC class I expression or changes in the tumor-associated antigen epitopes have been shown to affect CTL responses in both cell lines and mouse models. Additional lines of evidence from patients prove that this phenomenon is clinically relevant and could impact immunotherapy treatment strategies. In many cancer types, downregulation of MHC class I was observed more frequently in metastases than primary tumors, suggesting an association between antigen presentation and clinical progression.[64] Defects in antigen processing machinery were also seen to correlate with disease progression.[65] It stands to reason that tumor cell loss of antigen presentation to CTLs would serve as a highly predictive prognostic marker for patients, and in fact, downregulation of either MHC class I or antigen transport components is associated with poor response to immunotherapy and survival in many cancers.[66] For example, in melanoma decreased expression of any of the proteins important for antigen presentation is significantly associated with worse survival (Fig. 2).

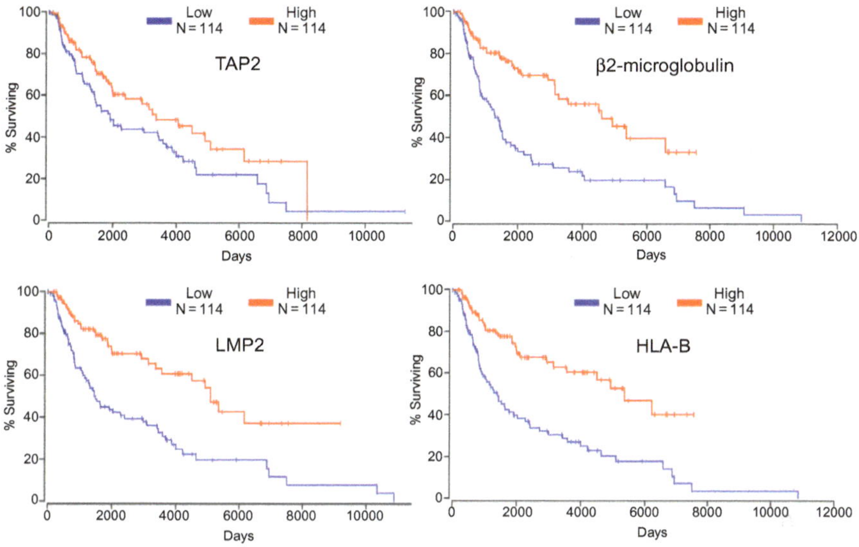

Fig. 2 Association of antigen presentation gene expression with survival in melanoma patients. Red line represents top 25th percentile and blue line represents bottom 25th percentile of mRNA expression. All differences were statistically significant. The results shown here are in whole or part based upon data generated by the TCGA Research Network and were analyzed using OncoLnc.[67]

3. Loss of immunogenicity

3.1 Regulation of immune cell activity

Cancer cells are very adept at adapting and developing new mechanisms for survival. Tumors that retain their antigenicity and are recognizable by the immune system can still escape elimination by employing a variety of other means to decrease their immunogenicity. In fact, the majority of tumor cells do express some form of tumor-associated antigen that T cells can respond to, which has led to the development of targeted immunotherapy strategies.[68] The failure of immune clearance of tumor cells in most cases of cancer is therefore attributable to factors that interfere with immune cell responses. The immune system requires extremely complex regulatory mechanisms to ensure that immune cells do not target self and only become activated when needed and for the proper duration. The two cell types important for mediating destruction of tumor, CD8+ CTLs and NK cells, have multiple activating and inhibiting receptors that modulate their function.

These receptors have collectively been termed immune checkpoints and play important roles in regulating self-tolerance and effector cell function.

T cell engagement of peptide antigens presented in the context of MHC class I *via* their TCRs is the main signal of activation for a T cell, but it is actually insufficient to initiate the proper signaling that leads to an effector response. T cells express many other costimulatory and coinhibitory receptors that determine cell fate. The primary and most characterized costimulatory signal comes from the interaction between the CD28 receptor on T cells and its ligands B7-1 or B7-2 on antigen presenting cells (APCs).[69] T cells also express other receptors with costimulatory functions, including ICOS, 4-1BB, OX40, CD27, HVEM, GITR, and CD2.[70] Importantly, binding of TCR to MHC class I in the absence of proper costimulatory signaling leads to the induction of anergy in the T cell.[71] T cell anergy is a mechanism for triggering tolerance and occurs normally in the periphery to prevent the expansion of autoreactive immune cells. While costimulatory signals are used to activate T cells, coinhibitory signaling exists as a way to turn off the immune response. Upon activation, T cells will upregulate the expression of several receptors that provide signaling to prevent a never-ending immune response. Researchers have been constantly identifying new inhibitory immune checkpoints for study, and some of the most characterized receptors include CTLA-4, PD-1, B7-H4, LAG3, TIM3, and TIGIT.[72] These receptors have a variety of mechanisms for negatively regulating T cell activation. CTLA-4 has high homology to CD28 but a much stronger binding affinity for B7-1 and B7-2, and when CTLA-4 becomes expressed following T cell activation it begins to outcompete CD28 for the costimulatory ligand, thereby dampening response.[73] Similarly, PD-1 binds its ligands PD-L1 and PD-L2, which are also members of the B7 family, on APCs to transmit signals that downregulate proliferation.

NK cells express some of the same activating and inhibitory receptors as T cells, such as CD27 and PD-1. However, in contrast to T cells, NK cells do not express receptors for specific antigens. Instead, NK cells express molecules that more generally interact with MHC class I. These activating and inhibitory receptors include NKG2 and the previously mentioned KIRs.[74] KIRs inhibit NK cell activation when they are bound to MHC class I and activate NK cells to lyse targets with aberrant expression of MHC class I. NKG2 family members that dimerize with CD94 can be either activating or inhibiting. NKG2D is an exception, instead homodimerizing and providing a strong costimulatory signal when bound to its ligands from the MIC and ULBP families that are found on stressed cells.[75] NK cells also express

CD16, also known as FcγRIIIA, in order to bind target cells coated in antibody and mediate antibody-dependent cellular cytotoxicity.[76]

In the context of tumor immunology, recent evidence suggests that tumors may co-opt the negative regulator pathways that are normally used to prevent autoimmune reactivity in order to thwart the anti-tumor response. Tumor cells have been found to upregulate ligands to inhibitory immune checkpoints, reduce the availability of costimulatory molecules, and provide various other inhibitory signals that prevent an effective immune response. Additionally, tumor cells can also avoid immune cell conjugation, resist immune effector cell-mediated apoptosis, or directly delete tumor-specific CTLs.

3.2 Upregulation of inhibitory immune checkpoints

The discovery that tumors use immune checkpoint pathways as a major mechanism of resistance to immune attack has revolutionized immunotherapy for patients. Seminal work from Dong et al. demonstrated that PD-L1 was highly expressed by many different cancer types and was responsible for inhibiting T cell responses.[77] Several mechanisms behind the upregulation of PD-L1 in tumor cells have been proposed. For some tumors, innate oncogenic signaling pathways were responsible for increases in PD-L1.[78,79] The majority of cancers lack these inherent drivers of PD-L1 and instead upregulate PD-L1 as an adaptive response to immune pressure. Production of IFNγ by tumor infiltrating lymphocytes has been found to induce PD-L1 expression in the tumor cells.[80] These findings suggest a feedback loop where immune effector cell attack causes tumor cells to express PD-L1 in order to suppress immune cell activity. Although PD-L1 has been the focus of many of these studies, others have shown that PD-L2 is also upregulated by some tumors and plays a similar role in inhibiting immune responses.[81] Further evidence for the importance of the PD-1/PD-L1 axis in tumor immune resistance comes from the characterization of tumor infiltrating lymphocytes (TILs). PD-1 is highly expressed by many TILs, and TILs with higher PD-1 levels produced less cytokines.[82] It has since been shown that PD-1 is the central regulator of the T cell exhaustion phenotype. T cell exhaustion has been defined as a dysfunctional state associated with the loss of effector functions, reduced proliferation, and high expression of inhibitory checkpoint receptors.[83] Originally identified to occur during chronic viral infections, T cell exhaustion is the result of chronic antigen stimulation. As TILs are continuously activated by tumor-associated antigens, they upregulate their expression of PD-1, while at the same time the tumor cells will upregulate PD-L1 in response to effector cytokines. The subsequent increases in PD-1/PD-L1 signaling at the site of the tumor ultimately exhausts the TIL populations.

Tumor cells also upregulate several other ligands for inhibitory immune checkpoint receptors. The inhibitory ligand from the B7 family, B7-H4, has been found to be highly expressed by breast and ovarian cancers.[84] Although the receptor on immune cells for B7-H4 has not been confirmed, animal studies supported B7-H4 as an inhibitory immune checkpoint.[85] More recent evidence suggests that B7-H4 expression by tumor-associated macrophages may be more important for immune suppression than its expression by the tumor cells themselves.[86] LAG3 is a homolog to CD4 that becomes expressed by activated T cells. MHC class II molecules are ligands for LAG3 and are upregulated by some cancers.[87] However, LAG3's role in inhibiting the immune response appears to be more related to its expression by T regulatory cells rather than its interaction with MHC class II. TIM3 is another receptor on T cells whose putative ligand, galectin 9, is upregulated in some cancers and this interaction has been hypothesized as a contributor to the inhibition of antitumor immunity.[88] Finally, the complex interactions of HVEM can modulate T cell behavior. When HVEM receptors on T cells bind their ligand LIGHT on APCs they send a costimulatory signal. However, another possible pairing is the BTLA receptor on T cells with HVEM present on the APCs, which leads instead to inhibitory signaling.[89] HVEM was found to be overexpressed in some cases of melanoma.[90] Taken together, studies on immune checkpoints indicate that tumor cells will upregulate several different ligands to inhibitory receptors in order to suppress T cell responses. Inhibitory checkpoints are summarized in Fig. 3.

3.3 Reduced costimulation and other inhibitory signals

Proper signaling during the first encounter between a T cell specific for a tumor antigen and a MHC class I molecule presenting that antigen is required to initiate a CTL anti-tumor response. Many of the positive costimulatory molecules are mainly expressed on professional antigen presenting cells. The lack of these molecules on tumor cells, especially members of the B7 family, led to the hypothesis that the induction of T cell anergy may occur during tumor development. Tumor cells that retain the ability to present recognizable antigens through MHC class I will cause sub-optimal TCR stimulation of T cells. Naïve T cells that receive TCR signaling without costimulation become anergic and unresponsive to future antigen

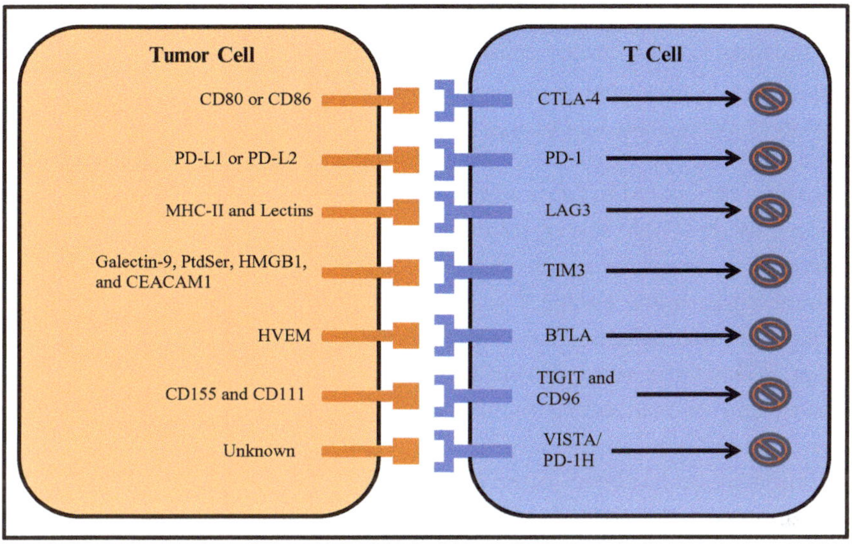

Fig. 3 Inhibitory immune checkpoints. Tumor cells (left) can express ligands for receptors expressed on T cells (right) that inhibit T cell anti-tumor activity.

exposure.[91] As opposed to T cell exhaustion, which develops over time from chronic antigen exposure, T cell anergy was hypothesized to develop early during initial encounters with antigen. Indeed, the induction of T cell anergy was shown to be an early event of tumor progression.[92] The importance of the lack of B7-1 to tumor immune escape was later demonstrated *in vivo* when melanoma cells transfected to express B7-1 were unable to grow in mouse models.[93]

Many other cell surface molecules play important roles in determining immune effector cell functions besides the immune checkpoints that have been implicated in tumor immune escape. Cell removal by phagocytosis is crucial for clearing old or unwanted cells without causing release of inflammatory or cytotoxic products. Phagocytes are activated depending on the balance of "eat me" and "don't eat me" signals expressed by the target cell. Calreticulin is a pro-phagocytic signal that, when present on the cell surface, is a marker of cell stress.[94] The DNA damage and dysregulation of cellular processes during neoplastic transformation are cell stress events that can induce calreticulin. Therefore, many cancer types express elevated levels of calreticulin.[95] CD47 is a protein in the immunoglobulin superfamily that inhibits phagocytosis by ligating SIRPα on phagocytes. CD47 is overexpressed in many cancer types, demonstrating that tumor cell manipulation of CD47 is a common immune evasion strategy to counteract calreticulin.[96] Phagocytosis is also a mechanism for engulfing apoptotic cells and for APCs to process cellular proteins to present as antigens to T cells. By upregulating CD47, tumor cells are able to interfere with these processes as well. Several additional possible functions for CD47 in cancer immunity have been described. CD47 is involved in regulating the immunosuppressive activity of the cytokine vascular endothelial growth factor in T cells[97] and is also implicated in facilitating cell migration and metastasis.[98] An *in vivo* genome-wide RNAi screen to identify tumor cell-based genes that regulate resistance to immune attack identified CD47 as one of the critical genes influencing immune escape.[99]

Another mechanistic class of inhibitory immune signaling employed by tumor cells is the secretion and production of tumor-derived antigens, ligands, and exosomes. In many cancer types, tumor-associated antigens have been detected in the serum of patients. Tumor antigen shedding into the bloodstream has been found to cause deletion of reactive thymocytes[100] and suppress tumor-specific CD8+ CTLs.[101] Tumors can secrete soluble forms of ligands to activating receptors on T and NK cells. Release of tumor-derived MIC ligands can bind NKG2D receptors on circulating cells away from the tumor site to induce endocytosis and degradation of the receptor, thereby interfering with the immune response.[102] In leukemia patients, the release of soluble BAG6 ligand for the activating receptor NKp30 on NK cells limited anti-tumor activity.[103] A relatively new concept in the field of tumor immunology is the tumor-derived exosome. Exosomes are membrane-bound microvesicles that are produced by both normal and malignant cells for distant cell–cell communication. Many tumors produce significant numbers of exosomes, which can contain immunosuppressive molecules.[104] Concordant with data from studies that examined soluble ligands, *in vitro* experiments indicated that tumor-derived exosomes contained NKG2D ligands that upon interaction with effector cells downregulated NKG2D expression.[105] These findings were replicated in prostate cancer patients.[106] The latest studies have identified subsets of T cell genes that may be modulated by tumor-derived exosomal contents.[107] There is now considerable interest in tumor-derived exosomes with researchers focusing on elucidating the exact mechanisms behind tumor-derived exosome production and whether they can be therapeutically targeted.

3.4 Death receptor signaling and loss of sensitivity

Activated CTLs and NK cells kill their targets *via* several mechanisms, one of which is the induction of apoptosis through the death receptor pathway.

The Fas receptor (a.k.a. the death receptor) is a member of the tumor necrosis factor (TNF) receptor family and is present on most cells to facilitate programmed cell death. Fas ligand (FasL) is upregulated on the surface of activated CTLs and NK cells in order to bind Fas on the target and signal the cell to begin apoptosis.[108] To escape the apoptotic signal, tumors have been demonstrated to downregulate their Fas expression or cause dysfunctional signaling by mutating members of the pathway.[109] Tumors can also produce and secrete a decoy receptor to FasL that will bind FasL on infiltrating lymphocytes to prevent interaction with their Fas.[110] Interestingly, tumors have been shown to also use death receptor signaling to subvert immune effector cells. Tumor cells can both decrease their expression of Fas and increase their expression of FasL. Tumor cells with elevated FasL were able to cause apoptosis of Fas expressing T cells in *in vitro* systems,[111] but the role of tumor FasL in immune evasion *in vivo* is not fully understood.[112] Even when tumors are signaled to undergo apoptosis by a functional death receptor pathway, additional mechanisms exist to prevent cell death. An array of studies has revealed that tumors block apoptosis by upregulating anti-apoptotic proteins and/or downregulating or mutating pro-apoptotic proteins.[113]

Anti-tumor immunity is promoted by the release of inflammatory cytokines such as IFNγ and TNF, which in addition to facilitating activation of immune effector cells, also sensitize tumor cells to killing. CRISPR screens have identified losses in IFNγ and TNF pathway genes that confer tumor cell resistance to CTL and NK-mediated killing.[114] These results confirmed earlier observations that tumors had deficits in JAK/STAT components of the IFN signaling pathway that rendered them unresponsive to IFNγ.[115,116] IFNγ is particularly important for inducing expression of MHC class I and the loss of IFN signaling contributes to losses in antigen expression. The other main mechanism for target cell killing is the release of granzyme B and perforin by activated immune cells. CTLs and NK cells conjugate to tumor cells and release these proteins from their cytotoxic granules. Perforins create pores in the cell membrane that facilitates granzyme B entry into the target cell, resulting in DNA fragmentation and apoptosis.[117] As a method of escape, tumor cells were shown to secrete serine protease inhibitors that inactivate granzyme B.[118]

For CTLs and NK cells to act on their tumor cell targets, proper binding and interaction between adhesion molecules must occur in addition to activating receptors. On one model system, both NK- and CTL-mediated lysis of tumor cells were dependent on the binding of leukocyte function antigen-1 (LFA-1) on effectors and intercellular adhesion molecule-1 (ICAM-1) on targets.[119] Previous research has shown that cellular adhesion molecules are often downregulated in cancer.[120] The decreased expression of these molecules is known to correlate with progression and helps facilitate metastasis.[121] Downregulation of adhesion molecules may represent yet another mechanism of resistance to immune attack. Tumor cells that developed resistance to ADCC-based mAb therapy by exhibiting reduced conjugation to NK cells were found to have significant decreases in cell surface expression of several adhesion molecules.[122] Future studies will determine if the phenomenon is replicable in other cell lines and with other immune effector cell populations.

3.5 Clinical relevance

The study of tumor loss of immunogenicity has been extremely impactful for the clinical management of cancer. For example, aberrant tumor cell expression of PD-L1 is associated with tumor immune escape; however elevated PD-L1 seems to correlate with worse prognosis only in some cancers.[123] In other cancer types, PD-L1 either did not correlate with prognosis or was actually linked to improved outcomes. It has been hypothesized that the link between increased PD-L1 on tumor cells and better prognosis may be due to PD-L1 expression representing an active host immune response to the tumor. Other inhibitory checkpoints on tumors such as PD-L2,[124] B7-H4,[125] galectin 9,[126] and HVEM[127] have been investigated as biomarkers of poor prognosis in many cancer types. Most of these studies have concluded that overexpression of inhibitory checkpoints alone does not serve as a robust marker of prognosis and instead must be combined with other measures such as number of TILs. Monitoring of immune checkpoint expression on circulating T cells has also been proposed as a way to track ongoing immune responses in patients.[128]

The targeting of inhibitory immune checkpoints by blocking monoclonal antibodies has revolutionized cancer therapy. Current strategies will be discussed in Section 5. Expression of immune checkpoints on both tumor and immune effector cells has been used as a predictor of immune checkpoint blockade therapeutic success.[129]

Other cell surface molecules and tumor-derived products also serve as markers of prognosis and sensitivity to treatment. High CD47 expression increases the incidence of metastasis[130] and is correlated with worse patient outcomes.[131] Tumors continually shed their antigens over the course of their growth. A thorough analysis of the antigen shedding process has linked serum levels of tumor-specific antigens with disease stage and prognosis.[132] Furthermore, detection of tumor antigens in the blood can serve as a biomarker useful in diagnosis, monitoring of progression, and monitoring of response to treatment.[133] Likewise, increased production of tumor-derived exosomes is significantly associated with poor prognosis and reduced survival.[134] There is excitement regarding using tumor-derived exosomes as diagnostic tools and measures of disease progression. Tumor-derived exosomes have complex contents that may serve as important biomarkers that are more specific than single antigens.[135] The expression profile of Fas/FasL in tumors correlates with progression, relapse, and survival.[136] Finally, expression signatures of genes in the TNF and IFN signaling pathways may predict outcomes in patients as well.[137,138] In summation, most of the mechanisms underlying tumor loss of immunogenicity can significantly alter the clinical course of the disease and may provide guidance in therapy choice.

4. Immunosuppressive microenvironment

4.1 The tumor microenvironment (TME)

So far, this review has mostly focused on mechanisms of immune escape that relate to changes in the direct interaction between tumor cells and tumor-specific immune effector cells. Another large category of strategies employed by tumors to avoid immune destruction involves the manipulation of the surrounding tumor microenvironment to become tumor promoting and immunosuppressive. There are already certain "immune privileged" sites in the body where access to immune cells is restricted. For example, the brain is surrounded by a specialized vascular structure known as the blood brain barrier that limits immune infiltration. Some forms of cancer such as glioblastoma can therefore avoid immune destruction by arising in, or metastasizing to, such immune privileged tissues.[139] The study of immune privileged sites and the mechanisms by which immune cells are excluded has shed light on this phenomenon's relevance in cancer, leading many to consider tumors as exhibiting similar features.[140]

Tumors do not develop as separate entities in the host and instead are more akin to an entire organ system. Tumor cells have many complex interactions with a variety of host cells near the tumor bed that leads to the creation of a tumor microenvironment. Non-transformed cells in the immediate vicinity of a developing tumor, such as fibroblasts, endothelial cells, pericytes, adipocytes, and lymphocytes are all affected by the tumor. Tumor cells mediate dynamic intercellular communication by producing cytokines, chemokines, growth factors, and enzymes that remodel the surrounding normal tissue. The non-transformed cells of the microenvironment become influenced by these factors to support tumor growth and are often tumor-promoting during all stages of carcinogenesis.[141] These non-malignant cells often comprise a large portion of the mass of primary tumors and function to sustain tumor growth. In addition to its tumorigenic properties, the tumor microenvironment is now also recognized as having immunomodulatory functions.[142] Several general processes occur in most tumors to reshape the surrounding tissue to be immunosuppressive. One of the most significant modifications tumors make to their environment is the recruitment of blood vessels. Angiogenesis is a physiological process where new blood vessels are created from existing ones and is vital to normal wound healing. Tumor angiogenesis is considered one of the hallmarks of cancer and is observed in most solid tumors. In order for solid tumors to progress pass their initial development stages, they must induce angiogenesis to obtain oxygen and nutrients to support continued cellular proliferation. Tumor cells produce and secrete proangiogenic factors such as vascular endothelial growth factor (VEGF) to initiate vasculature remodeling.[143] Tumor blood vessels are aberrant, with dysregulated structure, organization, and function compared to normal vasculature. Tumor angiogenesis results in a chaotic system of vessels that are leaky, inefficient, and have abnormal interactions with leukocytes.[144] Effector immune cells must be able to traffic to tissues to carry out an immune response. Unfortunately, the nature of the tumor vasculature restricts tumor-specific lymphocytes from extravasation into the tumor microenvironment. Tumor angiogenesis suppresses the expression of adhesion molecules on blood vessel endothelial cells that are important for interacting with immune cells.[145] Tumor endothelium thus acts as barrier that restricts access to leukocytes. However, certain immunosuppressive leukocyte subsets are commonly seen infiltrating tumors. This selective trafficking of immune cells is not fully understood, but in some cancers, the preferential recruitment of immunosuppressive cells has been demonstrated.[146] Tumor angiogenesis has additional immunosuppressive properties that include direct action on effector lymphocytes. Endothelial cells of tumor blood vessels are stimulated by tumor cytokines to express molecules that inhibit T cell activation such as PD-L1 and FasL.[147] The mural cells surrounding endothelial cells also have been shown to have immunosuppressive functions in the context of tumor angiogenesis.[148]

Tumor angiogenesis is inexorably linked with immunosuppression as many of the proangiogenic factors are also immunosuppressive.

Extensive study of the tumor microenvironment has revealed additional mediators of immunosuppression. Tumor cells produce and stimulate stromal cell release of immunosuppressive cytokines that inhibit tumor-specific effector cells if they manage to reach the tumor. Tumor cells also recruit or reprogram immune effectors into inhibitory variants in order to co-opt normal immunosuppressive mechanisms. These two common features of tumor microenvironments suggest universal immune escape strategies that can be therapeutically targeted.

4.2 Immunosuppressive cytokines

An immune response is dependent on more than just the integration of signaling of activating and inhibitory receptors on effector cells. Cytokines and chemokines in the local environment are critical regulators of immune cell functionality, and tumor cells often exploit them in order to suppress CTL and NK cell activity. It has been well established that tumor cells spontaneously release immunosuppressive cytokines.[149] These cytokines and chemokines include transforming growth factor beta (TGF-β), VEGF, prostaglandin E_2 (PGE_2), IL-10, and Th2 inflammatory mediators.

TGF-β signaling in cancer has both anti-tumor and tumorigenic properties and has been referred to as a "double-edged sword."[150] This is because TGF-β can arrest the cell cycle and suppresses early tumorigenesis. However, established tumors become resistant to those effects and produce TGF-β for inhibition of host-tumor immune responses. TGF-β is a critical regulator of chemokine release and promotes the proangiogenic switch. Moreover, TGF-β has been shown to directly inhibit the activity of CTLs, NK cells, and macrophages.[151] TGF-β signaling in CD8+ T cells also promotes their conversion to an IL-17 expressing phenotype that promotes tumor cell survival.[152] Additionally, TGF-β secreted by tumor cells acts on dendritic cells in the microenvironment to prevent their maturation.[153] Immature dendritic cells are unable to process and present antigens to T cells to activate them. By suppressing CTL activation and function, TGF-β is thought to play a predominant role in tumor growth *in vivo*, a supposition that has been confirmed in several mouse models.[154]

One of the most well studied of all the tumor-derived cytokines is VEGF. VEGF is the most prominent proangiogenic cytokine in the tumor microenvironment and was the first cytokine demonstrated to inhibit dendritic cell maturation.[155] As tumors grow large enough to require their own blood supply, they initially encounter a hypoxic environment. Hypoxic conditions drive the expression of transcription factors for VEGF in tumor cells. Subsequent VEGF signaling promotes angiogenesis, inflammation, and alterations in the development of immune progenitors. VEGF was identified to affect the maturation process of the immature granulocyte–macrophage progenitors. VEGF stimulates the recruitment of macrophages to the tumor microenvironment and is associated with their conversion to the immunosuppressive variant: tumor-associated macrophage (TAM).[156]

The COX2 gene is frequently overexpressed in cancer, with angiogenic pathway consequences.[157] The product of COX2 can be converted to PGE_2, a proinflammatory factor implicated in several immunosuppressive mechanisms. PGE_2 signaling in tumor cells is important for growth, angiogenesis, metastasis, and stem-cell like properties.[158] PGE_2 signaling in lymphocytes triggers their differentiation into an immunosuppressive Treg phenotype.[159] PGE_2 leads to increased IL-10 production in lymphocytes but decreased IL-12 secretion by macrophages.[160] IL-10 is a powerful cytokine with a myriad of effects on cell types present in the tumor microenvironment, and IL-12 has an important role in stimulating the cytotoxic activity of NK and CD8+ T cells. Like TGF-β and VEGF, IL-10 also compromises the function and maturation of dendritic cells.[161] IL-10 protects tumor cells from immune attack by also stimulating the loss of antigenicity. IL-10 has been shown to mediate MHC class I loss on tumor cells by inhibiting TAP1 and TAP2 gene expression.[162] Furthermore, IL-10 in combination with TGF-β is responsible for shifting the balance between Th1 and Th2 immune responses.[163] The Th1 immune response helps facilitate CTL activation and tumor cell lysis, and shifts to Th2 phenotypes abrogates those responses.[164] Tumor cells will release other factors that help initiate and maintain a Th2 phenotype. Tumor cell lysates contained IL-4, IL-5, and IL-9, which are all Th2 promoting inflammatory cytokines.[165]

In addition to cytokines, tumors produce enzymes that are immunosuppressive. Indoleamine 2,3-dioxygenase (IDO) is an enzyme that catalyzes the degradation of the amino acid tryptophan. Tumor cells will upregulate IDO production in response to IFNγ released by TILs.[166] T cells are preferentially sensitive to tryptophan starvation and IDO expression is linked with immunosuppression and conversion of T cells to inhibitory phenotypes.[167] Metabolism of amino acids and their substrates appears to be another powerful moderator of T cell function. The enzyme arginase-1 is also frequently upregulated in the tumor microenvironment and has similar effects on T cells as IDO.[168] Many of the cytokines and molecules released by tumor cells have tumor angiogenesis and growth enhancing properties on their own. For many, there is considerable overlap of their immunomodulatory effects so it is possible that their immunosuppressive functions are side effects rather than requirements for immune escape.

4.3 Recruitment of inhibitory immune cells

Tumor microenvironment cellular constituents include TILs, NK cells, macrophages, dendritic cells, and myeloid lineage cells. At first glance, the representation of immune effector cells in the tumor mass would suggest a productive immune response. However, most of the immune cells residing in the tumor microenvironment are functionally impaired in some manner. Many of the immune cell populations are converted to phenotypes that further impair immune system responses. Tumors are able to recruit Treg lymphocytes, TAMs, myeloid-derived suppressor cells (MDSCs), and cancer-associated fibroblasts (CAFs) to aid them in escape from immune recognition.

Certain populations of T cells were observed to have immunosuppressing regulatory functions in the 1970s.[169] Tregs are CD4+ T helper lymphocytes that also express FOXP3 and CD25. They were established as inhibitors of the anti-tumor immune response in animal experiments where their depletion significantly improved tumor-specific CTL function.[170] Several mechanisms for Treg immunosuppressive function have been reported. Tregs secrete TGF-β and IL-10 and also interact with other immune cells *via* their cytotoxic T-lymphocyte antigen 4 (CTLA4). The production of those cytokines and signaling *via* CTLA4 promotes a tolerogenic environment that propagates immunosuppression to other cell types.[171] Furthermore, Tregs express receptors with high affinity for cytokines that favor anti-tumor immunity and act as sinks to remove those cytokines from the environment.[172] Tregs are abundant in many tumor types, and become recruited to the tumor microenvironment in a process mediated by CCL22.[173]

MDSCs are a heterogeneous population of CD11b+ cells of either monocytic or granulocytic origins. They are stimulated to expand by granulocyte–macrophage colony-stimulating factor (GM-CSF) and possess the potential to differentiate into macrophages, dendritic cells, or neutrophils.[174] Recent evidence from a model of pancreatic cancer has implicated YAP as the primary driver of MDSC accumulation in the tumor microenvironment.[175] MDSCs are elevated in cancer patients and mouse models and have been linked to suppression of immune response.[176] MDSCs are known to inhibit the proliferation of T cells, block NK cell activation, and enhance the Th2 (tumor-promoting) phenotype. MDSCs also inhibit immune effector cell cytotoxicity by secreting arginase-1 and NOS2 into the tumor microenvironment, sequestering cysteine, or producing reactive oxygen species.[177,178] MDSCs can interfere with tumor cell antigen presentation, which allows tumors to avoid detection.[179] Another mechanism by which MDSCs facilitate tumor escape is by inducing the downregulation of L-selectin in T cells.[180] Without L-selectin on the surface, T cells are unable to home to the tumor. MDSCs also play a key role in determining the balance of macrophage phenotypes. M1 macrophages reinforce Th1 populations to promote anti-tumor immunity whereas M2 phenotypes are considered to be TAMs. In the presence of IL-10, interaction of MDSCs with M1 macrophages causes downregulation of IL-12 and MHC class II expression effectively shutting down antigen presentation and profoundly limiting immunosurveillance by the immune system.[181]

The inflammatory and perturbed environment present early in tumorigenesis attracts macrophages. These initial responders exhibit the classical M1 phenotype that secretes a large amount of proinflammatory cytokines that stimulate CTL responses. However, during tumor progression, prolonged exposure of the macrophages to tumor-derived TGF-β and IL-10 drives the macrophages toward the M2 phenotype commonly referred to as TAMs.[182] TAMs assist tumors by producing growth factors, proangiogenic molecules, and T cell inhibitory cytokines.[183] TAMs are abundant in the tumor milieu and are most often located in areas of hypoxia, suggesting that VEGF is a chemoattractant.

Normal resident fibroblasts only become activated in response to tissue injury, where they differentiate into myofibroblasts. In the tumor microenvironment, there is a high number of myofibroblasts that have been termed CAFs.[184] CAFs can be generated from a number of precursors. However, the exact factors that tumors secrete to influence their development have not been elucidated. CAFs support tumor growth by producing growth factors such as EGF, FGF, and IGF.[185] CAFs also facilitate epithelial–mesenchymal transition (EMT) in nearby tumor cells by releasing TGF-β. Recent studies suggest that CAFs may have immunosuppressive functions related to their secretion of fibroblast activation protein (FAP).[186]

4.4 Clinical relevance

Constituents of the tumor microenvironment have a significant effect on patient outcomes. Some of the deadliest cancers, such as pancreatic ductal adenocarcinoma, have relatively dense immunosuppressive environments that have rendered many therapeutic strategies ineffective. Even with the advent of immunotherapy, tumors that are considered immunologically "cold" due to the effects of the tumor microenvironment are less likely to respond to treatment. A strong piece of evidence for the powerful effects of the tumor microenvironment came from a study of T cell responses in mice bearing late stage tumors. Systemic T cells from these mice had normal effector functions when used in *in vitro* and *in vivo* assays, suggesting that T cell dysfunction only occurs at the site of the tumor.[187] Both tumor-derived cytokines and tumor-associated immune cell variants are useful prognostic indicators in patients. High concentrations of serum TGF-β are commonly observed in patients and are markers of disease progression.[188] Likewise, increased VEGF,[189] IL-10,[190] and IDO[191] are all associated with poor survival. Certain immune cell variants have complex roles as biomarkers. Increased numbers of Tregs in the tumor were initially reported as correlating with decreased survival. However, some studies stated that Tregs numbers were indicative of better outcomes. Considerations of the other immune cell populations revealed that the ratio of CD8+ T cells to Tregs was the most accurate prognostic maker.[192] MDSCs consist of many different subsets each with their own set of cell surface expression profiles, which complicates their use as a marker. In any event, studies have shown that the presence of certain MDSC populations correlate with poor prognosis in some cancer types.[193] More robust preclinical and clinical evidence exists to support the link between an abundance of TAMs and worse outcome.[194] Finally, while certain CAF produced molecules such as FAP have been correlated to patient prognosis in some studies, the lack of defined signatures for CAFs has made it difficult to evaluate their prognostic potential.[195]

5. Conclusion

Tumors cells must overcome or bypass a large number of mechanisms that have evolved in vertebrates to prevent malignant transformation. Until relatively recently, the contributions of the immune system to tumor prevention and control were under-appreciated by the scientific community. Based on extensive experimental studies in animal models and patients, there is now widespread agreement that the immune system is able to recognize tumors as aberrant growths and mount anti-tumor immune responses. However, tumors develop a myriad of tools to counteract host anti-tumor immunity. The main mechanisms for resistance to immune attack that has been identified are the loss of antigenicity, loss of immunogenicity, and construction of an immunosuppressive microenvironment. The discovery and characterization of these mechanisms are implicit confirmation that immunosurveillance does exist and that cancers must continually develop means of evading immune-mediated destruction. Although tumor immune evasion strategies are being used as promising new therapeutic targets, it should also be considered that the anti-tumor response may play a limited role when confronted with a rapidly growing and mutating heterogeneous tumor. A fundamental immune escape strategy may simply be the rapid proliferation and overwhelming of effector immune cells. Additionally, the cancer stem cell (CSC) theory proposes that tumor growth is driven by a small subset of tumor cells with stem cell like properties. CSCs have been identified in many cancer types and it has been suggested that their plasticity confers them resistance to both the immune system and conventional therapy.[196,197] Subscribers of the CSC theory argue that to eliminate tumors, the CSC subpopulation must be targeted in order to prevent recurrence. Since CSCs may express unique antigens ongoing studies aim to develop immunotherapy regimens that can specifically target these cells.[198] However, there still remains debate about the nature of CSCs and how identify them as well as the competing clonal evolution model of cancer.[199] Ultimately, either model of cancer development suggests that, in order to eliminate an entire tumor, treatments must address different tumor cell populations. Immunotherapies have shown promising results in the clinic and represent a promising therapeutic option able to target such diverse tumor cells.

The finding that tumor cells expressed tumor-specific antigens was the first major step in establishing immunotherapy as a viable treatment option. Based on those results, several approaches to prompting an anti-tumor immune response were explored. Tumor cells can evade host immunity by becoming poor antigen presenters, thus vaccination with tumor-derived antigens was hypothesized as an effective stimulant to tumor-specific effector cells. Unfortunately, most cancer therapeutic vaccines have been unsuccessful and only one has been FDA approved so far.[200] It has been suggested that presentation of tumor antigens by dendritic cells may be a critical component for determining the anti-tumor response and therefore dendritic cell-based vaccine strategies may improve outcomes.[201] Building on the fact that tumor-specific antigens existed, the isolation of CTLs that recognized those antigens from patients led to the development of adoptive T cell therapy. Obtaining a patient's own T cells for expansion, stimulation, and re-transplantation has proved to be a powerful weapon against some cancers in the clinic.[202] The other therapeutic modality has seen widespread success is the use of targeted monoclonal antibodies. Recombinant antibody technology has allowed for the creation of a vast number of antibodies that target tumor-specific antigens or proteins. Antibodies bind target cells and engage the Fc receptors on effector cells, leading to proper activation of an immune response even in the absence of correct antigen presentation on tumor cells. Unfortunately, tumor cells can become resistant to monoclonal antibodies by employing the mechanisms underlying the loss of antigenicity to downregulate or eliminate the targets of the antibodies.[203] Researchers are currently exploring more advanced antibodies using engineering strategies to create bi-specific variants that recruit immune cells and to include Fc portions with more robust activation activity. In order to more directly address tumor mechanisms behind the loss of antigenicity, several methods for counteracting the loss of MHC class I are currently undergoing trials. These include gene therapy to restore defective antigen presentation genes, HDAC inhibitors to stop epigenetic modification of the antigen presentation process, and interferon-based therapy to induce MHC class I expression.[204]

Perhaps the most successful new cancer treatment strategy has been the introduction of immune checkpoint blockade. In some cases tumors retain their antigenicity but block the immune response by upregulating inhibitory checkpoint ligands such as PD-L1. The first immune checkpoint to be targeted was CTLA4, which is expressed exclusively on T cells. Significant increases in overall survival in many patients led to the rapid approval of anti-CTLA4 antibodies.[205] The proof of concept that immune checkpoint blockade could potentiate long lasting tumor remission led to a massive increase in interest in tumor immune escape mechanisms. Anti-PD-1 therapy has demonstrated that immune checkpoint blockade not only overcomes tumor loss of immunogenicity, but it can also reinvigorate and restore the functionality of exhausted T cells.[206] Furthermore, immune checkpoint blockade also inhibits immunosuppressive cells such as Tregs and relieves their obstruction. Anti-CD47 antibodies have also shown great promise in preclinical studies and are currently being tested in patients.[207] However, targeting death receptor signaling has proved difficult and so far there have been few successes.[208] Resistance to all of these antibody strategies still arises in a large proportion of patients and those mechanisms are currently the subject of concentrated investigation.

Tumor microenvironment composition can vary greatly between cancer types and from patient to patient. However, common features of immunosuppressive cell populations suggest that targeting the TME is a viable strategy. Indeed, targeted approaches aimed at eliminating Tregs, MDSCs, and TAMs have had some successes.[209–211] The most successful application of targeting the tumor microenvironment has been anti-angiogenic therapy. Anti-VEGF antibodies were originally developed with the intention of blocking tumor angiogenesis and thus starving solid tumors of nutrients. Upon closer inspection, one of the main mechanisms behind therapeutic efficacy was the removal of the immunosuppressive functions of angiogenic cytokines. In addition, the restoration of normalized vasculature facilitated better immune infiltration and anti-tumor response.[212] Anti-VEGF therapy has been FDA approved in the treatment of several cancers and is currently being investigated in combination with many immunotherapies. The targeting of other soluble mediators in the tumor microenvironment has been an attractive option and strategies to block TGF-β, IDO, and prostaglandin E_2 are in development. For the so-called "immune privileged" tumors, whose microenvironments are dominated by immunosuppressive mediators, the targeting of the microenvironment may open the door for other immunotherapy options. Future treatment design will likely incorporate a multi-step process where initial targeting of immunosuppressive cells and cytokines paves the way for immune stimulating regimens such as immune checkpoint blockade.

The immune system plays a pivotal role in tumor biology. Tumors are under constant assault by the host during their development, which shapes their evolution of immune escape mechanisms. Immunotherapy has been an extremely successful addition to the clinical arsenal in the fight against cancer. Still, many patients either do not respond or quickly become resistant to current immunotherapy regimens. This reflects the ability of tumors to continue to adapt to immune selection pressure at all stages of development. Combinations of immunotherapy with more traditional chemotherapy or surgery are promising. Additional study of immune escape mechanisms and immunotherapy resistance mechanisms will be needed to inform future treatment approaches. Finally, characterization of each patient's unique tumor immune profile will help further personalize cancer therapy and improve clinical outcomes.

References

1. Hanahan D, Weinberg R. The hallmarks of cancer. *Cell.* 2000;100(1):57–70.
2. Ehrlich P. Ueber den jetzigen stand der karzinomforschung. *Ned Tijdschr Geneeskd.* 1909;5:273–290.
3. Old L, Boyse E. Immunology of experimental tumors. *Annu Rev Med.* 1964;15(1): 167–186.

4. Burnet M. Cancer—a biological approach: I. The processes of control. II. The significance of somatic mutation. *Br Med J.* 1957;1(5022):779–786.
5. Thomas L. Cellular and humoral aspects of the hypersensitive states: a symposium at the New York Academy of Medicine. *JAMA.* 1959;170(7):883.
6. Stutman O. Tumor development after 3-methylcholanthrene in immunologically deficient athymic-nude mice. *Science.* 1974;183(4124):534–536.
7. Pardoll D. Does the immune system see tumors as foreign or self? *Annu Rev Immunol.* 2003;21(1):807–839.
8. Dighe A, Richards E, Old L, Schreiber R. Enhanced in vivo growth and resistance to rejection of tumor cells expressing dominant negative IFNγ receptors. *Immunity.* 1994;1(6):447–456.
9. van den Broek M. Decreased tumor surveillance in perforin-deficient mice. *J Exp Med.* 1996;184(5):1781–1790.
10. Shankaran V, Ikeda H, Bruce A, et al. IFNγ and lymphocytes prevent primary tumour development and shape tumour immunogenicity. *Nature.* 2001;410(6832):1107–1111.
11. Vesely M, Kershaw M, Schreiber R, Smyth M. Natural innate and adaptive immunity to cancer. *Annu Rev Immunol.* 2011;29(1):235–271.
12. Reuschenbach M, von Knebel Doeberitz M, Wentzensen N. A systematic review of humoral immune responses against tumor antigens. *Cancer Immunol Immunother.* 2009;58(10):1535–1544.
13. Clark W, Elder D, Guerry D, et al. Model predicting survival in stage I melanoma based on tumor progression. *J Natl Cancer Inst.* 1989;81(24):1893–1904.
14. Clemente C, Mihm M, Bufalino R, Zurrida S, Collini P, Cascinelli N. Prognostic value of tumor infiltrating lymphocytes in the vertical growth phase of primary cutaneous melanoma. *Cancer.* 1996;77(7):1303–1310.
15. Zhang L, Conejo-Garcia J, Katsaros D, et al. Intratumoral T cells, recurrence, and survival in epithelial ovarian cancer. *N Engl J Med.* 2003;348(3):203–213.
16. Galon J. Type, density, and location of immune cells within human colorectal tumors predict clinical outcome. *Science.* 2006;313(5795):1960–1964.
17. Fridman W, Dieu-Nosjean M, Pagès F, et al. The immune microenvironment of human tumors: general significance and clinical impact. *Cancer Microenviron.* 2012;6(2):117–122.
18. Bindea G, Mlecnik B, Tosolini M, et al. Spatiotemporal dynamics of intratumoral immune cells reveal the immune landscape in human cancer. *Immunity.* 2013;39(4):782–795.
19. Schreiber R, Old L, Smyth M. Cancer immunoediting: integrating immunity's roles in cancer suppression and promotion. *Science.* 2011;331(6024):1565–1570.
20. Dunn G, Bruce A, Ikeda H, Old L, Schreiber R. Cancer immunoediting: from immunosurveillance to tumor escape. *Nat Immunol.* 2002;3(11):991–998.
21. DuPage M, Mazumdar C, Schmidt L, Cheung A, Jacks T. Expression of tumour-specific antigens underlies cancer immunoediting. *Nature.* 2012;482(7385):405–409.
22. Hanahan D, Weinberg R. Hallmarks of cancer: the next generation. *Cell.* 2011;144(5):646–674.
23. Williams A, Peh C, Elliott T. The cell biology of MHC class I antigen presentation. *Tissue Antigens.* 2002;59(1):3–17.
24. Rock K, Gramm C, Rothstein L, et al. Inhibitors of the proteasome block the degradation of most cell proteins and the generation of peptides presented on MHC class I molecules. *Cell.* 1994;78(5):761–771.
25. Rock K, Goldberg A. Degradation of cell proteins and the generation of MHC class I-presented peptides. *Annu Rev Immunol.* 1999;17(1):739–779.
26. Townsend A. The epitopes of influenza nucleoprotein recognized by cytotoxic T lymphocytes can be defined with short synthetic peptides. *Cell.* 1986;44(6):959–968.

27. Ljunggren H, Kärre K. In search of the 'missing self': MHC molecules and NK cell recognition. *Immunol Today*. 1990;11:237–244.
28. Zilber L. Specific tumor antigens. *Adv Cancer Res*. 1958;5:291–329.
29. Ilyas S, Yang J. Landscape of tumor antigens in T cell immunotherapy. *J Immunol*. 2015;195(11):5117–5122.
30. DeVita V, Hellman S, Rosenberg S. *Biological Therapy of Cancer*. Philiadelphia, PA: J.B. Lippincott; 1995.
31. Hellstrom KE. Studies on isoantigenic variation in mouse lymphomas. *J Natl Cancer Inst*. 1960;25(2):237–269.
32. Moller E. Quantitative studies of the sensitivity of normal and neoplastic mouse cells to the cytotoxic action of isoantibodies. *J Exp Med*. 1962;115(3):527–553.
33. Garrido F, Cabrera T, Concha A, Glew S, Ruiz-Cabello F, Stern P. Natural history of HLA expression during tumour development. *Immunol Today*. 1993;14(10):491–499.
34. Hui K. Re-expression of major histocompatibility complex (UMHC) class I molecules on malignant tumor cells and its effect on host-tumor interaction. *Bioessays*. 1989;11(1): 22–26.
35. Cabrera T, Fernandez M, Sierra A, et al. High frequency of altered HLA class I phenotypes in invasive breast carcinomas. *Hum Immunol*. 1996;50(2):127–134.
36. Koopman L, Corver W, van der Slik A, Giphart M, Fleuren G. Multiple genetic alterations cause frequent and heterogeneous human histocompatibility leukocyte antigen class I loss in cervical cancer. *J Exp Med*. 2000;191(6):961–976.
37. Garrido F, Ruiz-Cabello F, Cabrera T, et al. Implications for immunosurveillance of altered HLA class I phenotypes in human tumours. *Immunol Today*. 1997;18(2):89–95.
38. Hicklin D, Wang Z, Arienti F, Rivoltini L, Parmiani G, Ferrone S. beta2-Microglobulin mutations, HLA class I antigen loss, and tumor progression in melanoma. *J Clin Investig*. 1998;101(12):2720–2729.
39. Branch P, Bicknell D, Rowan A, Bodmer W, Karran P. Immune surveillance in colorectal carcinoma. *Nat Genet*. 1995;9(3):231–232.
40. Cabrera T, Concha A, Ruiz-Cabello F, Garrido F. Loss of HLA heavy chain and beta2-microglobulin in HLA negative tumours. *Scand J Immunol*. 1991;34(2):147–152.
41. Garrido F. HLA and cancer. *Tissue Antigens*. 1996;47(5):361–363.
42. Blanchet O, Bourge J, Zinszner H, Tatari Z, Degos L, Paul P. DNA binding of regulatory factors interacting with MHC-class-I gene enhancer correlates with MHC-class-I transcriptional level in class-I-defective cell lines. *Int J Cancer*. 1991;47(S6): 138–145.
43. van't Veer L, Beijersbergen R, Bernards R. N-myc suppresses major histocompatibility complex class I gene expression through down-regulation of the p50 subunit of NF-κB. *EMBO J*. 1993;12(1):195–200.
44. Dovhey S, Ghosh N, Wright K. Loss of interferon-gamma inducibility of TAP1 and LMP2 in a renal cell carcinoma cell line. *Cancer Res*. 2000;60(20):5789–5796.
45. Campoli M, Ferrone S. HLA antigen changes in malignant cells: epigenetic mechanisms and biologic significance. *Oncogene*. 2008;27(45):5869–5885.
46. Garcia-Lora A, Martinez M, Algarra I, Gaforio J, Garrido F. MHC class I-deficient metastatic tumor variants immunoselected by T lymphocytes originate from the coordinated downregulation of APM components. *Int J Cancer*. 2003;106(4):521–527.
47. Algarra I, Cabrera T, Garrido F. The HLA crossroad in tumor immunology. *Hum Immunol*. 2000;61(1):65–73.
48. Miguel Real L, Cabrera T, Collado A, et al. Expression of HLA G in human tumors is not a frequent event. *Int J Cancer*. 1999;81(4):512–518.
49. York I, Rock K. Antigen processing and presentation by the class I major histocompatibility complex. *Annu Rev Immunol*. 1996;14(1):369–396.

50. Restifo N. Identification of human cancers deficient in antigen processing. *J Exp Med.* 1993;177(2):265–272.
51. Singal D, Ye M, Ni J, Snider D. Markedly decreased expression of TAP1 and LMP2 genes in HLA class I-deficient human tumor cell lines. *Immunol Lett.* 1996;50(3): 149–154.
52. Leone P, Shin E, Perosa F, Vacca A, Dammacco F, Racanelli V. MHC class I antigen processing and presenting machinery: organization, function, and defects in tumor cells. *J Natl Cancer Inst.* 2013;105(16):1172–1187.
53. Setiadi A, David M, Seipp R, Hartikainen J, Gopaul R, Jefferies W. Epigenetic control of the immune escape mechanisms in malignant carcinomas. *Mol Cell Biol.* 2007;27(22): 7886–7894.
54. Rotem-Yehudar R. LMP-associated proteolytic activities and TAP-dependent peptide transport for class 1 MHC molecules are suppressed in cell lines transformed by the highly oncogenic adenovirus 12. *J Exp Med.* 1996;183(2):499–514.
55. Johnsen A, Templeton D, Sy M, Harding C. Deficiency of transporter for antigen presentation (TAP) in tumor cells allows evasion of immune surveillance and increases tumorigenesis. *J Immunol.* 1999;163(8):4224–4231.
56. Dierssen J, de Miranda N, Ferrone S, et al. HNPCC versus sporadic microsatellite-unstable colon cancers follow different routes toward loss of HLA class I expression. *BMC Cancer.* 2007;7(1):33.
57. Arshad N, Cresswell P. Tumor-associated calreticulin variants functionally compromise the peptide loading complex and impair its recruitment of MHC-I. *J Biol Chem.* 2018;293(25):9555–9569.
58. Boon T, De Plaen E, Lurquin C, et al. Identification of tumour rejection antigens recognized by T lymphocytes. *Cancer Surv.* 1992;3:23–37.
59. Cormier J, Hijazi Y, Abati A, et al. Heterogeneous expression of melanoma-associated antigens and HLA-A2 in metastatic melanoma in vivo. *Int J Cancer.* 1998;75(4): 517–524.
60. Jäger E, Ringhoffer M, Karbach J, Arand M, Oesch F, Knuth A. Inverse relationship of melanocyte differentiation antigen expression in melanoma tissues and CD8+ cytotoxic-T-cell responses: evidence for immunoselection of antigen-loss variants in vivo. *Int J Cancer.* 1996;66(4):470–476.
61. Tanaka Y, Tevethia M, Kalderon D, Smith A, Tevethia S. Clustering of antigenic sites recognized by cytotoxic T lymphocyte clones in the amino terminal half of SV40 T antigen. *Virology.* 1988;162(2):427–436.
62. Seung S. A tumor escape variant that has lost one major histocompatibility complex class I restriction element induces specific CD8+ T cells to an antigen that no longer serves as a target. *J Exp Med.* 1993;178(3):933–940.
63. Hopper-Borge E, Nasto R, Ratushny V, Weiner L, Golemis E, Astsaturov I. Mechanisms of tumor resistance to EGFR-targeted therapies. *Expert Opin Ther Targets.* 2009;13(3):339–362.
64. Kageshita T, Hirai S, Ono T, Hicklin D, Ferrone S. Down-regulation of HLA class I antigen-processing molecules in malignant melanoma. *Am J Pathol.* 1999;154(3): 745–754.
65. Seliger B, Maeurer M, Ferrone S. Antigen-processing machinery breakdown and tumor growth. *Immunol Today.* 2000;21(9):455–464.
66. Restifo N, Marincola F, Kawakami Y, Taubenberger J, Yannelli J, Rosenberg S. Loss of functional beta2-microglobulin in metastatic melanomas from five patients receiving immunotherapy. *J Natl Cancer Inst.* 1996;88(2):100–108.
67. Anaya J. OncoLnc: linking TCGA survival data to mRNAs, miRNAs, and lncRNAs. *PeerJ Comput Sci.* 2016;2:e67.

68. Van den Eynde B, Scott A. Tumor antigens. In: *Encyclopedia of Immunology*. Elsevier; 1998:2424–2431.
69. Green J, Noel P, Sperling A, et al. Absence of B7-dependent responses in CD28-deficient mice. *Immunity*. 1994;1(6):501–508.
70. Chen L, Flies D. Molecular mechanisms of T cell co-stimulation and co-inhibition. *Nat Rev Immunol*. 2013;13(4):227–242.
71. Schwartz R. T cell anergy. *Annu Rev Immunol*. 2003;21(1):305–334.
72. Driessens G, Kline J, Gajewski T. Costimulatory and coinhibitory receptors in anti-tumor immunity. *Immunol Rev*. 2009;229(1):126–144.
73. Rudd C. The reverse stop-signal model for CTLA4 function. *Nat Rev Immunol*. 2008;8(2):153–160.
74. Pegram H, Andrews D, Smyth M, Darcy P, Kershaw M. Activating and inhibitory receptors of natural killer cells. *Immunol Cell Biol*. 2010;89(2):216–224.
75. Ho E, Carayannopoulos L, Poursine-Laurent J, et al. Costimulation of multiple NK cell activation receptors by NKG2D. *J Immunol*. 2002;169(7):3667–3675.
76. Yeap W, Wong K, Shimasaki N, et al. CD16 is indispensable for antibody-dependent cellular cytotoxicity by human monocytes. *Sci Rep*. 2016;6(1):34310.
77. Dong H, Strome S, Salomao D, et al. Erratum: tumor-associated B7-H1 promotes T-cell apoptosis: a potential mechanism of immune evasion. *Nat Med*. 2002;8(8):793–800.
78. Parsa A, Waldron J, Panner A, et al. Loss of tumor suppressor PTEN function increases B7-H1 expression and immunoresistance in glioma. *Nat Med*. 2006;13(1):84–88.
79. Marzec M, Zhang Q, Goradia A, et al. Oncogenic kinase NPM/ALK induces through STAT3 expression of immunosuppressive protein CD274 (PD-L1, B7-H1). *Proc Natl Acad Sci U S A*. 2008;105(52):20852–20857.
80. Taube J, Anders R, Young G, et al. Colocalization of inflammatory response with B7-H1 expression in human melanocytic lesions supports an adaptive resistance mechanism of immune escape. *Sci Transl Med*. 2012;4(127):127ra37.
81. Rosenwald A, Wright G, Leroy K, et al. Molecular diagnosis of primary mediastinal B cell lymphoma identifies a clinically favorable subgroup of diffuse large B cell lymphoma related to Hodgkin lymphoma. *J Exp Med*. 2003;198(6):851–862.
82. Ahmadzadeh M, Johnson L, Heemskerk B, et al. Tumor antigen-specific CD8 T cells infiltrating the tumor express high levels of PD-1 and are functionally impaired. *Blood*. 2009;114(8):1537–1544.
83. Pauken K, Wherry E. Overcoming T cell exhaustion in infection and cancer. *Trends Immunol*. 2015;36(4):265–276.
84. Simon I, Zhuo S, Corral L, et al. B7-H4 is a novel membrane-bound protein and a candidate serum and tissue biomarker for ovarian cancer. *Cancer Res*. 2006;66(3):1570–1575.
85. Yi K, Chen L. Fine tuning the immune response through B7-H3 and B7-H4. *Immunol Rev*. 2009;229(1):145–151.
86. Kryczek I, Zou L, Rodriguez P, et al. B7-H4 expression identifies a novel suppressive macrophage population in human ovarian carcinoma. *J Exp Med*. 2006;203(4):871–881.
87. Pardoll D. The blockade of immune checkpoints in cancer immunotherapy. *Nat Rev Cancer*. 2012;12(4):252–264.
88. Zhu C, Anderson A, Schubart A, et al. The Tim-3 ligand galectin-9 negatively regulates T helper type 1 immunity. *Nat Immunol*. 2005;6(12):1245–1252.
89. Cai G, Freeman G. The CD160, BTLA, LIGHT/HVEM pathway: a bidirectional switch regulating T-cell activation. *Immunol Rev*. 2009;229(1):244–258.
90. Derré L, Rivals J, Jandus C, et al. BTLA mediates inhibition of human tumor-specific CD8+ T cells that can be partially reversed by vaccination. *J Clin Investig*. 2010;120(1):157–167.

91. Schwartz R, Mueller D, Jenkins M, Quill H. T-cell clonal anergy. *Cold Spring Harb Symp Quant Biol.* 1989;54(0):605–610.
92. Staveley-O'Carroll K, Sotomayor E, Montgomery J, et al. Induction of antigen-specific T cell anergy: an early event in the course of tumor progression. *Proc Natl Acad Sci U S A.* 1998;95(3):1178–1183.
93. Townsend S, Allison J. Tumor rejection after direct costimulation of CD8+ T cells by B7-transfected melanoma cells. *Science.* 1993;259(5093):368–370.
94. Chao M, Majeti R, Weissman I. Programmed cell removal: a new obstacle in the road to developing cancer. *Nat Rev Cancer.* 2011;12(1):58–67.
95. Chao M, Jaiswal S, Weissman-Tsukamoto R, et al. Calreticulin is the dominant pro-phagocytic signal on multiple human cancers and is counterbalanced by CD47. *Sci Transl Med.* 2010;2(63):63ra94.
96. Chao M, Weissman I, Majeti R. The CD47–SIRPα pathway in cancer immune evasion and potential therapeutic implications. *Curr Opin Immunol.* 2012;24(2): 225–232.
97. Kaur S, Chang T, Singh S, et al. CD47 signaling regulates the immunosuppressive activity of VEGF in T cells. *J Immunol.* 2014;193(8):3914–3924.
98. Chao M, Tang C, Pachynski R, Chin R, Majeti R, Weissman I. Extranodal dissemination of non-Hodgkin lymphoma requires CD47 and is inhibited by anti-CD47 antibody therapy. *Blood.* 2011;118(18):4890–4901.
99. Shuptrine C, Ajina R, Fertig E, et al. An unbiased in vivo functional genomics screening approach in mice identifies novel tumor cell-based regulators of immune rejection. *Cancer Immunol Immunother.* 2017;66(12):1529–1544.
100. Lauritzsen G, Hofgaard P, Schenck K, Bogen B. Clonal deletion of thymocytes as a tumor escape mechanism. *Int J Cancer.* 1998;78(2):216–222.
101. Höchst B, Diehl L. Antigen shedding into the circulation contributes to tumor immune escape. *OncoImmunology.* 2012;1(9):1620–1622.
102. Groh V, Wu J, Yee C, Spies T. Tumour-derived soluble MIC ligands impair expression of NKG2D and T-cell activation. *Nature.* 2002;419(6908):734–738.
103. Reiners K, Topolar D, Henke A, et al. Soluble ligands for NK cell receptors promote evasion of chronic lymphocytic leukemia cells from NK cell anti-tumor activity. *Blood.* 2013;121(18):3658–3665.
104. Ichim T, Zhong Z, Kaushal S, et al. Exosomes as a tumor immune escape mechanism: possible therapeutic implications. *J Transl Med.* 2008;6(1):37.
105. Clayton A, Mitchell J, Court J, Linnane S, Mason M, Tabi Z. Human tumor-derived exosomes down-modulate NKG2D expression. *J Immunol.* 2008;180(11):7249–7258.
106. Lundholm M, Schröder M, Nagaeva O, et al. Prostate tumor-derived exosomes down-regulate NKG2D expression on natural killer cells and CD8+ T cells: mechanism of immune evasion. *PLoS One.* 2014;9(9):e108925.
107. Muller L, Mitsuhashi M, Simms P, Gooding W, Whiteside T. Tumor-derived exosomes regulate expression of immune function-related genes in human T cell subsets. *Sci Rep.* 2016;6(1):20254.
108. Nagata S. Apoptosis by death factor. *Cell.* 1997;88(3):355–365.
109. French L, Tschopp J. Defective death receptor signaling as a cause of tumor immune escape. *Semin Cancer Biol.* 2002;12(1):51–55.
110. Pitti R, Marsters S, Lawrence D, et al. Genomic amplification of a decoy receptor for Fas ligand in lung and colon cancer. *Nature.* 1998;396(6712):699–703.
111. Strand S, Hofmann W, Hug H, et al. Lymphocyte apoptosis induced by CD95 (APO-1/Fas) ligand-expressing tumor cells—a mechanism of immune evasion? *Nat Med.* 1996;2(12):1361–1366.
112. Walker P, Saas P, Dietrich P. Tumor expression of Fas ligand (CD95L) and the consequences. *Curr Opin Immunol.* 1998;10(5):564–572.
113. Fulda S. Tumor resistance to apoptosis. *Int J Cancer.* 2009;124(3):511–515.

114. Kearney C, Vervoort S, Hogg S, et al. Tumor immune evasion arises through loss of TNF sensitivity. *Sci Immunol.* 2018;3(23):eaar3451.
115. Dunn G, Sheehan K, Old L, Schreiber R. IFN unresponsiveness in LNCaP cells due to the lack of JAK1 gene expression. *Cancer Res.* 2005;65(8):3447–3453.
116. Rodríguez T, Méndez R, Del Campo A, et al. Distinct mechanisms of loss of IFN-gamma mediated HLA class I inducibility in two melanoma cell lines. *BMC Cancer.* 2007;7(1):34.
117. Ackerman M, Nimmerjahn F. *Antibody Fc.* Elsevier, Inc; 2014.
118. Medema J, de Jong J, Peltenburg L, et al. Blockade of the granzyme B/perforin pathway through overexpression of the serine protease inhibitor PI-9/SPI-6 constitutes a mechanism for immune escape by tumors. *Proc Natl Acad Sci U S A.* 2001;98(20): 11515–11520.
119. Boyer M, Orchard P, Gorden K, Anderson P, Mclvor R, Blazar B. Dependency on intercellular adhesion molecule recognition and local interleukin-2 provision in generation of an in vivo CD8+ T-cell immune response to murine myeloid leukemia. *Blood.* 1995;85(9):2498–2506.
120. Budinsky A, Brodowicz T, Wiltschke C, et al. Decreased expression of ICAM-1 and its induction by tumor necrosis factor on breast-cancer cells in vitro. *Int J Cancer.* 1997;71(6):1086–1090.
121. Wimmenauer S, Keller H, Ruckauer K, et al. Expression of CD44, ICAM-1 and N-CAM in colorectal cancer. Correlation with the tumor stage and the phenotypical characteristics of tumor-infiltrating lymphocytes. *Anticancer Res.* 1997;17(4A): 2395–2400.
122. Aldeghaither D, Zahavi D, Murray J, et al. A mechanism of resistance to antibody-targeted immune attack. *Cancer Immunol Res.* 2019;7(2):230–243.
123. Dyck L, Mills K. Immune checkpoints and their inhibition in cancer and infectious diseases. *Eur J Immunol.* 2017;47(5):765–779.
124. Jung H, Jeong D, Ji S, et al. Overexpression of PD-L1 and PD-L2 is associated with poor prognosis in patients with hepatocellular carcinoma. *Cancer Res Treat.* 2017;49(1):246–254.
125. Wu L, Deng W, Yu G, et al. B7-H4 expression indicates poor prognosis of oral squamous cell carcinoma. *Cancer Immunol Immunother.* 2016;65(9):1035–1045.
126. Thijssen V, Heusschen R, Caers J, Griffioen A. Galectin expression in cancer diagnosis and prognosis: a systematic review. *Biochim Biophys Acta—Rev Cancer.* 2015;1855(2): 235–247.
127. Lan X, Li S, Gao H, et al. Increased BTLA and HVEM in gastric cancer are associated with progression and poor prognosis. *OncoTargets Ther.* 2017;10:919–926.
128. Gros A, Robbins P, Yao X, et al. PD-1 identifies the patient-specific CD8+ tumor-reactive repertoire infiltrating human tumors. *J Clin Investig.* 2014;124(5): 2246–2259.
129. Maleki Vareki S, Garrigós C, Duran I. Biomarkers of response to PD-1/PD-L1 inhibition. *Crit Rev Oncol Hematol.* 2017;116:116–124.
130. Uluckan O, Becker S, Deng H, et al. CD47 regulates bone mass and tumor metastasis to bone. *Cancer Res.* 2009;69(7):3196–3204.
131. Majeti R, Chao M, Alizadeh A, et al. CD47 is an adverse prognostic factor and therapeutic antibody target on human acute myeloid leukemia stem cells. *Cell.* 2009;138(2):286–299.
132. Herlyn M, Rodeck U, Koprowski H. Shedding of human tumor-associated antigens in vitro and in vivo. *Adv Cancer Res.* 1987;49:189–221.
133. Zusman I. Soluble tumor-associated antigens in cancer detection, prevention and therapy. *Med Sci Monit.* 2004;10(12):ra317–324.

134. Silva J, Garcia V, Rodriguez M, et al. Analysis of exosome release and its prognostic value in human colorectal cancer. *Genes Chromosomes Cancer*. 2011;51(4):409–418.
135. Whiteside T. The potential of tumor-derived exosomes for noninvasive cancer monitoring. *Expert Rev Mol Diagn*. 2015;15(10):1293–1310.
136. de Carvalho-Neto P, Santos M, de Carvalho M, et al. FAS/FASL expression profile as a prognostic marker in squamous cell carcinoma of the oral cavity. *PLoS One*. 2013;8(7):e69024.
137. Rodríguez-Berriguete G, Sánchez-Espiridión B, Cansino J, et al. Clinical significance of both tumor and stromal expression of components of the IL-1 and TNF-α signaling pathways in prostate cancer. *Cytokine*. 2013;64(2):555–563.
138. Duarte C, Willey C, Zhi D, et al. Expression signature of IFN/STAT1 signaling genes predicts poor survival outcome in glioblastoma multiforme in a subtype-specific manner. *PLoS One*. 2012;7(1):e29653.
139. Mrass P, Weninger W. Immune cell migration as a means to control immune privilege: lessons from the CNS and tumors. *Immunol Rev*. 2006;213(1):195–212.
140. Joyce J, Fearon D. T cell exclusion, immune privilege, and the tumor microenvironment. *Science*. 2015;348(6230):74–80.
141. Hanahan D, Coussens L. Accessories to the crime: functions of cells recruited to the tumor microenvironment. *Cancer Cell*. 2012;21(3):309–322.
142. Gajewski T, Schreiber H, Fu Y. Innate and adaptive immune cells in the tumor microenvironment. *Nat Immunol*. 2013;14(10):1014–1022.
143. Mittal K, Ebos J, Rini B. Angiogenesis and the tumor microenvironment: vascular endothelial growth factor and beyond. *Semin Oncol*. 2014;41(2):235–251.
144. Nagy J, Chang S, Shih S, Dvorak A, Dvorak H. Heterogeneity of the tumor vasculature. *Semin Thromb Hemost*. 2010;36(03):321–331.
145. Dirkx A, Oude Egbrink M, Kuijpers M, et al. Tumor angiogenesis modulates leukocyte-vessel wall interactions in vivo by reducing endothelial adhesion molecule expression. *Cancer Res*. 2003;63(9):2322–2329.
146. Nummer D, Suri-Payer E, Schmitz-Winnenthal H, et al. Role of tumor endothelium in CD4+CD25+ regulatory T cell infiltration of human pancreatic carcinoma. *J Natl Cancer Inst*. 2007;99(15):1188–1199.
147. Motz G, Coukos G. The parallel lives of angiogenesis and immunosuppression: cancer and other tales. *Nat Rev Immunol*. 2011;11(10):702–711.
148. Mulligan J, Young M. Tumors induce the formation of suppressor endothelial cells in vivo. *Cancer Immunol Immunother*. 2009;59(2):267–277.
149. Chouaib S, Asselin-Paturel C, Mami-Chouaib F, Caignard A, Blay J. The host–tumor immune conflict: from immunosuppression to resistance and destruction. *Immunol Today*. 1997;18(10):493–497.
150. Akhurst R, Derynck R. TGF-β signaling in cancer—a double-edged sword. *Trends Cell Biol*. 2001;11(11):S44–S51.
151. Yang L, Pang Y, Moses H. TGF-β and immune cells: an important regulatory axis in the tumor microenvironment and progression. *Trends Immunol*. 2010;31(6):220–227.
152. Nam J, Terabe M, Kang M, et al. Transforming growth factor beta subverts the immune system into directly promoting tumor growth through interleukin-17. *Cancer Res*. 2008;68(10):3915–3923.
153. Gabrilovich D. Mechanisms and functional significance of tumour-induced dendritic-cell defects. *Nat Rev Immunol*. 2004;4(12):941–952.
154. Wojtowicz-Praga S. Reversal of tumor-induced immunosuppression by TGF-beta inhibitors. *Invest New Drugs*. 2003;21(1):21–32.

155. Gabrilovich D, Chen H, Girgis K, et al. Production of vascular endothelial growth factor by human tumors inhibits the functional maturation of dendritic cells. *Nat Med.* 1996;2(10):1096–1103.
156. Johnson B, Osada T, Clay T, Lyerly H, Morse M. Physiology and therapeutics of vascular endothelial growth factor in tumor immunosuppression. *Curr Mol Med.* 2009;9(6):702–707.
157. Roelofs H, te Morsche R, van Heumen B, Nagengast F, Peters W. Over-expression of COX-2 mRNA in colorectal cancer. *BMC Gastroenterol.* 2014;14:1.
158. Majumder M, Xin X, Liu L, Girish G, Lala P. Prostaglandin E2 receptor EP4 as the common target on cancer cells and macrophages to abolish angiogenesis, lymphangiogenesis, metastasis, and stem-like cell functions. *Cancer Sci.* 2014;105(9):1142–1151.
159. Sharma S, Yang S, Zhu L, et al. Tumor cyclooxygenase-2/prostaglandin E2–dependent promotion of FOXP3 expression and CD4 + CD25 + T regulatory cell activities in lung cancer. *Cancer Res.* 2005;65(12):5211–5220.
160. Huang M, Stolina M, Sharma S, et al. Non-small cell lung cancer cyclooxygenase-2-dependent regulation of cytokine balance in lymphocytes and macrophages: up-regulation of interleukin 10 and down-regulation of interleukin 12 production. *Immunology.* 1998;58(6):1208–1216.
161. de Smedt T, van Mechelen M, De Becker G, Urbain J, Leo O, Moser M. Effect of interleukin-10 on dendritic cell maturation and function. *Eur J Immunol.* 1997;27(5): 1229–1235.
162. Asadullah K. Interleukin-10 therapy—review of a new approach. *Pharmacol Rev.* 2003;55(2):241–269.
163. Maeda H, Shiraishi A. TGF-beta contributes to the shift toward Th2-type responses through direct and IL-10-mediated pathways in tumor-bearing mice. *J Immunol.* 1996;156(1):73–78.
164. Knutson K, Disis M. Tumor antigen-specific T helper cells in cancer immunity and immunotherapy. *Cancer Immunol Immunother.* 2005;54(8):721–728.
165. Yan H, Pickup M, Pang Y, et al. Gr-1 + CD11b+ myeloid cells tip the balance of immune protection to tumor promotion in the premetastatic lung. *Cancer Res.* 2010;70(15):6139–6149.
166. Spranger S, Spaapen R, Zha Y, et al. Up-regulation of PD-L1, IDO, and T_{regs} in the melanoma tumor microenvironment is driven by CD8+ T cells. *Sci Transl Med.* 2013;5(200):200ra116.
167. Prendergast G. Immune escape as a fundamental trait of cancer: focus on IDO. *Oncogene.* 2008;27(28):3889–3900.
168. Bronte V, Zanovello P. Regulation of immune responses by L-arginine metabolism. *Nat Rev Immunol.* 2005;5(8):641–654.
169. Gershon R, Kondo K. Cell interactions in the induction of tolerance: the role of thymic lymphocytes. *Immunology.* 1970;18(5):723–737.
170. Shimizu J, Yamazaki S, Sakaguchi S. Induction of tumor immunity by removing CD25 +CD4 + T cells: a common basis between tumor immunity and autoimmunity. *J Immunol.* 1999;163(10):5211–5218.
171. Campbell D, Koch M. Treg cells: patrolling a dangerous neighborhood. *Nat Med.* 2011;17(8):929–930.
172. Gattinoni L, Finkelstein S, Klebanoff C, et al. Removal of homeostatic cytokine sinks by lymphodepletion enhances the efficacy of adoptively transferred tumor-specific CD8 + T cells. *J Exp Med.* 2005;202(7):907–912.
173. Shevach E. Fatal attraction: tumors beckon regulatory T cells. *Nat Med.* 2004;10(9): 900–901.
174. Gabrilovich D, Nagaraj S. Myeloid-derived suppressor cells as regulators of the immune system. *Nat Rev Immunol.* 2009;9(3):162–174.

175. Murakami S, Shahbazian D, Surana R, et al. Yes-associated protein mediates immune reprogramming in pancreatic ductal adenocarcinoma. *Oncogene*. 2016;36(9):1232–1244.
176. Burkholder B, Huang R, Burgess R, et al. Tumor-induced perturbations of cytokines and immune cell networks. *Biochim Biophys Acta—Rev Cancer*. 2014;1845(2): 182–201.
177. Talmadge J. Pathways mediating the expansion and immunosuppressive activity of myeloid-derived suppressor cells and their relevance to cancer therapy. *Clin Cancer Res*. 2007;13(18):5243–5248.
178. Srivastava M, Sinha P, Clements V, Rodriguez P, Ostrand-Rosenberg S. Myeloid-derived suppressor cells inhibit T-cell activation by depleting cystine and cysteine. *Cancer Res*. 2009;70(1):68–77.
179. Lu T, Ramakrishnan R, Altiok S, et al. Tumor-infiltrating myeloid cells induce tumor cell resistance to cytotoxic T cells in mice. *J Clin Investig*. 2011;121(10):4015–4029.
180. Hanson E, Clements V, Sinha P, Ilkovitch D, Ostrand-Rosenberg S. Myeloid-derived suppressor cells down-regulate L-selectin expression on CD4+ and CD8+ T cells. *J Immunol*. 2009;183(2):937–944.
181. Ostrand-Rosenberg S, Sinha P, Beury D, Clements V. Cross-talk between myeloid-derived suppressor cells (MDSC), macrophages, and dendritic cells enhances tumor-induced immune suppression. *Semin Cancer Biol*. 2012;22(4):275–281.
182. Sica A, Schioppa T, Mantovani A, Allavena P. Tumour-associated macrophages are a distinct M2 polarised population promoting tumour progression: potential targets of anti-cancer therapy. *Eur J Cancer*. 2006;42(6):717–727.
183. Ono M. Molecular links between tumor angiogenesis and inflammation: inflammatory stimuli of macrophages and cancer cells as targets for therapeutic strategy. *Cancer Sci*. 2008;99(8):1501–1506.
184. Sugimoto H, Mundel T, Kieran M, Kalluri R. Identification of fibroblast heterogeneity in the tumor microenvironment. *Cancer Biol Ther*. 2006;5(12):1640–1646.
185. Spaeth E, Dembinski J, Sasser A, et al. Mesenchymal stem cell transition to tumor-associated fibroblasts contributes to fibrovascular network expansion and tumor progression. *PLoS One*. 2009;4(4):e4992.
186. Kraman M, Bambrough P, Arnold J, et al. Suppression of antitumor immunity by stromal cells expressing fibroblast activation protein-α. *Science*. 2010;330(6005): 827–830.
187. Radoja S, Rao T, Hillman D, Frey A. Mice bearing late-stage tumors have normal functional systemic T cell responses in vitro and in vivo. *J Immunol*. 2000;164(5):2619–2628.
188. Gorsch S, Memoli V, Stukel T, Gold L, Arrick B. Immunohistochemical staining for transforming growth factor β1 associates with disease progression in human breast cancer. *Cancer Res*. 1992;52(24):6949–6952.
189. Saito H, Tsujitani S, Ikeguchi M, Maeta M, Kaibara N. Relationship between the expression of vascular endothelial growth factor and the density of dendritic cells in gastric adenocarcinoma tissue. *Br J Cancer*. 1998;78(12):1573–1577.
190. Wang H, Wang L, Chi P, et al. High level of interleukin-10 in serum predicts poor prognosis in multiple myeloma. *Br J Cancer*. 2016;114(4):463–468.
191. Brandacher G. Prognostic value of indoleamine 2,3-dioxygenase expression in colorectal cancer: effect on tumor-infiltrating T cells. *Clin Cancer Res*. 2006;12(4): 1144–1151.
192. Facciabene A, Motz G, Coukos G. T-regulatory cells: key players in tumor immune escape and angiogenesis. *Cancer Res*. 2012;72(9):2162–2171.
193. Tian T, Gu X, Zhang B, et al. Increased circulating CD14(+)HLA-DR-/low myeloid-derived suppressor cells are associated with poor prognosis in patients with small-cell lung cancer. *Cancer Biomark*. 2015;15(4):425–432.

194. Bingle L, Brown N, Lewis C. The role of tumour-associated macrophages in tumour progression: implications for new anticancer therapies. *J Pathol.* 2002;196(3): 254–265.
195. Paulsson J, Micke P. Prognostic relevance of cancer-associated fibroblasts in human cancer. *Semin Cancer Biol.* 2014;25:61–68.
196. Batlle E, Clevers H. Cancer stem cells revisited. *Nat Med.* 2017;23(10):1124–1134.
197. Codd A, Kanaseki T, Torigo T, Tabi Z. Cancer stem cells as targets for immunotherapy. *Immunology.* 2017;153(3):304–314.
198. Maccalli C, Volontè A, Cimminiello C, Parmiani G. Immunology of cancer stem cells in solid tumours. A review. *Eur J Cancer.* 2014;50(3):649–655.
199. Wang T, Shigdar S, Gantier M, et al. Cancer stem cell targeted therapy: progress amid controversies. *Oncotarget.* 2015;6(42):44191–44206.
200. Ogi C, Aruga A. Clinical evaluation of therapeutic cancer vaccines. *Hum Vaccin Immunother.* 2013;9(5):1049–1057.
201. Klein C, Bueler H, Mulligan R. Comparative analysis of genetically modified dendritic cells and tumor cells as therapeutic cancer vaccines. *J Exp Med.* 2000;191(10): 1699–1708.
202. Yee C, Thompson J, Byrd D, et al. Adoptive T cell therapy using antigen-specific CD8+ T cell clones for the treatment of patients with metastatic melanoma: in vivo persistence, migration, and antitumor effect of transferred T cells. *Proc Natl Acad Sci U S A.* 2002;99(25):16168–16173.
203. Schmitz K, Brugger W, Weiss B, Kaiserling E, Kanz L. Clonal selection of CD20-negative non-Hodgkin's lymphoma cells after treatment with anti-CD20 antibody rituximab. *Br J Haematol.* 1999;106(2):571–572.
204. Lampen M, van Hall T. Strategies to counteract MHC-I defects in tumors. *Curr Opin Immunol.* 2011;23(2):293–298.
205. Mellman I, Coukos G, Dranoff G. Cancer immunotherapy comes of age. *Nature.* 2011;480(7378):480–489.
206. Callahan M, Wolchok J. At the bedside: CTLA-4- and PD-1-blocking antibodies in cancer immunotherapy. *J Leukoc Biol.* 2013;94(1):41–53.
207. Liu J, Wang L, Zhao F, et al. Pre-clinical development of a humanized anti-CD47 antibody with anti-cancer therapeutic potential. *PLoS One.* 2015;10(9):e0137345.
208. Micheau O, Shirley S, Dufour F. Death receptors as targets in cancer. *Br J Pharmacol.* 2013;169(8):1723–1744.
209. Byrne W, Mills K, Lederer J, O'Sullivan G. Targeting regulatory T cells in cancer. *Cancer Res.* 2011;71(22):6915–6920.
210. Melancon M, Lu W, Huang Q, et al. Targeted imaging of tumor-associated M2 macrophages using a macromolecular contrast agent PG-Gd-NIR813. *Biomaterials.* 2010;31(25):6567–6573.
211. Gabrilovich D, Ostrand-Rosenberg S, Bronte V. Coordinated regulation of myeloid cells by tumours. *Nat Rev Immunol.* 2012;12(4):253–268.
212. Vasudev N, Reynolds A. Anti-angiogenic therapy for cancer: current progress, unresolved questions and future directions. *Angiogenesis.* 2014;17(3):471–494.

CHAPTER FOUR

TNFR2-expressing $CD4^+Foxp3^+$ regulatory T cells in cancer immunology and immunotherapy

Jiang He[a,b,†], Ruixin Li[a,†], Yibo Chen[a], Yuanjia Hu[a], Xin Chen[a,*]
[a]State Key Laboratory of Quality Research in Chinese Medicine, Institute of Chinese Medical Sciences, University of Macau, Macau, China
[b]Investment Banking, Shenzhen Rhino Star Information Co. Ltd., Shenzhen, China
*Corresponding author: e-mail address: xchen@um.edu.mo

Contents

1. Introduction 102
2. Ovarian cancer 103
3. Lung cancer 105
4. Acute myelocytic leukemia (AML) 105
5. Lymphoma 106
6. Colon cancer 106
7. Cervical intraepithelial neoplasia (CIN) and cervical cancer 107
8. Hepatocellular carcinoma (HCC) 108
9. Closing remarks 109
Acknowledgments 113
Conflict of interest 113
References 113

Abstract

$CD4^+Foxp3^+$ regulatory T cells (Tregs) represent a major cellular mechanism in tumor immune evasion. Elimination of Treg activity has become a strategy to devise an effective tumor immunotherapy. We reported that TNF receptor type II (TNFR2), one of two receptors transducing TNF biological activity, is preferentially expressed by the most suppressive subset of Tregs. By interaction with TNFR2, TNF plays a decisive role in the activation, expansion and phenotype stability of Tregs. We also found that highly suppressive TNFR2-expressing Tregs appear to be tumor-associated Tregs. This finding has been supported by recent studies in mouse tumor models and in cancer patients. In this chapter, published data revealing the important role of $TNFR2^+$ Tregs in tumor development and metastasis in different tumor types are reviewed and analyzed.

† These authors have contributed equally to this work.

Progress in Molecular Biology and Translational Science, Volume 164
ISSN 1877-1173
https://doi.org/10.1016/bs.pmbts.2019.03.010

The therapeutic potential of targeting TNF-TNFR2 interaction as means to eliminate Treg activity, and consequently to enhance anti-tumor immune responses, also is discussed.

1. Introduction

$CD4^+Foxp3^+$ regulatory T cells (Tregs) are potent immunosuppressive cells that play a major role in establishment and maintenance of immunosuppressive tumor microenvironments.[1,2] The accumulation of Tregs in tumor tissues hampers naturally occurring and therapeutically induced antitumor immune responses.[3] The increased prevalence of Tregs correlates with poor clinical prognosis in patients with lung cancer,[4] ovarian cancer,[5] breast cancer,[6] colorectal cancer,[7] pancreatic cancer[8] and other malignancies. Elimination of tumor-infiltrating Tregs, or suppression of their function, has become a strategy to enhance anti-tumor immune responses.[9,10]

Tumor necrosis factor (TNF) is a pleiotropic cytokine involved in diverse physiological and pathological processes, including cell growth, cell death, inflammatory and autoimmune responses, and tumor progression and metastasis. The discovery of TNF dates back to 120 years ago when Dr. William Coley made "Coley's mixed toxins" from the production of cultured *Streptoccocus pyogenes* and Gram-negative endotoxin-producing *Serratia marcasens* to treat sarcoma.[11] The term "tumor necrosis factor" was proposed in several follow up studies of "Coley's mixed toxins."[12] In 1975, Carswell and colleagues reported the identification of TNF from human serum, which was responsible for the induction of tumor necrosis in different mouse models.[13] In 1984, Aggarwal and colleagues reported the gene cloning of TNF.[14] Subsequent studies showed that, by binding to two distinct receptors, namely TNF receptor type I (TNFR1) and TNFR2, TNF plays various roles *via* sophisticated signal transduction.[15] Unlike the ubiquitous expression of TNFR1 on almost all cell types, TNFR2 expression is restricted to certain cell types, including immune cells, certain neural cells, and endothelial cells.[16] TNFR2 is predominantly expressed on $CD4^+Foxp3^+$ regulatory T cells (Tregs), and the expression of TNFR2 defines the maximally suppressive subset of mouse and human Tregs.[17–19] We (Xin Chen and Joost J. Oppenheim) for the first time reported that the interaction of TNF and TNFR2 promotes the activation, expansion and phenotype stability of Tregs, and enhanced immunosuppressive effect and expression of TNFR2 on Tregs.[20–24] Therefore, in addition to a marker of activated Tregs, TNFR2 expression by Tregs has important functional implications.[25,26] Our findings have been supported and substantiated by independent studies from other groups.[27–32] There is now increasing evidence that the decisive role of TNF-TNFR2 interaction in Treg activity can be harnessed therapeutically in the treatment of human diseases.[33,34]

We also found that highly suppressive TNFR2$^+$ Treg cells accumulated in the tumor microenvironment in Lewis Lung cancer (LLC) and 4T1 mouse models.[17,35] The concept that TNFR2-expressing Tregs are attributable to the immunosuppressive tumor environment is now supported by an increasing number of studies in human tumor patients.[7,36,37] Consequently, targeting TNFR2$^+$ Tregs has emerged as a novel strategy for boosting anti-tumor immune responses.

In addition to the interactions discussed above, it has been reported that TNF–TNFR2 interactions are involved in the accumulation and the suppressive effect of myeloid-derived suppressor cells (MDSCs).[38–40] Moreover, TNFR2 was reported to be expressed by multiple tumor types, including breast cancer,[41] cervical cancer,[42] colon cancer,[43] Hodgkin lymphoma,[44] myeloma,[45] renal carcinoma,[46] ovarian cancer,[47] skin cancer,[48] esophageal squamous cell carcinoma[49] and lymphoma[50] and was proposed as an oncoprotein.[51] We note these roles of TNF–TNFR2 interaction but detailed discussion of them is beyond scope of this review. We now discuss TNF–TNFR2 interactions in Treg-mediated immunosuppression in ovarian cancer, lung cancer, acute myelocytic leukemia (AML), colon cancer, cervical intraepithelial neoplasia (CIN) and cervical cancer, and hepatocellular carcinoma (HCC).

2. Ovarian cancer

Ovarian cancer typically presents in postmenopausal women as abdominal pain and distension[52] and was the fifth leading cause of cancer death in females in the United States in 2015.[53] Accumulation of Tregs has long been known to contribute to ovarian tumor progression and is correlated with reduced survival of patients.[54] Rosekeila et al. reported that the level of soluble TNFR2 (sTNFR2) in the malignant ovarian neoplasms was significantly higher than that in the benign ovarian neoplasms, and it was associated with tumor differentiation grades 2 and 3.[55] Govindaraj et al. reported a substantially higher proportion of $CD4^+CD25^{hi}TNFR2^+$ T cells, in mononuclear cell (MNCs) fractions of tumor ascites, as compared with that in peripheral blood mononuclear cells (PBMCs) from ovarian cancer patients or from healthy donors. Furthermore, TNFR2 expression (assessed by mean fluorescence intensity (MFI)) on $CD25^+$ T cells in tumor ascites, especially on these $CD25^{hi}$ T cells, was markedly increased as compared with that in PBMCs from ovarian cancer patients or from healthy donors.[56] Meanwhile, there was no difference in the proportion of $CD4^+$ $CD25^{hi}$ T cells in MNCs in the tumor ascites, or in PBMCs either from cancer patients or from healthy donors, indicating $CD4^+CD25^{hi}$ $TNFR2^+$ T cells with elevated expression of TNFR2 preferentially accumulated in the tumor microenvironment. Furthermore, depletion of $CD25^{hi}TNFR2^+$, but not $CD25^{hi}TNFR2^-$ T cells, resulted in the increased proliferation of Teffs and production of IFN-γ. The proliferation of Teffs was markedly suppressed by adding back the depleted $CD25^{hi}TNFR2^+$ T cells from tumor ascites. $CD25^{hi}TNFR2^+$ T cells appeared to be more suppressive than $CD25^{hi}TNFR2^-$ T cells, and $CD25^{hi}TNFR2^+$ T cells from tumor ascites were even more suppressive as compared with those from PBMCs. In line with the suppressive activity, the expression of immunosuppressive molecules CD39, CD73, GARP, and TGF-β was also up-regulated on $CD25^{hi}Foxp3^+TNFR2^+$ T cells, as compared with $CD25^{hi}Foxp3^+$ $TNFR2^-$ T cells from patients and healthy donors. Kampan et al. also reported that $TNFR2^+$ Tregs presented in ascites of ovarian cancer patients were highly suppressive cells and had the capacity to suppress $TNFR2^+$ Teffs, which were relatively more resistant to inhibition by regular Tregs.[57] Thus, targeting $TNFR2^+$ Tregs would presumably be effective in enhancing anti-tumor immune responses in ovarian cancer patients. To this end, antagonistic antibodies (Abs) targeting TNFR2 have been developed. Such antibodies preferentially killed tumor-residing Tregs from ovarian cancer patients, as compared with its cytotoxic effect on T cells from normal peripheral blood.[47] This effect of antagonistic TNFR2 Abs should be verified by *in vivo* study. Interestingly, blockade of IL-6, but not TNF, decrease the frequency of $TNFR2^+$ Tregs in ascites in advanced ovarian cancer.[57] This result suggests that the accumulation of $TNFR2^+$ Tregs is attributable to IL-6. Moreover, antagonistic agents of IL-6 may be useful for reducing $TNFR2^+$ Treg activity in ovarian cancer patients. This possibility should be studied further.

3. Lung cancer

Lung cancer is the number one cause of cancer-related deaths worldwide. The 5-year survival of patients with lung cancer varies from 4% to 17%, depending on stage and regional differences.[58] The proportion of Tregs is increased in the peripheral blood of lung cancer patients, an observation that has diagnostic and prognostic value.[59–61] We reported that highly suppressive TNFR2^{+} Tregs accumulated in tumor environments in mouse model of Lewis lung carcinoma.[17] In human lung cancer patients, in line with the increased proportion of Tregs in CD4^{+} cells of peripheral blood (PB), the proportion of TNFR2^{+} Tregs in PB CD4^{+} was also increased as compared with that in healthy donor.[62] TNFR2^{+} Tregs in PB of lung cancer patients expressed high levels of Ki67, a nuclear antigen presents only in replicating cells,[63] indicating that the increased number of TNFR2^{+} Treg cells resulted from proliferation of pre-existing Tregs, rather than from conversion of naïve CD4 T cells. Furthermore, TNFR2^{+} Tregs in PB of lung cancer expressed high levels of CTLA-4, which was consistent with their highly suppressive capacity in the inhibition of IFN-γ production from CD8 T cells.[61] More importantly, the proportion of TNFR2^{+} Tregs was positively correlated with lymphatic invasion, distant metastasis, and clinical stage (stages I + II vs stages III + IV).[61] Zhang et al. also reported that higher TNFR2 expression in human non-small cell lung cancer (NSCLC) leads to a more advanced clinical stage and shorter survival.[64] Therefore, the expression level of TNFR2 on Tregs in PB of lung cancer patients can be an indicator of poor prognosis of patients with lung cancer.

4. Acute myelocytic leukemia (AML)

AML is a hematopoietic malignancy of bone marrow, blood, and other tissues.[65] The long-term survival of patients with AML is poor as most patients relapse despite achieving temporary remission.[66] Studies consistently show that the frequency of Tregs is increased in patients and is correlated with poor clinical outcomes.[67–69] Govindaraj et al. and Wang et al. reported that the frequency of TNFR2^{+} Tregs in PB CD4^{+} T cells and expression levels of TNFR2 on Tregs from newly diagnosed AML patients was markedly higher than those in healthy donors or in patients with complete remission.[36,70] Govindaraj et al. further showed that the frequency of TNFR2$^+$ Tregs was 25-fold higher than TNFR2$^-$ Tregs, indicating TNFR2$^+$ Tregs were dominant in Treg subsets in AML patients.[37] These TNFR2$^+$ Tregs expressed higher levels of the immunosuppressive molecules CTLA-4, CD73 and replicating maker Ki67, and produced more immunosuppressive cytokines, including IL-10 and TGF-β, suggesting a highly suppressive and proliferative phenotype.[36] The expression of CXCR4 was positively correlated with expression of TNFR2 on Tregs, and the capacity of migration of TNFR2$^+$ Tregs to CXCL12/SDF-1 was higher as compared with TNFR2$^-$ Tregs. Azacitidine (a hypomethylating agent) and panobinostat (a histone deacetylase inhibitor, HDACi) have synergistic effects in AML treatment. This therapeutic effect may be associated with the reduction of TNFR2$^+$Foxp3$^+$ Tregs and increased levels of IL-2 and IFN-γ.[36] *In vitro* treatment with panobinostat also downregulated TNFR2 expression on Tregs.[36] Lenalidomide, another agent used for AML treatment, was also observed to reduce the expression of TNFR2 on Tregs. The levels of TNFR2 expression were associated with responsiveness to lenalidomide treatment, as the TNFR2 expression on CD4$^+$ cells was markedly lower in non-relapse patients as compared with patients with disease relapse.[37]

5. Lymphoma

Torry and colleagues recently reported that Tregs and tumor cells in patients with Sézary syndrome (SS), a rare form of cutaneous T-cell lymphoma, expressed high levels of TNFR2 that are associated with late stage and aggressive lymphoma with a poor prognosis.[71] The lymphocytes from patients with late stage SS treated with TNFR2 antagonists *in vitro* resulted in the death of TNFR2-expressing Tregs and tumor cells, and consequently restored the normal numbers of the CD26$^-$ population of lymphocytes.[71] Furthermore, treatment with TNFR2 antagonists also caused the rapid expansion of Teff cells and restored a normal ratio of Treg/Teff.[71] Thus, TNFR2 antagonism may have beneficial therapeutic potential on end-stage SS patients.

6. Colon cancer

Colon cancer was estimated to rank as the third most frequent cause of cancer-related death in 2018 among both male and female patients, although the incidence rates declined annually by about 2–3% from 2005 through

2014 due to increased numbers of colonoscopies.[53] It has been reported that the number of Tregs is higher in human colon cancers than in surrounding unaffected mucosa.[72] CD103-expressing effector Tregs expressed higher levels of TNFR2, as compared with CD103^{-} Tregs, in spleen and TILs in a mouse model of CT26 colon cancer.[73] In line with this, the level of TNF in the tumor microenvironment was increased in CT26 tumor-bearing mice. Blockade of TNF-TNFR2 signaling, either by an anti-TNFR2 antibody or by a soluble TNFR2 fusion protein (sTNFR2-Fc), efficiently inhibited TNF-induced expansion of CD103^{+} Tregs *in vitro*. Further, TNF-pretreated Tregs showed more potent immunosuppressive capacity when co-transferred with tumor-specific CD8 T cells in a mouse model of colon cancer.[73] Furthermore, blockade of TNF-TNFR2 signaling by sTNFR2-Fc after cyclophosphamide treatment potently inhibited tumor growth with decreased the number of CD103^{+} Tregs in spleen and in draining lymph nodes.[73] Clinical studies also have shown that, in patients with colorectal cancer (CRC), the proportion of CD45RA^{-}Foxp3^{hi} effector Tregs was markedly increased in PB and that these Tregs expressed a higher level of TNFR2 as well as of CTLA-4 and CCR5.[74] The level of TNF in the serum of patients with CRC was also increased, as compared with healthy donors. Similarly, TNF was able to expand CD45RA^{-} Foxp3^{hi} effector Tregs *in vitro* and this effect was abrogated by sTNFR2-Fc.[74] Our group showed that the treatment with TNFR2-blocking Ab M861 markedly enhanced the therapeutic effect of CpG ODN in a mouse CT26 colon tumor model and this effect was associated with reductions in the numbers of TNFR2^{+} Tregs and increases in the numbers of IFN-γ^{+}CD8^{+} CTLs in the tumor environment.[74] Williams et al. reported that anti-mouse TNFR2 antibodies TR75–54.7 and TR75–89 inhibited tumor growth in mouse CT26 tumor model and promoted the infiltration of CD8 CTLs into tumor.[75] It was also shown that clone TR75–54 was an antagonistic antibody that had both *in vitro* and *in vivo* effect in the blockade of TNFR2 signaling.[76] These results lend support to the idea that blockade of TNFR2 is able to enhance anti-tumor immune responses in mouse colon cancer model.

7. Cervical intraepithelial neoplasia (CIN) and cervical cancer

Cervical cancer was the second leading cause of cancer death in women aged 20–39 years in the United States in 2015.[53] The ratio of

CD8$^+$ T cells/Tregs has proved to be a single variable independent prognostic factor in cervical cancer by multivariate statistical analysis.[77] Zhang et al. reported that the proportion of peripheral TNFR2$^+$ Tregs in patients with CIN and cervical cancer was higher as compared with that in healthy donors.[78] A dramatic increase of TNFR2$^+$ Tregs in TILs was observed in patients with cervical cancer, as compared with that in the PB of those patients.[78] Importantly, the level of peripheral TNFR2$^+$ Tregs was inversely correlated with cancer stage.[78]

8. Hepatocellular carcinoma (HCC)

HCC was the fifth-ranked cause of cancer-related death in male patients in 2018. Tregs are associated with the invasiveness of HCC and are a promising independent predictor of recurrence and survival in HCC patients.[79] Chang et al. reported that CD103$^+$ Tregs expressed a higher level of TNFR2, as compared with CD103$^-$ Tregs, in spleen and tumor in a mouse model of hepatocellular carcinoma (inoculated with BNL cells).[73] In line with this, the level of TNF in tumor microenvironments was increased in BNL tumor-bearing mice. Blockade of TNF–TNFR2 interactions, either by an anti-TNFR2 antibody or by a soluble TNFR2 fusion protein (sTNFR2-Fc), efficiently inhibited TNF-induced expansion of CD103$^+$ Tregs *in vitro*.[73] Further, Tregs pre-treated with TNF were more potent in the inhibition of co-transferred tumor-specific CD8$^+$ T cells in this model.[73] Blockade of TNF–TNFR2 interaction by sTNFR2-Fc after cyclophosphamide treatment markedly inhibited tumor growth, accompanied by a decrease in the number of CD103$^+$ Tregs in spleen and drainage lymph nodes.[73] Clinical studies showed that the proportion of CD45RA$^-$ FoxP3^{hi} effector Tregs was markedly increased in PB CD4 T cells in patients with HCC.[80] These effector Tregs expressed higher levels of TNFR2, as well as CTLA-4 and CCR5. Intriguingly, the level of TNF in the serum of patients with HCC was also increased as compared with healthy donors.[73] Therefore, high levels of TNFR2 are consistently expressed by CD103$^+$ effector Tregs in a mouse model of HCC and by CD45RA$^-$FoxP3^{hi} effector Tregs in patients with HCC, presumably resulting from the stimulation of elevated levels of TNF.[22] This study suggests that blockade of interactions between TNF and TNFR2 could be a promising treatment for HCC.

9. Closing remarks

Targeting Tregs for cancer immunotherapy has been intensively studied for decades. The strategy of therapeutic modulation of Tregs includes treatment with anti-CD25, anti-CTLA-4, anti-GITR, anti-TIM3, anti-LAG-3, or anti-OX40 antibodies, as well as low-dose chemotherapy.[81,82] However, the efficacies and safety of these treatments are controversial. There is now compelling evidence, as discussed in this chapter, that TNFR2$^+$ Tregs accumulate in TILs (lung cancer and cervical cancer) or tumor ascites (ovarian cancer) and are increased in peripheral blood (AML and cervical cancer). These tumor-associated Tregs possess more potent immunosuppressive activity and activated phenotype (ovarian cancer, lung cancer, and AML). Beneficial therapeutic effects of blockade of TNF–TNFR2 interactions have been observed in colon cancer and HCC. These published data clearly indicate that targeting of TNFR2 is a novel strategy to eliminate activity of tumor-associated Tregs and consequently boost anti-tumor immune responses (as summarized in Table 1).

In addition to promoting the growth of primary tumors, several studies also showed that TNFR2$^+$ Tregs contributed to tumor metastasis. For example, in an experimental liver metastasis model, the incidence of metastasis was reduced significantly in TNFR2$^{-/-}$ mouse inoculated with the murine colon (MC-38) and lung (H-59) carcinoma cells, while the number of Tregs and MDSCs accumulated in livers were decreased concomitantly.[39] It was also shown that the treatment with exogenous TNF promoted tumor metastasis to the lungs in a mouse model of melanoma, accompanied by an increase of Tregs in lung tissues.[83] In contrast, deficiency of TNFR2 on T cells completely abrogated TNF-induced expansion of Tregs and consequently reduced the growth and metastasis of B16F10 melanoma.[83] Therefore, TNFR2$^+$ Tregs also play a major role in tumor metastasis, presumably by inhibition of host immune surveillance capacity.

Global inhibition of TNF with the anti-TNF antibody infliximab has been shown to enhance the efficacy of oncolytic adenovirus activity in mouse ovarian cancer.[84] Treatment with infliximab was also found to stabilize the disease in patients with advanced tumor.[85] The beneficial effect of some anti-TNF therapy in tumor-bearing mice or patients is presumably mediated, at least partially, by the blockade of TNF–TNFR2 interactions and consequently elimination of Treg activity. If this is the case, the clinical

Table 1 Role of TNFR2$^+$ Tregs in cancer immunology and immunotherapy.

Tumor type	Species	TNFR2$^+$ Tregs: implications for cancer immunology and immunotherapy	References
Ovarian cancer	Human	• Higher proportion of CD4$^+$CD25^{hi}TNFR2$^+$ T cells in mononuclear cells (MNCs) of tumor ascites • The expression of immunosuppressive molecules CD39, CD73, GARP, and TGF-β was up-regulated on CD25^{hi}Foxp3$^+$TNFR2$^+$ T cells	47
Lung cancer	Human	• Increased proportion of TNFR2$^+$ Tregs in CD4 cells of PB from lung cancer patients • Higher proportion of TNFR2$^+$ cells in Tregs as compare with Teffs • TNFR2$^+$ Tregs was positively correlated with lymphatic invasion, distant metastasis, and clinical stage (stage I + II vs stage III + IV) • TNFR2$^+$ Tregs in PB of lung cancer patients were more proliferative, active, and suppressive	17,62
	Mouse	• The proportion of TNFR2$^+$ cells in TIL CD4$^+$CD25$^+$ cells was greater than 70%, higher than in CD4$^+$CD25$^+$ cells in the peripheral lymphoid organs	
Acute Myeloid leukemia	Human	• Higher proportion of TNFR2$^+$ Tregs in PB CD4$^+$ T cells of newly diagnosed AML patients • Expressed higher levels of immunosuppressive molecules CTLA-4 and CD73, produced more immunosuppressive cytokines IL-10 and TGF-β • The migration capacity of TNFR2$^+$ Tregs was enhanced • Combination treatment of azaci^{t}idine (a hypomethylating agent) and panobinostat (a histone deacetylase inhibitor, HDACi) selectively reduced TNFR2$^+$ Tregs, accompanied by the down-regulation of Foxp3 and CTLA-4 expression on TNFR2$^+$ Tregs • Panobinostat and lenalidomide inhibited the expression of TNFR2 on Tregs	36,70
Sézary syndrome	Human	• High expression of TNFR2 oncogene on SS tumor cells and on Treg cells • TNFR2 antagonist killed SS tumor cells, restored CD26$^-$ subpopulation • TNFR2 antagonist reduced the number of Treg and ratio of Treg/Teff	71

Colon cancer	Human	• The number of Tregs was higher in colon cancer tissues than in surrounding unaffected mucosa • Increased proportion of $CD45RA^-FoxP3^{hi}$ effector Tregs in peripheral blood, with expression of high levels of CTLA-4, CCR5, and TNFR2	73–75
	Mouse	• $CD103^+$ effector Tregs expressed a higher level of TNFR2, as compared with $CD103^-$ Tregs in spleen and TILs in mouse CT26 colon cancer • The level of TNF in tumor microenvironment was increased in CT26 tumor-bearing mice • Blockade of TNF-TNFR2 signaling by anti-TNFR2 antibody or by a soluble TNFR2 fusion protein (sTNFR2-Fc) inhibited TNF-induced expansion of $CD103^+$ Tregs *in vitro* • Blockade of TNF-TNFR2 signaling by sTNFR2-Fc after cyclophosphamide treatment Inhibit tumor growth Decreased number of $CD103^+$ Tregs in spleen and drainage lymph nodes • Combination treatment with CpG ODN and a TNFR2-blocking Ab Reduced the proportion of TIL $TNFR2^+$ Tregs Increased of TIL $IFN\text{-}\gamma^+CD8^+$ CTLs Inhibited the growth of CT26 colon cancer • Mice treated with TNFR2 mAb TR75–54.7 or TNFR2 mAb TR75–89 resulted in complete regression of CT26 tumor	
Cervical intraepithelial neoplasia and cervical cancer	Human	• Percentage of peripheral $TNFR2^+$ Tregs elevated in patients with cervical intraepithelial neoplasia and cervical cancer • The proportion of $TNFR2^+$ Tregs in the PB of patients with cervical cancer was higher than that in healthy donors • The proportion of $TNFR2^+$ Tregs was markedly increased in tumor-infiltrating lymphocytes • Expanded $TNFR2^+$ Tregs population was correlated with cancer stage	78

Continued

Table 1 Role of $TNFR2^+$ Tregs in cancer immunology and immunotherapy.—cont'd

Tumor type	Species	$TNFR2^+$ Tregs: implications for cancer immunology and immunotherapy	References
Hepatocellular carcinoma	Human	• Increased proportion of $CD45RA^- Foxp3^{hi}$ effector Tregs in PB, with high expression of CTLA-4, CCR5, and TNFR2, in patients with HCC	73
	Mice	• $CD103^+$ Tregs expressed higher levels of TNFR2 as compared with $CD103^-$ Tregs in spleen and tumor in a mouse model of hepatocellular carcinoma (inoculated with BNL cells) • The level of TNF in tumor microenvironment was increased • Blockade of TNF-TNFR2 signaling by anti-TNFR2 antibody or by a soluble TNFR2 fusion protein (sTNFR2-Fc) inhibited TNF-induced expansion of $CD103^+$ Tregs *in vitro* • Blockade of TNF-TNFR2 signaling by sTNFR2-Fc after cyclophosphamide treatment Inhibited tumor growth Decreased number of $CD103^+$ Tregs in spleen and drainage lymph nodes	
Hepatic metastasis	Mouse	• Metastasis was markedly reduced in $TNFR2^{-/-}$ mouse inoculated with the murine colon (MC-38) and lung (H-59) carcinoma cells • Treg accumulation was decreased in $TNFR2^{-/-}$ tumor-bearing mice • Treatment with TNFR2 antisense oligodeoxynucleotides inhibited hepatic metastasis in WT mice	39
Lung metastasis	Mouse	• Metastasis of B16F10-Luc melanoma and numbers of Tregs in lungs was decreased in immune cell-restricted deficiency of TNF or TNFR2	83

efficacy of TNFR2-targeting agent should be markedly better than TNF-targeting agents. This idea has spurred the development of TNFR2-targeting agents in recent years.[86] In addition to biologics, our recent research showed that natural products may contain small molecules with the potential to specifically bind to TNFR2 and block TNF–TNFR2 interactions.[87] Deficiency of major TNFR2 signaling components in mice reduces Treg number.[88] Indeed, small molecules may inhibit Treg activity by acting on TNFR2 signaling pathways.[89] This highlights the importance of further understanding TNFR2 signaling pathways in Tregs. Since TNFR2 has been appreciated as an emerging target in cancer treatment,[51] we expect more potent and more specific TNFR2-targeting agents will be developed or identified, and their anti-tumor potential will be evaluated in appropriate mouse tumor models and then in human patients.

Taken together, increasing evidence supports the idea that $TNFR2^+$ Tregs are attractive targets for cancer immunotherapy. TNFR2-targeting therapeutic agents, such as anti-TNFR2 biologics or small molecule TNFR2 inhibitors, should be identified and developed. The efficacy and safety of TNFR2-targeting agents in cancer treatment should be evaluated. Moreover, combination therapy with TNFR2-targeting agents and other immune stimulators, such as checkpoint inhibitors, may be a safer and more effective tumor immunotherapy regimen and thus should be a focus of future study.

Acknowledgments

This work was supported by University of Macau under Grants MYRG2016-00023-ICMS-QRCM and MYRG2017-00120-ICMS; and the Science and Technology Development Fund of Macao S.A.R. (FDCT) under grant 014/2015/A1 and 201/2017/A3.

Conflict of interest

The authors declare no conflict of interest.

References

1. Wing JB, Tanaka A, Sakaguchi S. Human FOXP3+ regulatory T cell heterogeneity and function in autoimmunity and cancer. *Immunity*. 2019;50(2):302–316.
2. Togashi Y, Shitara K, Nishikawa H. Regulatory T cells in cancer immunosuppression—implications for anticancer therapy. *Nat Rev Clin Oncol*. 2019. https://doi.org/10.1038/s41571-019-0175-7. Epub ahead of print.
3. Nishikawa H, Sakaguchi S. Regulatory T cells in tumor immunity. *Int J Cancer*. 2010;127(4):759–767.
4. Tao H, et al. Prognostic potential of FOXP3 expression in non-small cell lung cancer cells combined with tumor-infiltrating regulatory T cells. *Lung Cancer*. 2012;75(1): 95–101.

5. Leffers N, et al. Prognostic significance of tumor-infiltrating T-lymphocytes in primary and metastatic lesions of advanced stage ovarian cancer. *Cancer Immunol Immunother.* 2009;58(3):449–459.
6. Shou J, et al. Worse outcome in breast cancer with higher tumor-infiltrating FOXP3+ Tregs: a systematic review and meta-analysis. *BMC Cancer.* 2016;16:687.
7. Lin YC, et al. Activated but not resting regulatory T cells accumulated in tumor microenvironment and correlated with tumor progression in patients with colorectal cancer. *Int J Cancer.* 2013;132(6):1341–1350.
8. Tang Y, et al. An increased abundance of tumor-infiltrating regulatory T cells is correlated with the progression and prognosis of pancreatic ductal adenocarcinoma. *PLoS One.* 2014;9(3), e91551.
9. Munn DH, Sharma MD, Johnson TS. Treg destabilization and reprogramming: implications for cancer immunotherapy. *Cancer Res.* 2018;78(18):5191–5199.
10. Liu C, Workman CJ, Vignali DAA. Targeting regulatory T cells in tumors. *FEBS J.* 2016;283(14):2731–2748.
11. Balkwill F. Tumour necrosis factor and cancer. *Nat Rev Cancer.* 2009;9(5):361–371.
12. Wiemann B, Starnes CO. Coley's toxins, tumor necrosis factor and cancer research: a historical perspective. *Pharmacol Ther.* 1994;64(3):529–564.
13. Carswell EA, et al. An endotoxin-induced serum factor that causes necrosis of tumors. *Proc Natl Acad Sci U S A.* 1975;72(9):3666–3670.
14. Marmenout A, et al. Molecular cloning and expression of human tumor necrosis factor and comparison with mouse tumor necrosis factor. *Eur J Biochem.* 1985;152(3):515–522.
15. Kalliolias GD, Ivashkiv LB. TNF biology, pathogenic mechanisms and emerging therapeutic strategies. *Nat Rev Rheumatol.* 2015;12:49.
16. Faustman DL, Davis M. TNF receptor 2 and disease: autoimmunity and regenerative medicine. *Front Immunol.* 2013;4:478.
17. Chen X, et al. Cutting edge: expression of TNFR2 defines a maximally suppressive subset of mouse CD4+CD25+FoxP3+ T regulatory cells: applicability to tumor-infiltrating T regulatory cells. *J Immunol.* 2008;180(10):6467–6471.
18. Chen X, et al. Co-expression of TNFR2 and CD25 identifies more of the functional CD4+FOXP3+ regulatory T cells in human peripheral blood. *Eur J Immunol.* 2010;40(4):1099–1106.
19. Chen X, et al. Expression of costimulatory TNFR2 induces resistance of CD4+FoxP3-conventional T cells to suppression by CD4+FoxP3+ regulatory T cells. *J Immunol.* 2010;185(1):174–182.
20. Chen X, et al. Interaction of TNF with TNF receptor type 2 promotes expansion and function of mouse CD4+CD25+ T regulatory cells. *J Immunol.* 2007;179(1):154–161.
21. Zaragoza B, et al. Suppressive activity of human regulatory T cells is maintained in the presence of TNF. *Nat Med.* 2016;22(1):16–17.
22. Hamano R, et al. TNF optimally activatives regulatory T cells by inducing TNF receptor superfamily members TNFR2, 4-1BB and OX40. *Eur J Immunol.* 2011;41(7): 2010–2020.
23. Chen X, et al. TNFR2 is critical for the stabilization of the CD4+Foxp3+ regulatory T. cell phenotype in the inflammatory environment. *J Immunol.* 2013;190(3): 1076–1084.
24. Chen X, Oppenheim JJ. TNF-alpha: an activator of CD4+FoxP3+TNFR2+ regulatory T cells. *Curr Dir Autoimmun.* 2010;11:119–134.
25. Chen X, Oppenheim JJ. Resolving the identity myth: key markers of functional CD4+FoxP3+ regulatory T cells. *Int Immunopharmacol.* 2011;11(10):1489–1496.
26. Chen X, Oppenheim JJ. The phenotypic and functional consequences of tumour necrosis factor receptor type 2 expression on CD4(+) FoxP3(+) regulatory T cells. *Immunology.* 2011;133(4):426–433.

27. Kleijwegt FS, et al. Critical role for TNF in the induction of human antigen-specific regulatory T cells by tolerogenic dendritic cells. *J Immunol.* 2010;185(3):1412–1418.
28. Ablamunits V, Bisikirska B, Herold KC. Acquisition of regulatory function by human CD8(+) T cells treated with anti-CD3 antibody requires TNF. *Eur J Immunol.* 2010;40(10):2891–2901.
29. Grinberg-Bleyer Y, et al. Pathogenic T cells have a paradoxical protective effect in murine autoimmune diabetes by boosting Tregs. *J Clin Invest.* 2010;120(12):4558–4568.
30. Housley WJ, et al. Natural but not inducible regulatory T cells require TNF-alpha signaling for in vivo function. *J Immunol.* 2011;186(12):6779–6787.
31. Okubo Y, et al. Homogeneous expansion of human T-regulatory cells via tumor necrosis factor receptor 2. *Sci Rep.* 2013;3:3153.
32. Atretkhany K-SN, et al. Intrinsic TNFR2 signaling in T regulatory cells provides protection in CNS autoimmunity. *Proc Natl Acad Sci U S A.* 2018;115(51):13051–13056.
33. Chen X, Oppenheim JJ. Contrasting effects of TNF and anti-TNF on the activation of effector T cells and regulatory T cells in autoimmunity. *FEBS Lett.* 2011;585(23): 3611–3618.
34. Chen X, Oppenheim JJ. Therapy: paradoxical effects of targeting TNF signalling in the treatment of autoimmunity. *Nat Rev Rheumatol.* 2016;12(11):625–626.
35. Chen X, et al. Effective chemoimmunotherapy with anti-TGFbeta antibody and cyclophosphamide in a mouse model of breast cancer. *PLoS One.* 2014;9(1), e85398.
36. Govindaraj C, et al. Reducing TNF receptor 2^+ regulatory T cells via the combined action of azacitidine and the HDAC inhibitor, panobinostat for clinical benefit in acute myeloid leukemia patients. *Clin Cancer Res.* 2014;20(3):724–735.
37. Govindaraj C, et al. Lenalidomide-based maintenance therapy reduces TNF receptor 2 on CD4 T cells and enhances immune effector function in acute myeloid leukemia patients. *Am J Hematol.* 2014;89(8):795–802.
38. Hu X, et al. Transmembrane TNF-alpha promotes suppressive activities of myeloid-derived suppressor cells via TNFR2. *J Immunol.* 2014;192(3):1320–1331.
39. Ham B, et al. TNF receptor-2 facilitates an immunosuppressive microenvironment in the liver to promote the colonization and growth of hepatic metastases. *Cancer Res.* 2015;75(24):5235–5247.
40. Zhao X, et al. TNF signaling drives myeloid-derived suppressor cell accumulation. *J Clin Invest.* 2012;122(11):4094–4104.
41. Yang F, Zhao N, Wu N. TNFR2 promotes adriamycin resistance in breast cancer cells by repairing DNA damage. *Mol Med Rep.* 2017;16(3):2962–2968.
42. Singhal P, et al. Identification of genetic variants in TNF receptor 2 which are associated with the development of cervical carcinoma. *Biomarkers.* 2016;21(7):665–672.
43. Hamilton KE, et al. Cytokine induction of tumor necrosis factor receptor 2 is mediated by STAT3 in colon cancer cells. *Mol Cancer Res.* 2011;9(12):1718–1731.
44. Nakayama S, et al. Expression of tumour necrosis factor-alpha and its receptors in Hodgkin lymphoma. *Br J Haematol.* 2014;167(4):574–577.
45. Rauert H, et al. TNFR1 and TNFR2 regulate the extrinsic apoptotic pathway in myeloma cells by multiple mechanisms. *Cell Death Dis.* 2011;2, e194.
46. Wang J, Al-Lamki RS. Tumor necrosis factor receptor 2: its contribution to acute cellular rejection and clear cell renal carcinoma. *Biomed Res Int.* 2013;2013:821310.
47. Torrey H, et al. Targeting TNFR2 with antagonistic antibodies inhibits proliferation of ovarian cancer cells and tumor-associated Tregs. *Sci Signal.* 2017;10(462). pii: eaaf8608.
48. Arnott CH, et al. Expression of both TNF-alpha receptor subtypes is essential for optimal skin tumour development. *Oncogene.* 2004;23(10):1902–1910.
49. Yang D, et al. Clinical significance of tumor necrosis factor receptor 2 in middle and lower thoracic esophageal squamous cell carcinoma. *Oncol Lett.* 2018;16(3):2971–2978.

50. Ungewickell A, et al. Genomic analysis of mycosis fungoides and Sezary syndrome identifies recurrent alterations in TNFR2. *Nat Genet.* 2015;47(9):1056–1060.
51. Chen X, Oppenheim JJ. Targeting TNFR2, an immune checkpoint stimulator and oncoprotein, is a promising treatment for cancer. *Sci Signal.* 2017;10(462). pii: eaal2328.
52. Jayson GC, et al. Ovarian cancer. *Lancet.* 2014;384(9951):1376–1388.
53. Siegel RL, Miller KD, Jemal A. Cancer statistics, 2018. *CA Cancer J Clin.* 2018;68(1): 7–30.
54. Curiel TJ, et al. Specific recruitment of regulatory T cells in ovarian carcinoma fosters immune privilege and predicts reduced survival. *Nat Med.* 2004;10(9):942–949.
55. Nomelini RS, et al. TNF-R2 in tumor microenvironment as prognostic factor in epithelial ovarian cancer. *Clin Exp Med.* 2018;18(4):547–554.
56. Govindaraj C, et al. Impaired Th1 immunity in ovarian cancer patients is mediated by TNFR2+ Tregs within the tumor microenvironment. *Clin Immunol.* 2013;149(1): 97–110.
57. Kampan NC, et al. Interleukin 6 present in inflammatory ascites from advanced epithelial ovarian cancer patients promotes tumor necrosis factor receptor 2-expressing regulatory T cells. *Front Immunol.* 2017;8:1482.
58. Hirsch FR, et al. Lung cancer: current therapies and new targeted treatments. *Lancet.* 2017;389(10066):299–311.
59. Erfani N, et al. Increase of regulatory T cells in metastatic stage and CTLA-4 over expression in lymphocytes of patients with non-small cell lung cancer (NSCLC). *Lung Cancer.* 2012;77(2):306–311.
60. Hasegawa T, et al. Prognostic value of peripheral and local forkhead box P3(+) regulatory T cells in patients with non-small-cell lung cancer. *Mol Clin Oncol.* 2014;2(5):685–694.
61. Wolf AM, et al. Increase of regulatory T cells in the peripheral blood of cancer patients. *Clin Cancer Res.* 2003;9(2):606–612.
62. Yan F, et al. Expression of TNFR2 by regulatory T cells in peripheral blood is correlated with clinical pathology of lung cancer patients. *Cancer Immunol Immunother.* 2015;64(11):1475–1485.
63. Soares A, et al. Novel application of Ki67 to quantify antigen-specific in vitro lymphoproliferation. *J Immunol Methods.* 2010;362(1–2):43–50.
64. Zhang YW, et al. Expression of tumor necrosis factor receptor 2 in human non-small cell lung cancer and its role as a potential prognostic biomarker. *Thorac Cancer.* 2019;10(3): 437–444.
65. Dohner H, Weisdorf DJ, Bloomfield CD. Acute myeloid leukemia. *N Engl J Med.* 2015;373(12):1136–1152.
66. Shlush LI, et al. Tracing the origins of relapse in acute myeloid leukaemia to stem cells. *Nature.* 2017;547(7661):104–108.
67. Ustun C, et al. Regulatory T cells in acute myelogenous leukemia: is it time for immunomodulation? *Blood.* 2011;118(19):5084–5095.
68. Szczepanski MJ, et al. Increased frequency and suppression by regulatory T cells in patients with acute myelogenous leukemia. *Clin Cancer Res.* 2009;15(10):3325–3332.
69. Shenghui Z, et al. Elevated frequencies of $CD4^{+}$ $CD25^{+}$ $CD127^{lo}$ regulatory T cells is associated to poor prognosis in patients with acute myeloid leukemia. *Int J Cancer.* 2011;129(6):1373–1381.
70. Wang M, et al. Increased regulatory T cells in peripheral blood of acute myeloid leukemia patients rely on tumor necrosis factor (TNF)-alpha-TNF receptor-2 pathway. *Front Immunol.* 2018;9:1274.
71. Torrey H, et al. Targeted killing of TNFR2-expressing tumor cells and Tregs by TNFR2 antagonistic antibodies in advanced Sezary syndrome. *Leukemia.* 2018; https://doi.org/10.1038/s41375-018-0292-9. Epub ahead of print.

72. Le Gouvello S, et al. High prevalence of Foxp3 and IL17 in MMR-proficient colorectal carcinomas. *Gut*. 2008;57(6):772–779.
73. Chang LY, et al. Blockade of TNF-alpha signaling benefits cancer therapy by suppressing effector regulatory T cell expansion. *Oncoimmunology*. 2015;4(10), e1040215.
74. Nie Y, et al. Blockade of TNFR2 signaling enhances the immunotherapeutic effect of CpG ODN in a mouse model of colon cancer. *Sci Signal*. 2018;11(511). pii: eaan0790.
75. Williams GS, et al. Phenotypic screening reveals TNFR2 as a promising target for cancer immunotherapy. *Oncotarget*. 2016;7(42):68278–68291.
76. Sheehan KC, et al. Monoclonal antibodies specific for murine p55 and p75 tumor necrosis factor receptors: identification of a novel in vivo role for p75. *J Exp Med*. 1995; 181(2):607–617.
77. Jordanova ES, et al. Human leukocyte antigen class I, MHC class I chain-related molecule a, and CD8+/regulatory T-cell ratio: which variable determines survival of cervical cancer patients? *Clin Cancer Res*. 2008;14(7):2028–2035.
78. Zhang T, et al. Aberrant frequency of TNFR2(+) Treg and related cytokines in patients with CIN and cervical cancer. *Oncotarget*. 2018;9(4):5073–5083.
79. Gao Q, et al. Intratumoral balance of regulatory and cytotoxic T cells is associated with prognosis of hepatocellular carcinoma after resection. *J Clin Oncol*. 2007;25(18): 2586–2593.
80. Kalathil S, et al. Higher frequencies of GARP$^+$CTLA4$^+$Foxp3$^+$ T regulatory cells and myeloid-derived suppressor cells in hepatocellular carcinoma patients are associated with impaired T-cell functionality. *Cancer Res*. 2013;73(8):2435–2444.
81. Chaudhary B, Elkord E. Regulatory T cells in the tumor microenvironment and cancer progression: role and therapeutic targeting. *Vaccines (Basel)*. 2016;4(3). pii: E28.
82. Pere H, et al. Comprehensive analysis of current approaches to inhibit regulatory T cells in cancer. *Oncoimmunology*. 2012;1(3):326–333.
83. Chopra M, et al. Tumor necrosis factor receptor 2-dependent homeostasis of regulatory T cells as a player in TNF-induced experimental metastasis. *Carcinogenesis*. 2013;34(6): 1296–1303.
84. Salako MA, et al. Inhibition of the inflammatory cytokine TNF-alpha increases adenovirus activity in ovarian cancer via modulation of cIAP1/2 expression. *Mol Ther*. 2011; 19(3):490–499.
85. Brown ER, et al. A clinical study assessing the tolerability and biological effects of infliximab, a TNF-alpha inhibitor, in patients with advanced cancer. *Ann Oncol*. 2008;19(7):1340–1346.
86. Zou H, et al. Modulation of regulatory T cell activity by TNF receptor type II-targeting pharmacological agents. *Front Immunol*. 2018;9:594.
87. Shaikh F, et al. TNF receptor type II as an emerging drug target for the treatment of cancer, autoimmune diseases, and graft-versus-host disease: current perspectives and in silico search for small molecule binders. *Front Immunol*. 2018;9:1382.
88. Chen X, et al. IKKalpha is required for the homeostasis of regulatory T cells and for the expansion of both regulatory and effector CD4 T cells. *FASEB J*. 2015;29(2):443–454.
89. He T, et al. The p38 MAPK inhibitor SB203580 abrogates tumor necrosis factor-induced proliferative expansion of mouse CD4(+)Foxp3(+) regulatory T cells. *Front Immunol*. 2018;9:1556.

CHAPTER FIVE

Cancer stem cell immunology and immunotherapy: Harnessing the immune system against cancer's source

Roberto Ruiu, Lidia Tarone, Valeria Rolih, Giuseppina Barutello, Elisabetta Bolli, Federica Riccardo, Federica Cavallo*, Laura Conti
Department of Molecular Biotechnology and Health Sciences, Molecular Biotechnology Center, University of Torino, Torino, Italy
*Corresponding author: e-mail address: federica.cavallo@unito.it

Contents

1. Introduction 120
2. Cancer stem cells 121
   2.1 The cancer stem cell model: An overview 121
   2.2 Biological features of the cancer stem cells 126
   2.3 Methods for the identification, isolation and characterization of cancer stem cells 131
   2.4 Clinical implications of the cancer stem cell concept 135
3. The immunology of cancer stem cells 137
   3.1 Tumor antigens expressed by cancer stem cells 138
   3.2 Immune evasion of cancer stem cells 143
   3.3 Immune response and targeted immunotherapies to cancer stem cells 153
4. Conclusions and future perspectives 170
   4.1 Cancer stem cell immunology and immunotherapy in short 170
   4.2 Integration of cancer immunology in the CSC model 171
   4.3 Caveats associated with CSC immunology and possible solutions 173
Acknowledgments 175
References 175

Abstract

Despite recent advances in diagnosis and therapy having improved cancer outcome, many patients still do not respond to treatments, resulting in the progression or relapse of the disease, eventually impairing survival expectations. The limited efficacy of therapy is often attributable to its inability to affect cancer stem cells (CSCs), a small population of cells resistant to current radio- and chemo-therapies. CSCs are characterized by self-renewal and tumor-initiating capabilities, and function as a reservoir for the local and

Progress in Molecular Biology and Translational Science, Volume 164
ISSN 1877-1173
https://doi.org/10.1016/bs.pmbts.2019.03.008

distant recurrence of the disease. Therefore, new therapeutic approaches able to effectively target and deplete CSCs are urgently needed.

Immunotherapy is facing a renewed interest for its potential in cancer treatment, and the possibility of harnessing the immune system to target CSCs is being addressed by a new exciting research field.

In this chapter, we discuss the cancer stem cell model and illustrate CSC biological and molecular properties, critically addressing theoretical and practical issues linked with their definition and study. We then review the existing literature regarding the immunological properties of CSCs and the complex interplay occurring between CSCs and immune cells. Finally, we present up-to-date studies on CSC immunotargeting and its potential future perspective.

In conclusion, understanding the interplay between CSC biology and tumor immunology will provide a deeper understanding of the mechanisms that regulate CSC immunological properties. This will contribute to the design of new CSC-directed immunotherapeutic strategies with the potential of strongly improving cancer outcomes.

1. Introduction

Despite cancer outcomes dramatically improving through advances in diagnosis, understanding, and management of the disease, many patients still do not respond to treatments, resulting in the progression or relapse of the disease, finally leading to reduced survival expectations. The limited efficacy of therapy for some types of tumors is often attributable to the intrinsic complexity of the tumor. In fact, many therapies treat cancer as a homogenous disease, while it is now clear that tumors can be represented as complex ecosystems that contain different populations of non-tumor cells (immune, endothelial, hematopoietic, stromal) that infiltrate the tumor and that contribute to create a microenvironment that sustains the growth of the tumor cells providing nutrients and stimuli.[1] Moreover, tumor cells themselves display many genetic, phenotypic, and functional differences.[2] These differences determine overall intratumor heterogeneity, with the presence of regions characterized by different degrees of differentiation, proliferation, vascularization, inflammation and invasion within the same tumor.[2] Evidence suggests that heterogeneity within tumors results from an intrinsic cell hierarchy based on tumorigenic and differentiation potential, with a small population of undifferentiated tumorigenic cells at the apex of this hierarchy, the so-called cancer stem cells (CSCs). CSCs are endowed with the biological properties required for tumor progression and spreading, and can differentiate into non-tumorigenic cancer cells that comprise the largest part of the tumor but which have little or no capacity to contribute to cancer progression. Importantly, CSCs would represent a reservoir for the local and distant recurrence of the disease, by virtue of their resistance to radio- and chemotherapies, which enrich the CSC population by preferentially eliminating differentiated cancer cells.[3] Therefore, the efficacy of a therapeutic approach may rest on its ability to effectively target and deplete CSCs.

In recent years, immunotherapy has regained popularity among anticancer treatments. Importantly, tumor immunotargeting would circumvent some of the mechanisms implemented by CSCs to resist chemo- and radiotherapy. Nevertheless, CSCs could be diverging from more differentiated cancer cells with regard to antigenicity and immune modulating properties.[4] Therefore, the design of efficient immunotherapies able to target CSCs would rely on an in-depth knowledge of CSC immunology. In this chapter, we will review and discuss the existing literature regarding the immunological profiles of CSCs, as well as the different immunotherapeutic strategies that have been used to target these tumor-initiating cells.

2. Cancer stem cells

2.1 The cancer stem cell model: An overview

2.1.1 The origin of intratumor heterogeneity: Clonal evolution model and hierarchical model

Two separate models have been historically used to explain tumor heterogeneity and progression.[3] One attributes intratumor heterogeneity to genetic causes (clonal evolution, or stochastic, model), the other to non-genetic causes (hierarchical, or CSC, model). The clonal evolution model of cancer, postulated by Nowell,[5] asserts that the tumor arises from a single cell upon the appearance of multiple driver mutations, and that maintenance and progression of the tumor depends on acquired genetic variability within the original clone. This variability allows progressive selection of more aggressive subclones following the rules of Darwinian selection. Thus, distinct stochastic mutations characterizing distinct clones (all deriving from the same cell of origin) can provide a selective advantage to a clone with respect to another, thus allowing a continuous alternation of dominant clones that expand and permit the tumor perpetuation. An important assumption of the clonal evolution model is that every cell belonging to the same genetic clone displays the same tumorigenic potential.[6]

In the CSC model, also named hierarchical model, phenotypic heterogeneity is explained as a reflection of non-genetic diversification that exists in normal tissues. As all the normal cells in an individual share the same genotype (with the exception of adaptive immune cells), the phenotypic diversity of normal cells depends on developmental pathways which are required for the maintenance of the hierarchy of normal tissues, with the stem cells at the apex and differentiated cells at the base. In a similar manner, these non-genetic features contribute to a hierarchical organization of the tumor, where a population of CSCs is responsible for its long-term maintenance, while more differentiated tumor cells representing the largest part of the tumor lose their tumorigenic potential.[7] In this model, only cancer cells displaying stem cell-features are able to generate a new tumor, independently from their genetic background.[8] Thus, cancer cells sharing a common genetic background can exist in a CSC and a non-CSC state.[9]

These two models have been long considered as mutually exclusive. However, each view taken individually is insufficient to fully explain the heterogeneity observed within tumors. The clonal evolution model focuses on genetic heterogeneity without taking in consideration that single cells within a genetically homogeneous tumor subclone could possess functional diversity due to non-genetic determinants. Similarly, an important limitation of the hierarchical model is that it considers the tumor as genetically homogeneous and static, not accounting for the appearance of genetically different subclones during tumor progression.[10]

In more recent years, there is the tendency to combine these models and to consider their contribution to the intratumor heterogeneity and tumor maintenance as complementary. Integrating these two models could describe tumor evolution with more accuracy. CSCs may in fact be subjected to clonal selection, with a dominant CSC clone emerging upon the appearance of advantageous mutations.[8] In this "*integrated model*," acquiring new genetic mutations could give rise to genetically different lineages of CSCs that are heterogeneously dominant and malignant (Fig. 1).

2.1.2 Historical perspective of the cancer stem cell model

The concept that cancer cells have characteristics reminiscent of stem cells has its roots in the 19th century. This idea was first proposed by Rudolph Virchow, who noted the similarities between teratocarcinomas and embryonic tissues regarding their ability to proliferate and differentiate. His "embryonal rest hypothesis" postulated that cancers are caused by cells with properties similar to those of the early embryo.[11] Subsequently, Julius

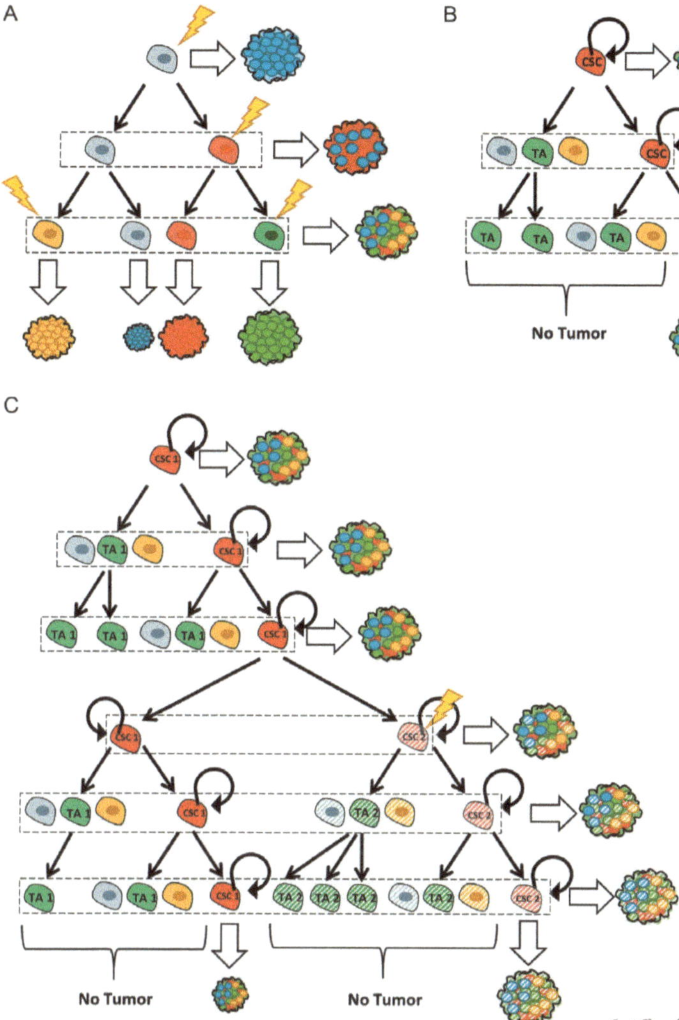

Fig. 1 The origin of intratumor heterogeneity: clonal evolution and hierarchical models. (A) Clonal evolution model. Tumor arises from a single cell upon the appearance of mutations. Accumulation of further genetic mutations results in a variety of cells with different genotype and thus phenotype. The best fitted cells are selected by Darwinian processes to generate clonal variants of the tumor. In this model, all cells in the dominant clones have similar tumorigenic capacity. (B) Cancer stem cell model. Tumor heterogeneity results from the existence of phenotypically diverse populations of different stages of cell maturation. In this model, only the CSC can generate a tumor, based on its self-renewal properties, unlimited proliferative potential and ability to generate both differentiated and transiently amplifying (TA) cells. (C) Integrated model. The appearance of mutations in the first CSC (CSC1) may lead to the generation of separate subclones of CSCs. Thus, a distinct and more aggressive CSC (CSC2) may arise and become dominant over CSC1, driving tumor progression.

Cohnheim extended this concept by speculating that embryonal remnants persist in adult organs, and that these residual embryonic cells could later be re-activated to form cancer.[11]

The modern idea that the same developmental programs that are at the basis of normal tissue organization are retained in cancer began with seminal studies on teratocarcinomas in the '60s. In 1960, Pierce and colleagues suggested that the majority of the cells within a teratocarcinoma are differentiated, and that the appearance of such differentiated cells occurs after an initial phase of proliferation, a phenomenon similar to that observed in normal embryo development (reviewed in Ref. 12). Later, in 1964, Pierce proved the clonal origin of mouse teratocarcinomas from single transplanted multipotent tumorigenic cells.[13] Thus, Pierce and colleagues formulated a general view in which tumors can be considered as a caricature of normal development in embryogenesis, where multipotent malignant cells propagate the cancer and give rise to more differentiated, benign populations of cells.[7,14] Early studies on blood tumors as well contributed to CSC research. Studies in the '60s and '70s showed the existence of a rare population of slow-cycling leukemia cells that were not susceptible to anti-proliferative therapies and were therefore responsible for recurrence. These slow-cycling leukemia cells were also responsible for the continuous generation of the proliferating fraction that eventually leads to post-mitotic leukemia cells, which represent the majority of leukemia blasts (reviewed in Ref. 10). In the same years, the heterogeneous potential of tumor cells to self-renew both *in vitro* and *in vivo* was reported. These studies showed that an extremely low percentage of cells had clonogenic potential and proposed a view of hierarchical organization of leukemia, as established for normal hematopoiesis (reviewed in Ref. 15).

However, it was not until the development of fluorescence-activated cell sorting techniques coupled with refined techniques of xenografting in immune-compromised mice that the hierarchical organization of leukemia was demonstrated. In the '90s, Dick and colleagues prospectively isolated human leukemia-initiating cells (L-IC) by sorting acute myeloid leukemia (AML) cells using the cell surface markers CD34 and CD38. It was shown that leukemia are cells able to engraft in an immunocompromised mouse were exclusively in the $CD34^{+}/CD38^{-}$ fraction.[16] Further studies established that AML is organized in a hierarchical manner with $CD34^{+}/CD38^{-}$ cells at the apex,[17] thus proving that AML is composed by heterogeneous populations in which only rare cells have tumorigenic potentials.

These initial studies on AML paved the way for the subsequent studies on CSCs in solid tumors. The first prospective isolation of CSCs in solid tumors was performed in human breast cancer over a decade ago by Al-Hajj and colleagues.[18] They showed that as few as 100 cells within the $CD44^+/CD24^-$ fraction were able to generate tumors in mice, while thousands of cells with alternative phenotypes were not. This tumorigenic subpopulation could be serially passaged, each time generating new tumors containing both $CD44^+/CD24^-$ tumor-initiating cells as well as the phenotypically heterogeneous populations of non-tumorigenic cells that were present in the initial tumor.[18] Since then, CSCs have been identified in many solid tumors.

2.1.3 What is a cancer stem cell? Concepts and definitions

Normal stem cells are commonly defined as undifferentiated cells that have the capability to generate the specialized cells of the tissue they reside in (differentiation), along with undifferentiated daughter stem cells with tissue-maintaining potential (self-renewal).[19] Similarly, the consensus definition of CSC (as established at the AACR Workshop on Cancer Stem Cells[20]) is "a cell within a tumor that possesses the capacity to self-renew and to cause the heterogeneous lineages of cancer cells that comprise the tumor" (Fig. 2).

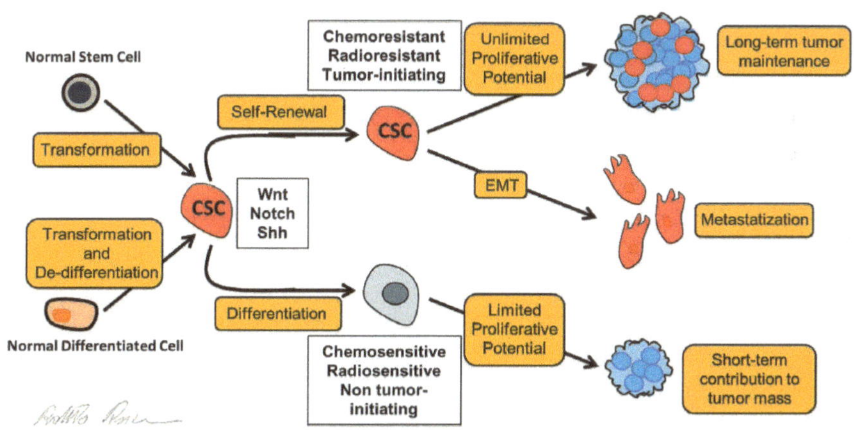

Fig. 2 Characteristics of CSCs. CSCs can derive from the malignant transformation of either a normal stem cell or a normal differentiated cell that acquires stem cell-like features. The activation of Notch, Shh and Wnt pathways endows CSCs with long-term self-renewal ability and the capacity to differentiate into the tumor bulk populations. The chemo- and radioresistant CSCs display an unlimited proliferative potential that maintains tumor growth, and can undergo EMT and give rise to metastasis. The differentiated cells, sensitive to chemo- and radiotherapy, display a limited proliferative potential and no tumor-initiating ability.

Therefore, CSCs can only be defined experimentally by their capability to generate a continuously growing tumor. For this reason, putative CSCs are operatively referred to as "tumor-initiating cells" or "tumorigenic cells."

It should be noted that the term "cancer stem cell" has led to some confusion over the years.[19] Some proposed that CSCs originate from normal tissue stem cells through the acquisition of transforming mutations, while others suggest that they arise from progenitor or differentiated cells that have reverted to a stem-like state, referring to a complex set of interactions including epigenetic modifications, intrinsic molecular pathways and transcription factors that regulate and maintain the stem cell form.[3] However, the CSC concept does not imply any specific hypothesis in this regard. CSCs may indeed arise from normal stem cells as a consequence of oncogenic mutations, but it is also conceivable that progenitors or more differentiated cells can acquire the self-renewal ability that characterizes CSCs through multiple mutagenic events.[20] In conclusion, any cell that acquires genetic or epigenetic modifications restoring the capability to self-renew and suppressing the terminal differentiation program could potentially act as a CSC of a given cancer.[21,22] The same confusion is caused by the term "tumor-initiating cell."[19] A tumor-initiating cell is not necessarily the cell that received the first oncogenic hits in the patient (the cell of origin of cancer), but simply refers to the cell that has the ability to initiate a new tumor when transplanted into a host.[20] As stated in the previous section (see "*integrated model*"), the stem-like status can be acquired by a cell at any given moment in the tumor development due to accumulation of genetic and epigenetic alterations.

2.2 Biological features of the cancer stem cells

2.2.1 Signaling pathways in cancer stem cells

Despite having in common the self-renewal and tumor-initiating capacity, not all CSCs necessarily share the same mechanisms that regulate these fundamental features. In fact, the intra- and inter-tumor heterogeneity characterizing different tumor types represents one of the major challenges in specifically defining CSCs from the phenotypic, genetic, functional and molecular points of view.[23] Nonetheless, it has been observed that gene expression profiles obtained from normal mammary stem cells can stratify breast cancers according to their molecular and biological features, and that high-grade breast cancers show enrichment in normal mammary stem cell profile due to a higher CSC content.[24] Furthermore, the activation of specific signaling pathways occurs frequently across different tumor types, and these pathways can contribute to the acquisition and maintenance of stemness features in CSCs. Indeed, the same pathways that are conserved in regulating normal stem cell self-renewal and tissue homeostasis, such as Notch, wingless-related integration site (Wnt) and Hedgehog (Hh) signaling pathways, appear to contribute to neoplastic progression when deregulated.[25] Other stem cell-related genes, such as Bmi1, Nanog, SOX2 and OCT4 have been frequently linked to the CSC phenotype.[23]

Wnt plays a critical role in highly proliferative tissues, such as the skin, intestine and hematopoietic system.[26] Wnt1 was first identified as an oncogene involved in breast cancer in mice infected by the mouse mammary tumor virus (MMTV), whose integration into the Wnt1 gene resulted in its ectopic activation. Activation of Wnt1, as well as constitutively activated β-catenin, has been shown to induce mammary tumor formation.[27,28] MMTV-Wnt-induced tumors are dependent on continuous Wnt signaling, which determines a progenitor-like signature in cancer cells.[29] In humans, Wnt signaling is activated in over 50% of breast cancer patients and is linked to reduced overall survival, with canonical Wnt ligands and receptors often upregulated in breast cancer despite the lack of somatic mutations.[29]

The Hh signaling pathway was initially described in Drosophila as a mediator of segmental patterning during embryonic development. The most studied vertebrate homolog of the Drosophila Hh gene is Sonic Hedgehog (Shh). Shh regulates proliferation, migration, and differentiation.[23] Abnormally active Shh signaling has been involved in different types of cancers and significantly contributes to the long-term self-renewal of breast CSCs.[30] A strong evidence for Shh implication in carcinogenesis has been provided by a study in which conditional overexpression of its downstream effector glioma-associated oncogene homolog GLI1 in the mouse mammary gland was sufficient to induce tumor formation.[31] GLI1 overexpression was accompanied by expansion of mammary progenitor cells and upregulation of genes involved in cell survival, proliferation and metastasis, indicating that the Shh signaling pathway is fundamental for initiation and propagation of breast CSCs.[23]

The Notch signaling pathway plays a role in controlling cell fate, proliferation and self-renewal during the development of most tissues.[32] Notch is a transmembrane receptor, and its ligands Delta-like (DLL) and Jagged are transmembrane proteins as well, thus ligand–receptor interactions occur through cell–cell contact. Activation of the Notch signaling pathway induces the expression of target genes including HES, PI3K,

AKT, NF-κB, PPAR, CyclinD1, p21 and p27.[23] The role of Notch in cancer was first described in mouse models. Integration of MMTV into the Notch4 or Notch1 genes results in constitutively activated forms of these receptors and is sufficient for the transformation of mouse mammary epithelial cells.[23] Expression of components belonging to the Notch signaling is common in breast cancer patients and relates to poor prognosis.[23]

2.2.2 Epithelial-to-mesenchymal transition and stem-like state

One of the most prominent processes linked to cancer cell stemness is the epithelial-to-mesenchymal transition (EMT), a multi-step process characterized by the reorganization of the cytoskeletal network and loss of cell junctions, resulting in the loss of epithelial polarity and acquisition of a mesenchymal-like phenotype.[33] During embryogenesis, EMT enables embryonic epithelial cells to become mesenchymal-like and travel to distant sites where new tissues and organs form. Similarly, during tumor progression, this EMT process is thought to be re-activated and ultimately facilitates tumor cell migration through the basement membrane, invasion into surrounding tissues, and entry into the blood stream. EMT is induced by humoral factors such as tumor growth factor (TGF)-β, epidermal growth factor (EGF), fibroblast growth factor (FGF) and Wnt and is mediated by transcription factors such as Snail, Slug, Twist1/2 and Zeb1/2,[34] which suppress the expression of E-cadherin and promote the loss of epithelial and adhesive features and the acquisition of invasive and migratory ability.[35] The expression of EMT-related transcription factors in primary tumors has been linked to tumor invasion, metastasis, and poor prognosis.[36] Besides these pro-invasive and metastatic functions, EMT has been linked to the promotion of stemness in cancer cells. The activation of EMT by TGF-β and the expression of various EMT-inducing transcription factors generate cells that express CSC markers and functional features, including tumor initiation in immunodeficient mice.[33,37] Furthermore, cellular changes induced by EMT have direct impact on stem cell signaling pathways. The loss of E-cadherin caused by EMT transcription factors leads to the release of β-catenin from cell membrane sequestration and allows its translocation into the nucleus, where it activates the canonical Wnt signaling.[38] Moreover, the disruption of epithelial polarity leads to the activation of YAP/TAZ, which are effectors of the Hippo pathway and favor the acquisition of CSC traits by increasing the chemo-resistance, tumorigenicity and metastatic capacity of cancer cells.[39,40]

2.2.3 Cancer stem cells and metastasis

A feature linked to EMT, and often attributed to CSCs, is the ability to generate distant metastasis (Fig. 2). Investigations of the clinical relevance of CSC marker expression in human tumors have suggested that a high percentage of $CD44^{+}/CD24^{-/low}$ cells in breast cancer may favor distant metastasis,[41] similarly to what has been shown for aldehyde dehydrogenase (ALDH) expression.[42] Cells isolated *via* stem cell marker expression from primary tumor samples or blood of breast cancer patients can generate metastasis when grafted in immunocompromised mice,[43] and a study on disseminating tumor cells in 50 bone marrow specimens of breast cancer patients showed that an average of 72% of the cells displayed CSC marker positivity, while the percentage in the primary tumors was less than 10%.[44] Although EMT promotes the migration of cancer cells from the primary tumor, it inhibits cell proliferation and interferes with the initiation of metastatic outgrowth.[43] Therefore, cancer cells that have undergone EMT must revert to an epithelial phenotype in order to initiate growth at the metastatic site, a process known as mesenchymal-to-epithelial transition (MET). Of note, it has been reported that CSCs can exist in both epithelial and mesenchymal states.[45] These two populations of CSCs appear to co-exist in the same tumor, with mesenchymal-like CSCs being mostly quiescent and localized at the tumor invasive front, while epithelial-like CSCs are proliferative and localized more centrally.[45] Therefore, it appears that plasticity of CSCs allows them to actuate transition between EMT and MET states, thus making them capable of both invading and disseminating, and growing at metastatic sites.

2.2.4 Resistance to conventional therapies

Chemo- and radiotherapy have represented the preferred treatments for cancer patients for the past decades, often being effective in reducing the tumor burden. The efficacy of these therapies is indeed assessed by efficient reduction in tumor mass. However, a minority of cells non-responding to therapies may still survive, representing the minimal residual disease that leads to local and distant recurrence.[46] In many cases, CSCs have been demonstrated to be more resistant to cytotoxic drugs compared to non-CSCs of the same tumor.[47] Along with chemo-resistance, CSCs of various histotypes exhibit greater resistance to radiation therapy[47] and small drug inhibitors[48] compared to non-CSCs (Fig. 2).

In general, CSC resistance to therapy can be ascribed to multiple mechanisms. Increased expression of multidrug efflux pumps belonging to the

ATP-binding cassette (ABC) transporter superfamily has been described in CSCs,[49] in particular multidrug resistance protein 1 (MRP1, ABCC1), P-glycoprotein (P-gp, MDR1, ABCB1), and breast cancer resistance protein (BCRP, ABCG2).[48] Enhanced expression of such proteins results in ATP-dependent efflux of a broad spectrum of cytotoxic drugs, including taxanes, anthracyclines, antimetabolites, as well as several tyrosine kinase inhibitors.[48]

Radiation therapy and many chemotherapeutic drugs such as DNA cross-linkers (platinum-based drugs) and topoisomerase inhibitors (anthracyclines) act by inducing DNA damage, which is lethal to cells if not repaired. CSCs can defend themselves from DNA-damaging treatments through multiple mechanisms, including regulation of the cell cycle, enhancement of DNA repair and prevention of DNA damage through scavenging of reactive oxygen species (ROS).[48]

As quiescent cells do not go through the cell cycle, they are not susceptible to conventional therapies that target the mitotic machinery of rapidly-dividing cells. As for normal adult stem cells, quiescence has been observed also for CSCs.[50] Besides passive resistance to therapy, also active DNA damage response (DDR) mechanisms have been shown to be highly upregulated by CSCs. These include nucleotide excision repair (NER), which is involved in intra-strand crosslinks repair, and homologous recombination (HR) pathway for the repair of double-strand breaks.[49]

Detoxification from ROS induced by chemo- and radiotherapy is also an important mechanism of resistance activated by CSCs. Most CSCs upregulate ALDH, which confers survival to therapy since it directly scavenges radiation or drug-induced free radicals generated by oxidative stress by producing the antioxidant NADPH.[51] Pharmacological inhibition of ALDH activity in therapy-resistant CSCs causes the accumulation of ROS at levels that are toxic for the cells, eventually leading to irreversible DNA damage and apoptosis.[52] Furthermore, ALDH activity has been correlated to the activation of pro-survival signaling pathways, including PI3K/AKT, as well as stem cell-associated pathways such as Notch.[48]

Anti-apoptosis mechanisms promote resistance to therapy either in a direct manner by blocking cell death, or in an indirect manner by providing enough time for DDR mechanisms to repair DNA damage induced by therapy.[49] Increased resistance to apoptosis has been linked to downregulation of death receptors by CSCs, thus providing resistance to the extrinsic pathway, as well as to upregulation of the members of the BCL2 family involved in negative regulation of the intrinsic pathway of apoptosis.[49]

2.3 Methods for the identification, isolation and characterization of cancer stem cells

Since normal stem cells and CSCs have some characteristics in common, it is possible to exploit some of them—such as expression of specific surface markers, increased activity of ABC transporters and ALDH, self-renewal and multipotency—to identify and isolate CSCs (Fig. 3).

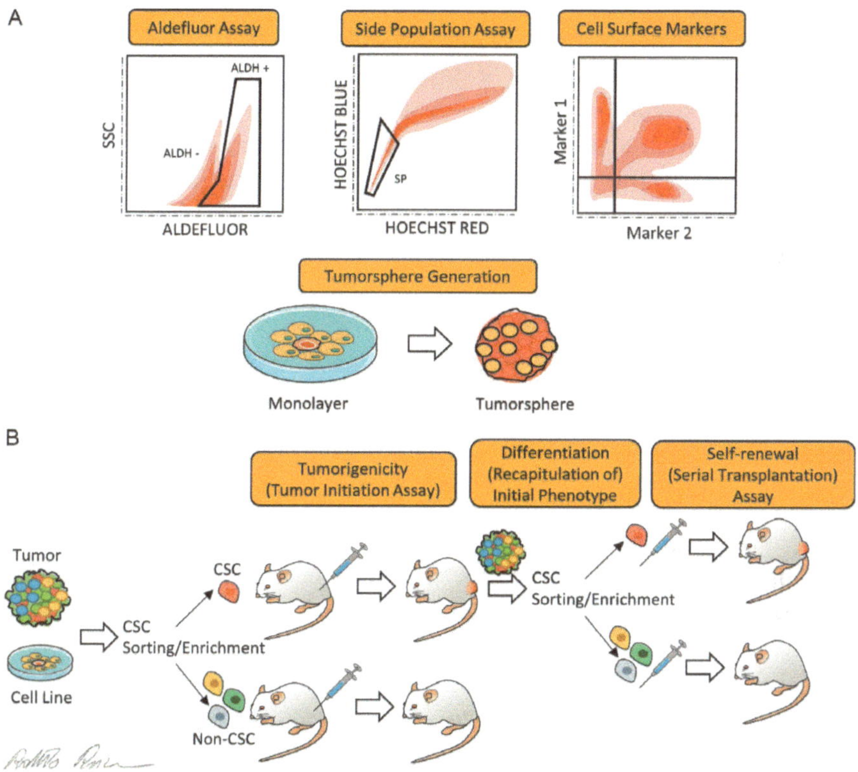

Fig. 3 Methods for the identification, isolation and characterization of CSCs. (A) CSCs can be identified and isolated by fluorescent activated cell sorting on the basis of (i) their enhanced ALDH activity, measured with the Aldefluor assay; (ii) Hoechst dye efflux, evaluated in the side population (SP) assay; and (iii) expression of specific cell surface markers. Moreover, their ability to generate floating tridimensional cell clones (tumorspheres) can be exploited to enrich the CSC population *in vitro*. (B) Methods for the *in vivo* validation of *bona fide* CSCs. A real CSC is endowed with tumor-initiating ability, self-renewal and the ability to histologically recapitulate the tumor of origin. To verify this, putative CSCs isolated from a tumor or cultured as tumorspheres are injected at low density in mice (tumor initiation assay); CSCs, but not differentiated cells, will give rise to a tumor that recapitulates the initial phenotype. CSCs explanted from this tumor will give rise to other tumors if re-implanted in new mice (serial transplantation assay).

2.3.1 *Methods commonly employed for the isolation and enrichment of CSCs*

They include: (i) sorting on the basis of cell surface markers expression; (ii) sorting of a side population (SP) based on Hoechst dye efflux; (iii) sorting on the basis of ALDH activity (Aldefluor assay); and (iv) generation of floating tridimensional cell clones (tumorspheres).[53]

2.3.1.1 Cell surface markers

CSCs in solid tumors can be identified and isolated through an extensive set of markers that are not uniformly expressed by the cells comprising the tumor and are often reflective of the developmental programs of the tissue of origin. The first studies on leukemia stem cells relied on the well understood differentiation hierarchy in the hematopoietic compartment. Thus, normal hematopoietic stem cell markers CD34 and CD38 were chosen to define leukemia stem cells.[16,17] In other cases, such as the CD44 and CD24 combination extensively used for human breast CSC isolation, markers were selected on the basis of their heterogeneous expression in breast tumor and not because they defined normal breast stem cells.[18] Although CSC isolation based on marker expression is still widely used, the lack of universal markers and the existence of contradictory results undermines its reliability. For example, the $CD44^{+}/CD24^{-}$ phenotype proposed by Al Haji et al. as characteristic of breast CSCs was not confirmed in all subsequent studies.[54] In particular, it has been proposed that the $CD44^{+}/CD24^{-}$ CSC phenotype could be associated to the basal-like breast cancer subtype,[55] and the finding that $CD44^{+}/CD24^{-}$ phenotype shows very little overlap with the $ALDH^{+}$ phenotype (that will be discussed later in this section) suggests that multiple CSC phenotypes could co-exist within the same tumor.[45,56,57] Furthermore, CD44 exists in several splice variants, with some of them (CD44v) better identifying breast CSCs compared to the standard isoform (CD44s).[58] Thus, studies using pan-CD44 antibodies may overestimate the CSC content of a tumor cell population.[55] Similarly, the use of CD133 (Prominin-1) to identify CSCs from a large number of tumor types, mainly from brain, colon and lung,[59] has recently become controversial because several studies reported the ability of $CD133^{-}$ cells to initiate tumors or, conversely, the inability of $CD133^{+}$ to recapitulate the initial tumor heterogeneity, suggesting the low reliability of CD133 as a CSC marker.[60]

2.3.1.2 ALDH activity

ALDH is broadly used as a functional marker in various tumor types. ALDH is a family of detoxifying enzymes responsible for the oxidation of intracellular aldehydes whose activity was initially found to be increased in hematopoietic and neural stem and progenitor cells. Increased ALDH activity in CSCs was first reported by Ginestier et al.,[61] which showed that $ALDH^+$ breast cancer cells had increased tumor-initiating ability in immunocompromised mice compared to $ALDH^-$ cells. ALDH activity can be detected in live cells through an enzymatic assay (Aldefluor) that was first developed by Jones and colleagues to isolate viable hematopoietic stem cells.[62]

2.3.1.3 "Side population"

The observation that a small population of murine cells from bone marrow aspirates did not accumulate the vital dye Hoechst 33343 was first described by Goodell et al.[63] The population capable of excluding this dye was named side population (SP) and was shown to be enriched in hematopoietic stem cells. Using flow cytometry, the SP has been isolated from different types of tumor and was associated to CSC features such as increased tumorigenicity and self-renewal.[64] The SP results from the extrusion of the dye from the cell through ABC transporters including MDR1, MRP1–5 and BRCP1.[64] However, the SP can represent differentiated cells with specialized function within organs, where ABC transporters play a role in protection against cytotoxic effects of toxins and xenobiotics.[64] Similarly, not all cancers contain SP cells, and SP could include cancer cells that are not CSCs.

2.3.1.4 Generation of floating tridimensional cell clones (tumorspheres)

The generation of spherical anchorage-independent clones as a means for propagating and analyzing stem cell populations emerged with the new millennium. Such floating tridimensional clones have been termed "spheres" or "spheroids," and can be obtained by both normal stem cells and CSCs. Originally, these floating spheroids were developed from normal neural stem cells and were termed "neurospheres." Each of them was the product of a single cell able to generate a sphere by clonal expansion.[65] In the following years, spheres were developed from a variety of solid tumors, including breast cancer,[66] under the same assumption that sphere-generation ability allows measuring self-renewal ability.[59]

"Tumorspheres" generically refers to spheres derived from cancer tissues or cell lines.[67] Methods for CSC isolation and expansion as spheres do not differ much between different cancer types.[67] Single cell suspensions are plated at low density in a specific serum-free medium supplemented with factors that favor stem cell growth, such as bFGF, insulin and EGF, in low-adherent conditions.[67] Since the ability to survive and grow in these conditions is peculiar of CSCs,[67] the "sphere-generation assay" gained popularity as a faster and cheaper *in vitro* alternative to the tumor initiation assay *in vivo* (that will be described in the next section). Readout of the "sphere-generation assay" consists in the evaluation of the number and size of the spheres. However, recent work has highlighted the high degree of phenotypical heterogeneity in tumorspheres from different cell lines,[68,69] while others reported that the amount of enrichment in CSCs that can be obtained through tumorsphere generation is cell line-dependent and that the conventional monolayer culture may maintain tumor-initiating phenotypes more effectively than the tumorspheres, depending on the cancer cell line.[70,71] Therefore, as described below, the use of the tumorsphere model should be complemented by *in vivo* assays.

2.3.2 Validation of the biological features of cancer stem cells

Independently from the isolation methodology used, affirming that a subpopulation of cells is a *bona fide* CSC population rests on the validation of the biological characteristics of CSCs, including the abilities to initiate a tumor, to self-renew and to histologically recapitulate the tumor of origin (Fig. 3).

2.3.2.1 Tumor-initiating ability

As stated previously in this chapter, the operative definition of CSC is that of tumor-initiating cell. In theory, grafting a single CSC should be enough to generate an entire tumor in a mouse model. Therefore, as mentioned earlier, an important assay for a prospectively isolated CSC population is to test its ability to recapitulate a tumor at low cell densities, and limiting dilution assays are useful to estimate the frequency of CSCs within a cell population. As stated previously, *in vitro* surrogate of tumor formation is considered the ability to form a sphere (sphere-generation assay) in serum-free medium or a colony in soft agar.[11]

2.3.2.2 Self-renewal

Normal stem cells are endowed with the ability to generate at least one cell with the same biological properties upon mitotic division. The same self-renewal ability can be experimentally demonstrated *in vivo* also for CSCs by serial transplantation of tumors. This assay represents the gold-standard for CSC assessment.[11] Following prospective isolation of CSCs from a bulk population, the putative CSCs are injected in a mouse to initiate a tumor. The tumor is subsequently removed and dissociated to re-isolate cells with prospective CSC phenotype, which are then re-injected into a new mouse to test again their tumor-initiating potential. If the isolated cells are *bona fide* CSCs, they should be able to maintain their tumor-initiating potential across subsequent passages of transplantation, while cells with alternative phenotypes should not. Also, the serial transplantation assay has its *in vitro* surrogate, represented by the assessment of the capability of the cells to form spheres or colonies in soft agar through multiple generations.[53]

2.3.2.3 Recapitulating tumor heterogeneity

Besides the ability of self-renewing, normal stem cells have the ability to generate a daughter cell that can differentiate into a more specialized, non-stem cell. Stem cells are thus defined as totipotent, pluripotent, multipotent or unipotent according to their potency of generating few or several differentiated cell types. Normal stem cells of adult tissues are usually multipotent, being able to regenerate multiple cell types composing the specific tissue they belong to. In the same way, CSCs should be able to differentiate and recapitulate the heterogeneity seen in the tumor of origin. This can be assessed *in vivo* by examining the intratumor heterogeneity in CSC-derived tumors in mice, or by determining the ability of the CSCs to differentiate *in vitro* in the presence of serum, either with or without other factors known to induce differentiation in the specific tissue type under study. Cells induced to differentiate *in vitro* lose both the expression of CSC markers and their tumorigenic potential.[11]

2.4 Clinical implications of the cancer stem cell concept

From a clinical perspective, the CSC concept has significant implications. Conventional anticancer approaches are in fact mainly directed against the bulk population of the tumor. As CSCs are endowed with features that render them resistant to chemo- and radiotherapy, these therapeutic strategies may fail in eradicating them, thus allowing this self-renewing tumor population to cause relapse. In breast cancer patients, it was shown that the cells composing tumor tissue remaining after endocrine therapy or chemotherapy, displayed a more pronounced CSC-related expression signature compared to pretreatment.[46] Using *in vivo* tracing of a tumor cell subpopulation in a genetically modified mouse model of glioblastoma, it was shown that not all cancer cells, but a restricted and rather quiescent cell population with CSC features, was responsible for the tumor re-growth following treatment with Temozolomide. Re-growth occurred through the production of highly proliferative transit amplifying tumor cells derived from these quiescent CSCs.[72] Similarly, experiments based on tracing a subpopulation of human breast cancer cells with CSC features through an *in vivo* fluorescent imaging system in xenograft models showed that fluorescent CSCs are not proliferating, but they expanded through increased proliferation following radiotherapy, thus repopulating the tumor.[73] In these works, the activation of a suicide gene in the tracked population led to ablation of the tumor, thus proving that the observed cells were effectively those leading to relapse. Besides enrichment of intrinsically resistant CSCs following therapy, evidence exists of chemo- and radiotherapy-induced conversion of non-stem cancer cells into a stem-like status. This reprogramming was shown to occur upon stress-induced re-expression of stemness transcription factors, including Sox2, OCT4 and Nanog, and the activation of CSC-related signaling pathways such NF-κB and Notch, finally resulting in the restoration of the tumorigenic capacity in these induced CSCs.[74–78]

If it is true that CSCs are responsible for therapy failure and tumor relapse and progression, then approaches targeting CSCs could potentially increase the long-term efficacy of currently available treatments (Fig. 4).

The potential of immunotherapy against CSCs has recently become an intriguing field of research. Immune effectors may succeed in overcoming some main chemo-resistance mechanisms adopted by CSCs, like their

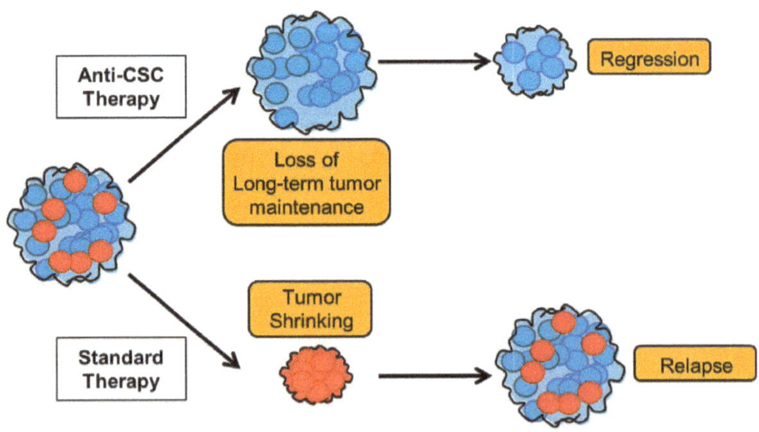

Fig. 4 Clinical implications of the CSC concept. Standard therapies mainly kill differentiated cells with limited proliferative potential, thus shrinking the tumor mass. However, since CSCs are resistant to these therapies, they will be spared and induce tumor relapse after treatment. By contrast, a therapy able to target CSCs may render the tumor unable to maintain itself or grow, eventually leading to tumor regression.

proliferative quiescence or the expression of drug-detoxifying enzymes and pumps.[79] Crucial issues under investigation are the expression of major histocompatibility complex (MHC) and antigen processing machinery (APM) molecules, of MHC-restricted and unrestricted antigens and of immunosuppressive or immunostimulatory molecules by CSCs, that will be taken into account in the next section.

3. The immunology of cancer stem cells

During cancer initiation and progression, tumor cells interact with and are shaped by a huge variety of accessory cells, which can either positively or negatively affect tumor evolution. Among them, cells belonging to the immune system are major players in this interaction. The concept that a physiological function of the immune system is to recognize and eliminate continuously arising transformed cells was proposed by Burnet and Thomas in the '50s with the name of *immunosurveillance*.[80] Many failed to experimentally validate this hypothesis in the subsequent decades, but recently better techniques and models contributed to establish the physiological relevance of cancer immunosurveillance in both mice and humans, giving new impetus to the research field of cancer immunology.[81]

Despite the pivotal role of immunosurveillance in counteracting tumor growth, it is now clear that the immune system's action is not limited to protection against tumor development, but can select for poorly immunogenic clones of tumor cells, thus indirectly promoting tumor growth.[82] This property is named cancer immunoediting and results from three subsequent phases described as the three Es: (i) **E**limination, which corresponds to the immunosurveillance, where innate and adaptive immune cells recognize arising transformed cells, destroying them before they can outgrow into an overt tumor; (ii) **E**quilibrium, where the immune system shapes the tumor through Darwinian selection, iteratively deleting more immunogenic tumor cells and selecting for tumor cells that are able to survive the immune attack; and (iii) **E**scape, where the immunologically shaped tumor expands in an uncontrolled manner and the immune system is no longer able to restrain it.[83]

Several tumor cell-autonomous and non-cell-autonomous mechanisms are involved in the process of immunoediting,[84] and more broadly in the tumor–microenvironment interaction.[85] Mechanisms through which tumor cells can escape immune attack can be thus divided into two broad categories: (i) intrinsic mechanisms, which are activated by the tumor cells themselves and (ii) extrinsic mechanisms, which are mediated by non-tumor cells. Scientific literature produced in the last decade supports the existence of differences in intrinsic mechanisms of immune evasion between CSCs and differentiated tumor cells, as well as a prominent role of CSCs in the interplay between cancer and the tumor microenvironment (TME), which is at the basis of the extrinsic mechanisms supporting tumor escape. Interestingly, immune evasion has been identified as an intrinsic property of normal stem cells.[86] This suggests that cancer cells with stem-like properties may evade immunosurveillance not only as a result of immunoediting but also, at least in some cases, as an intrinsic feature of their undifferentiated state. Understanding whether CSCs differentially regulate immune escape mechanisms compared to differentiated cancer cells would help to better define efficient anti-CSC immunotherapies.

3.1 Tumor antigens expressed by cancer stem cells

The first experimental demonstration that tumor establishment can induce protective immune responses came from studies on transplantable tumors performed in the '50s and reviewed by Parish.[80] In inbred mice, a sarcoma was induced by painting the skin with the chemical carcinogen methylcholanthrene. When the tumor was resected and transplanted into other syngeneic mice, it was still able to grow. However, if transplanted back into the original mouse, the tumor was rejected. Furthermore, T cells recovered from a tumor-bearing mouse and transferred to a tumor-free mouse conferred protection against tumor challenge. These data point out that the tumor expresses antigenic epitopes that can become targetable by an effective anti-tumor T cell response. These tumor rejection antigens expressed by experimental murine tumors are usually specific for an individual tumor. In fact, transferring T cells from a tumor-bearing mouse protects a syngeneic mouse from challenge with cells derived from that same tumor, but not from a different one.[80]

Earlier classification of these tumor antigens distinguished between tumor-associated antigens (TAAs) and tumor-specific antigens (TSAs). The formers are expressed by both normal and cancer cells and are usually represented by molecular components of normal cells whose expression is deregulated in tumors. The latter are expressed exclusively by cancer cells and are usually the product of gene mutations (the so-called "neoantigens"). However, the modern classification takes into account the molecular structure and the source of the tumor antigens.[80]

At present, an exhaustive CSC antigenic profile is lacking. Nevertheless, a number of studies sought to compare the expression of given antigens between CSCs and the corresponding non-CSCs. In general, based on the distribution between CSCs and non-CSCs, tumor antigens can be defined as (i) *CSC-specific antigens*, which are expressed by CSCs but not by non-CSCs; (ii) *shared antigens*, which are expressed by both CSCs and non-CSCs; and (iii) *non-CSC antigens*, which are expressed only by non-CSCs but not by CSCs.[87]

Based on existing literature, it appears that tumor antigens are differently distributed in these three groups at least in part depending on the subclass they belong to (e.g., differentiation antigens, cancer/testis antigens, overexpressed antigens, viral antigens, neoantigens).[87,88] In this section, we will consider the different subclasses of tumor antigens and discuss whether they are expressed in a differential manner by CSCs and their non-CSC counterpart. Clearly, this has implications in the choice of the right immunotherapy, which should be efficient in targeting CSCs while sparing normal cells.

3.1.1 Neoantigens

The most common tumor antigens are products of mutated genes, and thus defined as neoantigens.[89] The transformation of normal cells toward malignant cells, indeed, requires one or more mutations in oncogenes and oncosuppressors, possibly leading to the generation of mutated antigens. Since normal cells do not express these mutated antigens, they are recognized as non-self and can elicit both $CD4^+$ and $CD8^+$ T cell activation. Besides oncogenes and oncosuppressors, random mutations mostly involve non-tumor-related genes that, if protein-coding, can anyway produce immunogenic mutated antigens. Since a high number of mutations often occurs in mutagen-induced cancers,[90] tumors such as UV-induced melanomas and cigarette smoke-induced lung cancers are often highly immunogenic.[91] As a consequence, patients with tumors carrying a high mutational burden are more likely to be responsive to inhibition of immune checkpoints, thanks to the presence of non-self-antigen/tumor-specific T cells.[92–94] Immune checkpoints are inhibitory signals that prevent activation—or promote inactivation—of T cells. Blocking these signals unleashes T cells, which then are free to attack and kill tumor cells. Although it is currently unknown if the mutation pattern of CSCs is different from that of more differentiated cancer cells from the same tumor, a study on colorectal cancer patients suggests that cancer cells grown in standard conditions or as tumorspheres express the same neoantigens and are equally susceptible to T cell targeting.[95]

3.1.2 Viral antigens

Other antigens recognized as non-self by the immune system are represented by products of oncogenic viruses, such as hepatoma-associated hepatitis C virus (HCV), B cell lymphoma-associated Epstein–Barr virus (EBV) or cervical carcinoma-associated human papilloma virus (HPV). Unlike products of mutated genes, which in many cases are specific for each patient since they are often the result of random mutations, antigens of oncogenic viruses are shared by many patients and can thus represent an ideal target for anti-tumor vaccines. HPV-induced cervical cancers, whose incidence has been strongly reduced by preventive vaccination programs,[96] represent a significant case and therapeutic anti-HPV-related cancer vaccines are under study in patients.[97] Interestingly, it has been suggested that cervical stem cells residing in the transformation zone of the cervix, and not the basal cells of the cervical epithelium, are those responsible for cancer initiation and maintenance following HPV infection.[98] Conversely, HCV infection has been shown to induce the acquisition of a hepatic progenitor-like phenotype and CSC traits in hepatoma cells.[99] In particular, the infection of liver-derived cells has shown to enhance the expression of the stem cell marker DCAMKL-1, which in turn favors the replication of HCV, and curing hepatoma cells from HCV infection abrogates the stem cell features previously acquired through HCV-induced de-differentiation.[99] Therefore, seminal evidence suggests that normal stem cells can generate tumors following infection and, conversely, that viral infection can lead to de-differentiation of cancer cells into a CSC phenotype. This has important implications for CSC-targeted therapies, as targeting the tumor-inducing viral antigens would be *per se* a CSC-targeted therapy. However, research in this context is scarce and more studies are needed to disclose the link between viral infections and the CSC phenotype.

3.1.3 Tumor-associated antigens

Besides neoantigens and viral antigens, proteins that are not mutated but that are abnormally expressed by tumor cells can also represent tumor antigens able to induce immune responses in tumor-bearing patients. As their expression levels in tumors increases, this elicits an immune response.[100] For this reason, non-mutated antigens have been studied in several preclinical and clinical studies as targets for anticancer vaccination.[101] This is the case of the so-called differentiation antigens, which are expressed by terminally-differentiated cells of a given tissue and are retained by tumor cells derived from the same tissue. However, differentiation antigens would not be an ideal target for CSC-directed immunotherapies, as CSCs are in an undifferentiated state and thus lack the expression of markers of terminal differentiation. This is the case for melanoma antigen recognized by T cells (MART)-1, a lineage-specific antigen commonly expressed in melanoma but whose expression is downregulated in melanoma stem cells,[102,103] and prostate-specific antigen (PSA), whose expression is common in prostate cancer but low in prostate CSCs.[104]

Another class of TAAs is represented by cancer/testis (CT) antigens, which are expressed in gametes and trophoblasts but not by normal somatic cells. These antigens, such as NY-ESO-1 and those belonging to the melanoma-associated antigen gene (MAGE) family, are re-expressed by many tumors. Contrary to differentiation antigens, it was revealed that several known CT antigens are preferentially expressed by CSCs compared to non-CSCs of different tumor types. Since some of them are already known to be target of anti-tumor T cells, they could be optimal targets for anti-CSC immunotherapy.[88]

Most CSC markers have been identified based on our knowledge of the stem cell hierarchy of the tissue from which the tumor originates (see Section 3.1). T cell-mediated immune responses targeting CSC markers such as ALDH, CD133, CEP55, and EpCam among others, have been described.[105] CSC markers can be considered as overexpressed antigens, i.e., antigens that are expressed at low levels by normal tissues but constitutively overexpressed by tumors as a consequence of their transformed phenotype. Therefore, most CSC markers are not tumor-specific antigens, as they are shared with normal tissues. As a consequence, their immunogenicity is generally limited due to the presence of tolerogenic T cells that recognize these antigens. The low immunogenicity of this type of TAAs has been recognized as one of the mechanisms at the basis of the low therapeutic efficacy of cancer vaccines.[106] Nevertheless, these CSC markers could represent potential targets for passive immunotherapy, such as antibody therapy targeting CD44 or CD133 (see Section 3.3.2). However, potential safety issues should be kept in consideration when utilizing these markers as therapeutic targets, as they are often shared with normal tissue stem cells, thus questioning their use as therapeutic targets.

Additionally, antigenicity of cancer can derive from altered glycosylation pattern of surface glycoproteins and glycolipids. This can result in the presence of novel tumor-specific carbohydrate or peptide epitopes that can elicit a humoral immune response. An example of such altered glycoproteins is represented by mucin (MUC)-1, a member of the mucin glycoprotein family whose expression is deregulated in breast cancer. As for other non-patient specific antigens, studies aimed at targeting MUC-1 with vaccines are ongoing.[107] Several studies have indicated that the maintenance of CSC stemness properties is supported by aberrant protein glycosylation. The majority of surface markers that identify the CSC population are glycoproteins whose altered glycosylation affects tumorigenic processes.[108] Since CSCs express different glycosylation patterns from normal cells and the other cancer cells, exploiting glycan modifications could have an impact on CSC identification and targeting for therapeutic purposes. The epithelial cell adhesion molecule, EpCAM, whose expression is associated with the stem-like phenotype of CSCs, is differentially glycosylated in healthy tissue compared to tumor cells. Indeed, the role of EpCAM hyperglycosylation has been reported to have an impact on its function in head and neck carcinomas.[109] The altered CD44 standard and its variant glycosylation are associated with metastatic properties of human tumors.[110] Moreover, the glycosylation status of CD133 was demonstrated to play a critical role in stem cell maintenance.[111] Therefore, the early detection of CSCs could be made possible through the identification of aberrantly glycosylated CSC markers or TAAs, providing the opportunity of developing new therapeutic approaches targeting CSCs such as glycopeptide-based vaccines.

Besides known antigens, several studies are aiming at finding new CSC-specific targets. For example, Morita and co-workers found that DNAJB8 is a good candidate for CSC-targeting immunotherapy in preclinical models. DNAJB8 is a CT antigen found to be preferentially expressed in a side population of kidney and colon cancer cells and its targeting showed significant anti-tumor effects compared to results obtained by immunizing against the shared antigen Survivin.[112,113] Similarly, olfactory receptor family 7 subfamily C member 1 (OLF7C1), whose overexpression on CSCs has been linked to a poor prognosis in colorectal cancer patients, has been successfully targeted in an *in vivo* murine model using OLF7C1-specific CTLs.[114]

3.1.4 Candidate antigens as immunotherapy targets

Besides the identification of many antigens of the different classes, as described before, with a different distribution between CSCs and non-CSCs, the questions remain: "which are the best cancer immunotherapy targets in general and for anti-CSC immunotherapy in particular?" First of all, it should be considered that most tumor antigens can derive from proteins with a non-pivotal role in the biology of the tumor. Therefore, even if immune responses directed against these antigens could achieve temporary tumor control, they may also lead to tumor escape by selecting those cells that have lost antigens. For this reason, targeting a protein that is required for the survival of tumor cells could be the most effective strategy, as this protein cannot be easily downregulated due to its indispensable role.[115] These molecules have been termed *oncoantigens*, i.e., tumor-associated molecules recognized by the immune system that have a causal role in cancer.[116]

In the light of the encouraging results showing that CSCs could be a better source of antigens compared to non-CSCs, oncoantigens expressed by CSCs may represent good targets for anticancer vaccination.[117] In this case, a CSC oncoantigen would be a protein required for the survival, self-renewal and tumor-initiating ability of the CSCs. Currently, no consensus exists on the expression of specific tumor antigens by CSCs. Despite the lack of studies depicting the whole antigenic profile of CSCs of different tumor types, it is known that CSCs can display different groups of tumor antigens that can be recognized by T cells, and these antigens belong to the different functional groups described for tumors in general, such as differentiation antigens, CT antigens and mutated antigens[118] (Section 3.1). Clearly, more studies in this regard are needed.

Another hurdle in the choice of the target antigen is the heterogeneity that can occur in the CSC population itself. Indeed, different papers often report the expression of different TAAs on CSCs from the same tumor type, which sometimes may be due to the different methods used to characterize these cells. Finally, it should be considered that, while destroying the sole CSC population by targeting CSC-specific oncoantigens should be sufficient to halt tumor progression, the intrinsic plasticity of a tumor might cause differentiated cancer cells to de-differentiate in CSCs. Therefore, targeting oncoantigens shared among CSCs and non-CSCs would be theoretically more promising, resulting in depletion of both groups of cells. Regarding non-CSC antigens, they do not seem to be suitable immunotherapy targets since their targeting will result in an enrichment in CSCs in a way similar to what is achieved using radio- and chemotherapy.

3.2 Immune evasion of cancer stem cells

3.2.1 MHC-I and APM downregulation

In addition to the loss of immunogenic antigens because of the selective pressure exerted by CTL or anti-tumor antibodies, resulting in the generation of antigen loss variants of the tumor,[119] tumor cells can downregulate the expression, or produce non-functional forms, of the APM and MHC-I molecules, thus avoiding recognition by T cells.[120] Despite the key role of

APM and MHC-I in T cell recognition, the expression of these molecules on CSCs has been studied only to a limited extent. Available data suggest that defects in antigen processing and presentation are frequent in CSCs of some cancer types. A comprehensive immune profiling in CSCs grown as tumorspheres and autologous non-CSCs isolated from glioblastoma (GBM) patients revealed that CSCs expressed MHC-I molecules, but at a lower level than differentiated GBM cells, and that interferon (IFN)-α and IFN-γ treatment upregulated MHC-I expression to a higher extent in differentiated cells than in CSCs.[121] Similarly, a frequent down-modulation of MHC-I was found in CSCs isolated from colorectal cancer (CRC) patients and grown as tumorspheres, compared with their autologous differentiated tumor cells.[122,123] In another study performed on melanoma patients, MHC-I expression was significantly reduced on CSCs compared with differentiated melanoma cells.[102] A published meta-analysis performed on different studies on prostate cancer suggests that downregulation of MHC-I molecules also is common in prostate CSCs as compared to differentiated prostate cancer cells.[86]

Evidence supports the idea that the downregulation of MHC-I in CSCs is not an artifact of *in vitro* cell growth. Indeed, immunohistochemical analysis performed on tumor samples obtained before and after primary chemotherapy on head and neck cancer patients revealed that there is an enrichment of cells with stem cell markers in relapsed tumors as compared to the original tumor, which is accompanied by a decrease in MHC-I positivity. Furthermore, patients with low MHC-I expression after relapse had statistically significantly poorer survival compared with patients with medium to high MHC-I expression.[124] The same group also reported that docetaxel-resistant prostate cancer cells are enriched in CSCs features and that negativity for MHC-I closely overlapped with CSCs since the MHC-I-negative subpopulation, but not the MHC-I-positive one, is highly enriched in tumor-initiating cells that sustain serial xenotransplantation and recapitulate the phenotypic heterogeneity of the primary tumor in NSG mice, in which immunosurveillance is lacking.[125] This suggests that lack of MHC-I expression is an intrinsic feature of CSCs rather than the result of clonal selection following immunosurveillance, reminiscent of the immune-evasive phenotype of normal stem cells.

Besides patient-derived CSCs, MHC-I downregulation in CSCs compared to differentiated cancer cells has also been observed in established cancer cell lines of different tumor types, including colon, pancreas, melanoma and breast cancer.[126,127] In particular, Busse et al. reported that MHC-I was expressed on the surface of all the cell lines used. However, in contrast to cells growing in adherent conditions, a significant downregulation or loss of surface MHC-I expression on spheres was reported. MHC-I expression was upregulated upon administration of IFN-γ in nine out of 10 cell lines growing as monolayer, but just in two growing as tumorspheres. Interestingly, when these spheres were transferred into standard culture conditions, cells regained expression of MHC-I molecules to the same level as observed on adherent bulk cells continuously cultured in standard conditions, suggesting that MHC-I expression can be restored upon differentiation.[126] This could have important implications, as pro-differentiation treatment could restore CSC susceptibility to T cell recognition.

Regarding the APM, defective expression of several APM molecules, including some of the constitutive proteasome and immunoproteasome subunits, transporter molecules (TAP) and chaperone molecules, was commonly found in both CSCs and differentiated cells from GBM patients, and IFN treatment could restore the expression of some of these proteins.[121] No significant differences in the expression of the standard proteasome subunits were observed in established cell lines, while spheres showed equal or higher mRNA expression levels of immunoproteasome and TAP1 and TAP2 compared to those cells growing as a monolayer. In this case, IFN-γ led to upregulation of immunoproteasome subunits and TAP1 and 2 in cells growing as monolayers or tumorspheres, while standard proteasome subunits were not affected by IFN-γ treatment.[126] Conversely, in cells from CRC patients, all the APM molecules investigated were downmodulated or absent in CSCs compared to differentiated CRC cells, and no or low upregulation of these molecules was observed upon treatment with IFN-α or IFN-γ.[123] An interesting proof of the lower CSC susceptibility to T cell recognition was provided by Morrison et al.[128] They used the murine TC-1 lung cancer cell line engineered to express the human papillomavirus 16 antigens E6/E7 to test the ability of CSCs to escape the immune response elicited by an E6/E7 peptide vaccine in syngeneic mice. TC-1 cells grown as tumorspheres showed reduced expression of MHC-I compared to cells grown as monolayer. *In vivo*, TC-1 CSCs were resistant to E6/E7 vaccine-mediated killing, resulting in shorter survival and fewer $CD8^+$ tumor-infiltrating T lymphocytes compared to mice challenged with non-CSCs. *In vitro*, CTL from vaccinated mice showed lower ability to lyse CSCs compared to non-CSCs. Following treatment with IFN-γ, CSCs and non-CSCs expressed similar levels of MHC-I, and this increased CTL-mediated tumor lysis and improved tumor-free survival in mice.[128]

Together, these results suggest that CSCs of different origin display a low efficiency or impairment of antigen processing and presentation, possibly resulting in lower susceptibility to CTL-mediated killing.

It is important to highlight that conflicting and controversial results have also been reported. Wu et al. reported that both $CD133^+$ and $CD133^-$ cells isolated from tumorspheres derived from human GBM expressed low levels or no MHC-I, and IFN-γ treatment appreciably upregulated expression of MHC-I on both,[129] suggesting that both type of GBM cells may be similarly resistant to immune surveillance, and that IFN-γ may partially restore their immunogenicity. Conversely, Wei et al. reported that $CD133^+$ cells isolated from GBM patient-derived tumorspheres expressed high levels of MHC-I, but also of programmed death-ligand 1 (PD-L1), speculating that a contact between CSCs and T cells would be inhibitory on immune cells.[130] Moreover, in contrast to what previously has been reported, Inoda et al. showed no differences in MHC-I expression between non-CSCs and CSCs isolated as a side population, and both non-CSCs and CSCs were efficiently killed by CTLs *in vitro* and *in vivo*.[131] Similarly, CSCs and differentiated cancer cells derived from prostate cancer of transgenic cancer-prone TRaMp mice and grown as tumorspheres expressed comparable levels of MHC-I that further increased upon administration of IFN-γ, and both were susceptible to CTL targeting *in vitro* and *in vivo*.[132]

The discrepancies in the immunological profile of CSCs could be partially explained by the different isolation methods and culture conditions used, as well as by the different tumor types investigated. Although CSCs are speculated to have the ability of immune evasion by means of MHC-I or APM downregulation, it is important to highlight that several studies reported that anti-CSC vaccination is effective in inducing the generation of T cells able to specifically target and kill CSCs, thus impairing tumor initiation, metastatization and relapse (see Section 3.2.1). Moreover, many of the studies here reported suggest that, when MHC-I or APM downregulation occurs in CSCs, the use of IFNs can restore their susceptibility to T cell recognition.

3.2.2 Immune checkpoints expression

Another intrinsic mechanism of immune escape is the engagement of inhibitory molecules on the T cells by tumor cells. Immunosuppressive pathways acting on T lymphocytes have the physiological role to prevent unwanted or uncontrolled immune responses against self-antigens. For this reason, these pathways are named immune checkpoints. However, these checkpoints can favor the immune escape of tumors when dysregulated.[133] This is the case, for example, of the checkpoint molecule programmed cell death 1 (PDCD1, or PD-1) expressed by activated T cells.[134] Its ligand PD-L1 is expressed by many tumors, thus when an activated T cell reaches the tumor microenvironment the PD-1/PD-L1 interaction leads to T cell inactivation. Another important inhibitory molecule is cytotoxic T-lymphocyte antigen 4 (CTLA-4) expressed by T lymphocytes, which competes with CD28 for the binding to B7-1 and B7-2 on antigen presenting cells (APCs). This competitive binding prevents the co-stimulatory signal normally provided by CD28/B7 interaction, resulting in T cell anergy. Di Tomaso reported that both CSCs and non-CSC from GBM cell lines are universally positive for PD-L1, which may contribute to the inefficient induction *in vitro* of T cell-mediated responses by both CSCs and non-CSCs and thus to the possible impairment of *in vivo* anticancer immune reactions.[121] Conversely, Wei et al. showed that PD-L1 was expressed in GBM CSCs and direct cell-to-cell contact between CSCs and T cells induced apoptosis in the latter.[130] Similarly, Schatton et al. reported that melanoma stem cells have down-regulated expression of PD-L1 compared to non-CSCs, while curiously expressing PD-1 and B7-2.[102] In primary human and xenografted head and neck squamous cell carcinoma (HNSCC), PD-L1 expression is higher on $CD44^+$ CSCs compared to $CD44^-$ non-CSCs. PD-L1 expression was mediated by constitutive signal transducer and activator of transcription 3 (STAT3) activation in CSCs and further increased by IFN-γ administration.[135] High expression of PD-L1 has also been reported on colorectal and breast CSCs,[136] while B7-H4, another member of the B7 family that negatively regulates T cell-mediated immunity,[137] was found to be preferentially expressed in non-dividing brain tumor cells and in a subpopulation of brain CSCs.[138] All these observations suggest that CSCs may have an increased immunosuppressive phenotype compared to more differentiated cancer cells.

Interestingly, some studies suggest that PD-L1 upregulation may be linked to EMT,[139,140] a feature closely linked to the stem-like phenotype of CSCs (see Section 2.2.2). Besides its strong link to the CSC phenotype, EMT has been proposed as a major mechanism of immune escape.[141] A correlation study in non-small-cell lung carcinoma (NSCLC) showed that an EMT-gene signature is associated with increased expression of a variety of immune inhibitory surface molecules (e.g., PD-L1/2, PD-1, TIM-3, LAG-3, B7-H3, BTLA, CTLA-4). Mechanistically, EMT and PD-L1 upregulation in NSCLC cells are linked by the transcription factor

ZEB-1 and by miR-200. ZEB-1 is a promoter of EMT and a transcriptional repressor of miR-200, the latter targeting PD-L1 transcript. Therefore, ZEB-1 expression indirectly upregulates PD-L1 in lung, thus contributing to immune escape and metastatization of cancer.[142]

In addition to its established role in the immune suppression, PD-L1 expression may have a cell-autonomous effect on cancer cells themselves. Transcriptomic meta-analysis on a large breast cancer patient dataset revealed a strong link between PD-L1 expression and cancer stemness score. Importantly, this study highlighted that PD-L1 is fundamental for the expression of the embryonic stem cell transcriptional factors OCT4A, Nanog and the stemness factor BMI1 through AKT signaling. Moreover, downregulation of PD-L1 impaired the self-renewal ability of breast CSCs *in vitro* and *in vivo*.[143] PD-L1 has also been shown to enhance proliferation ability of gastric cancer stem-like cells following interaction with PD-1.[144] A similar intrinsic function has been reported for CTLA-4. It was demonstrated that $ALDH^+$ melanoma stem cells express CTLA-4, which supports cell proliferation and suppresses apoptosis *in vitro*. Blocking of CTLA-4 was able to suppress self-renewal *in vitro* and tumorigenic ability *in vivo* by depleting $ALDH^+$ cells.[145] Overall, these studies highlight how PD-L1 or CTLA-4 targeting has the potential to affect CSC immunoevasive properties, as well as CSC biology, in a direct manner.

3.2.3 Immune evasion and tumor microenvironment remodeling through membrane-bound and soluble factors

Immune tolerance may also be induced through inhibition of T lymphocytes and other immune cell populations *via* the release of immunosuppressive molecules.[146] CSCs isolated from different solid tumors have been shown to release a wide range of immunomodulating cytokines and soluble factors including vascular endothelial growth factor (VEGF), growth/differentiation factor (GDF)-15, TGF-β, interleukin (IL)-4, IL-6, IL-10, IL-13, prostaglandin E2 (PGE2), galectin 3 (LGALS3), tumor necrosis factor (TNF), and others.[105]

Colon CSCs sustain tumor growth, resist apoptosis and acquire resistance to chemotherapy by producing IL-4, which acts in an autocrine manner through interaction with IL-4R, both preferentially expressed by CSCs compared to non-CSCs.[123,147] At the same time, the increased amount of membrane-bound IL-4 on CSCs can directly inhibit T cell proliferation in the co-culture of peripheral blood mononuclear cells (PBMCs) from either CRC patients or healthy donors through cell-to-cell contact.

Of note, the neutralization of IL-4 signaling could prevent this immunosuppressive mechanism, supporting activation and proliferation of T cells.[123]

Cancer cells can directly impair CTL function through the release of TGF-β.[148] Notably, the TGF-β signaling pathway is specifically activated in the $CD44^+$ fraction of breast cancer cells,[149] and members of the TGF-β superfamily and their receptors are also preferentially expressed by melanoma CSCs.[102] The action of TGF-β and other cytokines in cancer-related immunosuppression is not limited to a direct impact on immune cells, but can favor immune escape through the induction of EMT in cancer cells,[141] which is strictly linked to a stem-like state. It was shown that exposure to inflammatory cytokines enhanced EMT features, which resulted in multiple immune-regulatory effects in cancer cells. This eventually led to NK and T cell apoptosis, inhibition of lymphocyte proliferation and stimulation of regulatory T and B cells.[150] Furthermore, the acquisition of EMT traits in human breast cancer is associated with inhibition of CTL-mediated killing,[151] and in melanoma, EMT sustains metastasis not only by enhancing invasion but importantly by inducing multiple immunosuppression and immunoresistance mechanisms, including immunosuppressive cytokines, regulatory T cells, dysfunctional dendritic cells (DCs), and CTL resistance.[152] Therefore, a combination of treatments able to revert the EMT phenotype could positively affect the outcome of CSC-directed immunotherapies. In addition, many of the above-mentioned immunosuppressive factors can help recruit suppressive immune cells to the tumor or remodel the TME to establish an immunosuppressive environment that in turn promotes tumor growth and progression,[105] as discussed in the next section.

3.2.4 Immune evasion through extrinsic mechanisms

As opposed to cancer cell intrinsic mechanisms of immune escape, the extrinsic ones involve non-transformed cell populations, which modulate the TME toward immunosuppression.[153] Paradoxically, the immune system plays a pivotal role in promoting cancer cell escape from immunosurveillance.[154] In particular, cells belonging to the innate immune system can either induce anti-tumor immune responses or promote tumor growth and progression depending on their functional subtype.[155]

Among these cells, an important tumor-promoting role is played by alternatively-activated macrophages (M2) that, contrarily to M1, behave as if they were trying to heal a wound and not to kill a pathogen. Consequently, they secrete factors that promote tumor neoangiogenesis, such as VEGF, and that impair T cell function, such as IL-10, PGE_2, arginase and TGF-β.[156]

Other major players in immunosuppression include another population of myeloid cells, the myeloid-derived suppressive cells (MDSCs). Deriving from diverse ranges of myeloid precursors, MDSCs are recruited to inflammation sites, including the tumor microenvironment. Here, MDSCs directly suppress T cell proliferation and function by secreting several molecules, including IL-10 and PGE_2, and produce arginase and indoleamine 2,3-dioxygenase that catabolizes arginine and tryptophan, respectively, thus starving effector T cells that need these amino acids for their function. Furthermore, they induce the generation of Tregs, which support an immunosuppressive microenvironment.[157] Physiologically, Tregs have an important function in shutting down the immune response after the elimination of the pathogen to prevent autoimmunity. In the cancer setting, they prevent effector T cell activation by competing with them in interactions with APCs and in IL-2 uptake. Moreover, they secrete the immunosuppressive molecules TGF-β and IL-10.[158]

Tumor cells, particularly CSCs, secrete cytokines and growth factors that influence normal hematopoiesis, inducing the expansion of immunosuppressive myeloid cells, and recruit these immune suppressive cell populations to the tumor site.[159] For instance, through the release of TGF-β, IL-10 and IL-13, GBM CSCs can induce the differentiation of Tregs when cultured with $CD4^+$ T cells.[121] $ALDH^+$ pancreatic CSCs can up-modulate the immunosuppressive properties of tumor-associated macrophages (TAMs) compared to $ALDH^-$ cells, leading to impairment of T cell-mediated responses.[160] Glioma CSCs produce colony-stimulating factor (CSF)-1, TGF-β1, and macrophage inhibitory cytokine (MIC)-1. These factors polarize the microglia macrophages to an M2 phenotype, inhibit phagocytosis and induce the secretion of the immunosuppressive cytokines IL-10 and TGF-β1 by the microglia, and enhance the capacity of microglia to inhibit T cell proliferation.[130] In ovarian cancer, the chemokine CXCL12 is involved in sustaining CSC self-renewal and in the recruitment in the TME of MDSCs and suppressive DCs as well as in the differentiation of Tregs. The inhibition of CXCL12/CXCR4 signaling leads to the decrease of tumor growth and induction of anti-tumor immune responses.[161] Elevated levels of IL-6 and IL-8, among others, have been observed in chemotherapy-induced lung CSCs.[162] Chronic inflammation is mediated by IL-6 and IL-8, suggesting that CSCs can have critical role in inducing and maintaining a pro-tumorigenic inflammatory microenvironment, which in turn may regulate CSCs.[163]

Nevertheless, partially contrasting results have been reported. Di Tomaso et al. conducted a detailed *in vitro* analysis of the soluble factors produced by CSCs from GBM. They found that CSCs release high amounts of angiogenic factors such as VEGF, macrophage-chemoattractant protein-1 (MCP-1), eotaxin, macrophage inhibitory factor (MIF) and growth-related oncogene a (GROa). Nevertheless, the immunosuppressive cytokines IL-10 or IL-13 were not released by CSCs and TGF-β, IL-6 and IL-8 were downmodulated compared to non-CSCs.[121] Wei et al. reported that the cancer-initiating cells of GBM did not produce any appreciable IL-6, IL-10, or soluble Fas, but did produce TGF-β1 and the Treg chemokine attractant CCL-2.[130]

In summary, cytokines preferentially secreted by CSCs may contribute to create an immune suppressive microenvironment and to escape from CTL, even though a general CSC cytokine profile cannot be determined based on available data.

The contribution of the different immunosuppressive cell populations in promoting CSC maintenance and tumorigenesis is becoming clearer. Inflammatory cells take part in the generation of the CSC niche in both primary tumors and metastases, where they favor CSC self-renewal by both direct and indirect mechanisms.[164] Different research groups have demonstrated that TAMs fuel cancer cell stemness through several mechanisms.[154,165] TAMs secrete cytokines and growth factors, such as IL-6 and milk-fat globule-epidermal growth factor-VIII (MFG-E8), that induce STAT3 and Shh activation in CSCs, which promote their proliferation and expansion, increase drug resistance and protect from apoptosis.[165,166] Together with IL-1 and IL-8, these cytokines induce the activation of STAT3 and NF-κB pathways also in stromal cells, inducing additional cytokine production in a positive feedback loop.[167] Moreover, TAM-secreted TNF-α, IL-6 and TGF-β stimulate EMT in differentiated cancer cells, thus acting as a source of stem-like cells and promoting the metastatic process.[168] Besides cytokine production, TAMs act also through cell-to-cell contacts to promote CSC self-renewal.[169] Indeed, Lu et al. demonstrated that in breast CSCs, EMT induces the expression of CD90 and EphA4. CD90 anchors CSCs to TAMs, thus promoting EphA4 binding to its ligands Ephrins expressed on TAMs, eventually leading to NF-κB activation in CSCs.[170]

The interaction between CSCs and MDSCs is not fully understood, but new evidence supporting a role of MDSCs in the CSC niche has being provided. As with TAMs, in pancreatic and breast cancer MDSCs secrete high amounts of IL-6 and nitric oxide (NO), which cooperate to induce STAT3 and Notch activation in cancer cells, endowing them with stem-like properties.[171,172] Moreover, MDSCs can regulate miRNA expression in CSCs. Indeed, Cui and colleagues have demonstrated that MDSCs induce the expression of stem-related genes in ovarian cancer cells through the upregulation of miRNA-101, which represses the corepressor gene C-terminal binding protein-2 (CtBP2), a negative regulator of stemness genes.[173] Additionally, in uterine cancer, MDSCs express high amounts of cyclooxygenase-2 (COX2) and produce PGE_2, which enhance the stem-like properties of cancer cells.[174] Thus, MDSCs and TAMs cooperate to promote CSC induction, and under hypoxic conditions MDSCs can even differentiate to give rise to immunosuppressive TAMs.[159]

Although TAMs and MDSCs are the main immunosuppressive cells in the tumor microenvironment, other myeloid cell populations concur to CSC promotion. Indeed, the microenvironment and CSCs themselves confer tolerogenic and immunosuppressive properties to DCs and neutrophils. In colon cancer, tumor-associated DCs secrete high amounts of IL-10 and the chemokine CXCL1, which induce the expression of many oncogenes and stemness genes in cancer cells. Moreover, CXCL1 favors EMT and cell migration, thus promoting tumor metastatic spread.[175] Furthermore, DCs secrete CXCL12, whose receptor CXCR4 is expressed on tumor and stromal cells and exerts a double action, promoting CSC self-renewal and recruiting MDSCs and Treg in the tumor milieu.[161] Also neutrophils play a central role in the orchestration of breast cancer metastases. Indeed, neutrophils are recruited either directly by factors produced by tumor cells or indirectly by γδ T cell-produced IL-17 in the pre-metastatic niches in lungs. Here, they produce different leukotrienes that induce MAPK-dependent migration and proliferation of cancer cells.[176,177] Of notes, cigarette smoke can worsen this effect, since it induces a chronic inflammatory state that stimulates the generation of the neutrophil extracellular traps (NETs) in the lungs. NETs promote the remodeling of the extracellular matrix, leading to a modification of laminin that allows its binding to integrins expressed on dormant cancer cells. This leads to the induction of cancer cell proliferation *via* the FAK/ERK/MLCK/YAP signaling pathway, eventually transforming a dormant cancer cell into an overt metastasis.[178]

Besides the different effects exerted by the various immunosuppressive cells present in the TME, it is important to remember that CSCs and immunosuppressive cells are engaged in complex crosstalk. CSCs are able to stimulate their recruitment in tumors and upregulate their inhibitory properties by producing chemokines and growth factors such as CCL2, CSF1 and

G-CSF, eventually leading to downmodulation of effector T cell activation, further promoting tumor progression.[159,168] Obviously, the communication between CSCs and immune cells also involves other components of the TME, such as vascular endothelial cells, fibroblasts, mesenchymal stem cells, but these are beyond the scope of this chapter and have been extensively reviewed elsewhere.[164,168]

3.3 Immune response and targeted immunotherapies to cancer stem cells

Progress in understanding the immune system and in identifying tumor antigens has spurred researchers and physicians to develop a large variety of immune-mediated strategies to fight cancer. Anticancer immunotherapy can be defined as the induction or enhancement of an immune response to treat cancer. Anticancer immunotherapies are generally classified as "active" or "passive," depending on the ability to actively stimulate the host immune response against the tumors or to directly act as anti-tumor agents, respectively.[80]

Passive immunotherapy involves the transfer of immune effectors endowed with intrinsic anti-neoplastic activity, including tumor-specific T cells and antibodies, into patients. Passive immunization against tumors has a rapid onset of action but does not lead to long-lasting immunity.[80]

Tumor-specific monoclonal antibodies (mAbs) are the best characterized form of anticancer immunotherapy, and one the most widely employed in the clinic.[179] Anti-tumor mAbs specifically recognize molecules expressed on the surface of tumor cells and lead to their destruction by the same effector mechanisms working for microbe elimination, including opsonization and phagocytosis, activation of the complement system, and antibody-dependent cell-mediated cytotoxicity (ADCC). In addition, some mAbs may interfere with signaling pathways required for the progression and survival of tumor cells or may trigger death signals by binding to surface receptors. Finally, anti-tumor mAbs can be conjugated to toxins or radionuclides and promote the delivery of these cytotoxic agents specifically to the tumor.[180]

Adoptive cellular immunotherapy, also named adoptive cell transfer (ACT), is the transfer of immune cells that have anticancer reactivity into a tumor-bearing host. In ACT, usually blood or tumor-infiltrating lymphocytes of a patient are isolated and expanded *ex vivo* by culture in immunostimulatory cytokines such as IL-2 and are infused back into the patient. This treatment, often combined with systemic IL-2 administration, leads to tumor regression in some patients.[181] Another approach for adoptive therapy is the T cell receptor (TCR) gene therapy, in which tumor patient's T cells are transduced *in vitro* with genes encoding a TCR specific for a given tumor antigen, expanded, and then infused back into the patient.[182] The patient's T cells TCR can be also engineered to produce a chimeric antigen receptor (CAR), in which the tumor antigen-binding domain of an immunoglobulin is linked to one or more immunostimulatory signal transduction domains, thus making T cells capable of recognizing tumor antigen-expressing cells in an MHC-independent fashion.[183] Usually, the antigen-binding region (a single chain variable fragment—scFv—of a specific mAb) is linked to a spacer/linker portion, a transmembrane domain and a TCR signaling domain (the zeta chain of CD3). To increase the CAR-T cell efficacy, in the second and third generation of CAR-T the signaling domain of respectively one or more co-stimulatory receptors such as CD28, CD134 (OX40) or CD137 (4-1BB) is added in order to elicit full T cell activation, IL-2 production and proliferation.[184] There are many preclinical studies evaluating CAR-T cells targeting different CSC antigens,[185] and CAR-T cells targeting the CD19 antigen have been approved by FDA and EMA for use in some hematologic malignancies (see Section 3.3.3). Nevertheless, CAR-T cell therapy could present some limits that are mainly linked to off-tumor toxicity and cytokine release syndrome induced by the immune system stimulation after cell infusion.[184] Thus, the target antigen has to be selected very carefully and extensive preclinical studies have to be conducted before going into clinical trials.[185]

Unlike passive immunotherapy, active immunotherapy exerts anticancer effects only upon the engagement of the host immune system. The earliest attempts to boost anti-tumor immunity relied on nonspecific immune stimulation, such as local administration of inflammatory molecules (e.g., PAMP and DAMP) and immunostimulatory cytokines (e.g., G-CSF, GM-CSF, TNF-α, IFN-α, IL-2). Notably, as cytokines exert pleiotropic functions in the organism, their use is often associated with different degrees of toxicity.[186] Another way of non-specifically activating novel, or restoring a pre-existing, host immune response against tumors is by means of immunomodulatory mAbs. Immunomodulatory mAbs, such as immune checkpoint inhibitors (see Section 3.2.2), act by binding to, and thus modulating the activity of, molecular components—such as receptors and ligands—of the immune system.

In contrast to nonspecific immune stimulators, the most tumor-targeted form of active immunotherapy in cancer treatment is anti-tumor vaccination. Indeed, anticancer vaccines can be designed to target TSAs, thus sparing normal cells. Moreover, vaccination can provide long-lasting protection, possibly acting against tumor relapse. Depending on the formulation, vaccines can elicit a T cell-mediated response, a humoral response, or both.[187] Therefore, despite CSCs being immunoevasive, immunotherapies could be successful in the task of re-directing the immune system against the tumor. In particular, immunotherapies able to target CSCs would be particularly effective, as they are responsible for tumor metastatization and relapse. In the next section, we discuss studies aimed at specifically target CSCs through different immunotherapeutic approaches.

3.3.1 Vaccines targeting cancer stem cells

Early cancer vaccine studies found that mice could be immunized with irradiated tumor cells, which were not able to give rise to neoplastic lesions but could act as antigen source. An alternative approach is to use proteins or mRNAs derived from tumor lysates to load APCs, in particular DCs derived from the patient's PBMCs, *ex vivo* and then to re-inject them back into the host. *In vivo*, these activated DCs present the tumor antigens to T cells, eventually activating a tumor-specific immune response. The nature of the antigen (peptide, full protein or mRNA) can affect the resulting immune response, with full protein being able to activate both $CD8^+$ and $CD4^+$ T cells and mRNA coding for the antigen being able to enter only the MHC-I processing pathway and thus being limited to a $CD8^+$ response.[188]

In one of the first studies aiming at targeting CSCs with a vaccine, lysates from glioma tumorspheres were exploited to load DCs to be used as a vaccine in mice. The vaccination showed significant protective effects in animals challenged either with tumorspheres or differentiated cancer cell lines, while DCs loaded with lysates from differentiated cancer cells achieved only moderate protection against the challenge with differentiated cells and no protection against tumorspheres. The protection elicited by this CSC-based vaccine was associated with increased tumor infiltration by $CD4^+$ and $CD8^+$ T cells *in vivo* and increased lytic activity of splenocytes from vaccinated mice *in vitro*.[189] This study provided a first proof-of-concept on the higher efficacy of CSC-based DC vaccines compared to standard anticancer DC vaccines. These achievements were obtained in murine models, but it was subsequently demonstrated that human brain CSCs were capable of antigen processing and presentation, expressed functional MHC-I molecules and were susceptible to CTL-mediated lysis *in vitro* and *in vivo*.[190] Therefore, human brain CSCs represent valuable source of antigens capable of triggering effective T cell activation upon DC pulsing with apoptotic CSCs.[191]

Indeed, the authors showed that CSCs expressed common GBM-TAAs at higher levels compared to differentiated cancer cells, and since no great difference in antigen processing and presentation between the two cell types nor differences in DCs were detected, they suggested that superior immune responses after pulsing with CSCs rather than differentiated cells are likely due to improved target cell recognition by antigen-specific CTL.[191] Translating this approach *in vivo* in a rat gliosarcoma model, they confirmed that CSC-loaded DCs conferred significantly longer survival in rats challenged with an intracranial injection of unselected cancer cells compared to DCs loaded with differentiated cancer cells.[191]

The ability of eliciting a superior protective immune response by loading DCs with irradiated CSCs or lysates compared to differentiated cancer cells was also demonstrated for other types of cancer, including prostate cancer,[132] hepatocellular carcinoma,[192] melanoma[193,194] and squamous cell carcinoma (SCC).[194–196] In particular, the studies on melanoma and SCC showed that CSCs, selected for their high ALDH activity, were a better source of antigens compared to non-CSCs or unselected cancer cells, since CSC-loaded DCs had a better protective effect on mice by means of the induction of CSC-specific humoral and CTL responses.[194]

Notably, all these studies were done in a prophylactic setting, i.e., vaccination was performed prior to cancer cell challenge in healthy animals with fully functioning immune systems. Compared to prophylactic vaccination, therapeutic vaccination against established cancer is much more challenging, since the vaccine must face an immune system that has been curbed by tolerizing or polarizing mechanisms that supports the disease in a misguided attempt of self-tolerance.[197] Cancer vaccines would more likely reach clinical success when used for the prevention of recurrence after treatment for the primary tumor, when a functional immune system able to benefit from vaccination is re-established.[198] Since CSCs have been implicated in metastatic progression and local recurrence of disease, patients would benefit from anti-CSC vaccines in this context. As a proof-of-concept, it was demonstrated that DCs loaded with $ALDH^+$ cells significantly reduced local tumor relapse, inhibited spontaneous lung metastasis and prolonged host survival when administered following localized radiotherapy or surgical resection of established tumors, while such a prominent anti-tumor effect was not produced by DCs loaded with non-CSCs or unselected cancer population.[195,196] Notably, both humoral and cellular immune responses produced by CSC-loaded DCs were CSC specific, as demonstrated by binding and lysis assays *in vitro* and reduction of $ALDH^+$ cells *in vivo*, while those elicited by DCs loaded with non-CSCs or unselected cells were shifted toward the non-CSC population.[195,196] This suggests that CSCs express a different set of antigens compared to differentiated tumor cells and that treatment efficacy can be reached only by depleting CSCs. Also, decreased efficacy of DCs loaded with lysates from mixed population suggests that, since CSCs represent a minor population, CSC antigens would be underrepresented, thus preventing the mounting of a CSC-specific immune response able to afford an efficient anti-tumor effect. These concepts are summarized in Fig. 5.

These encouraging preclinical results prompted the translation of CSC-loaded DC vaccines into the clinical setting. The first clinical trial was performed on GBM patients. Seven patients, following standard chemotherapy, received intradermal injections of DCs loaded with RNA extracted from autologous neurospheres. The vaccination induced *in vitro* detectable immune responses in all patients, who had prolonged progression-free survival compared to a historical group of matched controls.[199] Other clinical trials followed in lung and pancreatic cancer settings, showing that an anti-CSC DC vaccine was able to induce both CSC-specific and CSC-nonspecific immune responses and with no untoward side effects, proving the vaccine was safe.[200,201] However, it should be pointed out that in these clinical studies no comparison with DCs loaded with non-CSCs or unsorted cells was performed, thus the question of whether the effects observed in preclinical models can be effectively reproduced in humans remains unanswered. In general, APC-based vaccines have been extensively tested in clinical trials and have proven to be safe and able to promote clinically significant tumor regression in some patients. Notably, the only Food and Drug Administration (FDA)-approved anticancer vaccine for human application is a DC-based vaccine, namely Sipuleucel-T, consisting of autologous DCs pulsed with prostatic acid phosphatase (PAP), which is present in 95% of prostate cancers, fused with GM-CSF. However, a major hurdle of the use of DC-based vaccines is their cost: the treatment with Sipuleucel-T is in the order of $100,000 per cycle of treatment.[202]

A large-scale and cost-effective approach for the CSC-targeted vaccines might be the use of genetic vaccines. They are based on the use of viral or plasmid DNA vectors, or alternatively of RNA-based vectors, to deliver *in vivo* the antigen-coding sequence.[203] These vectors can be administered to the subject to be vaccinated through many different routes (e.g., intradermally or intramuscularly). The resident APC cells take up the vectors and the coding sequence is then translated to produce the antigen of interest.

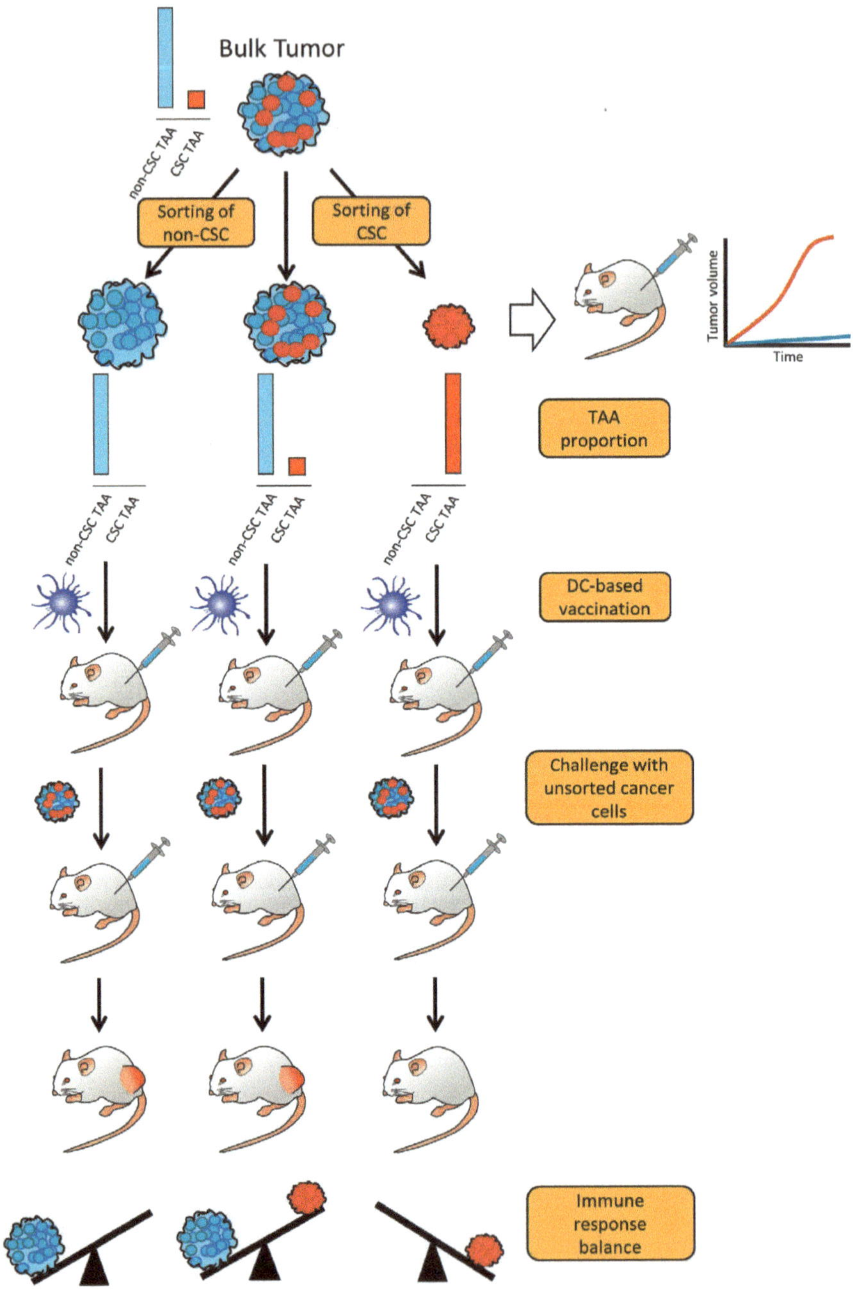

Fig. 5 See legend on opposite page

Genetic vaccines thus allow the endogenous production of the antigen of interest. As they consist in nucleic acid sequences, genetic vaccines are naturally endowed with the ability of simulating microbial infection, which contributes to induce inflammation at the injection site and to license DCs for antigen presentation to T cells. Moreover, the sequence carried by the vaccine can be manipulated to deliver multiple antigens or to produce antigens fused with immunostimulatory molecules, making genetic vectors a versatile tool for vaccination.[204]

A critical point for the development of genetic vaccines is the identification of the appropriate antigen(s) to be delivered by the vector, as discussed in Section 3.1. Tumors can be "shaped" by the immune system: the selective pressure of immune attack can destroy more immunogenic cells, leading to the expansion of less immunogenic clones. This is a problem also faced by anti-tumor vaccines, which may be able to eliminate the target antigen-expressing cells but leave other sub-populations untouched. Identification of novel candidate targets will be possible by evaluating differential expression between CSCs and non-CSCs, as we have performed for murine breast cancer,[205–207] since molecules upregulated in CSCs are more likely to have a functional role in their biology.

By comparing the transcriptional profile of CSC-enriched tumorspheres with that of the more differentiated parental cells in a mouse model of mammary cancer, we were able to identify the transmembrane amino-acid antiporter xCT as a candidate target for vaccination.[206] xCT is upregulated in different types of human cancers, and exerts its biological role mainly by protecting cancer cells from harmful ROS induced by microenvironmental stress and chemotherapy.[208] Different vaccine formulations were tested,

---

Fig. 5 CSC-based DC vaccines are more effective than standard anticancer DC vaccines. Within a heterogeneous bulk tumor mass, only CSCs are responsible for tumor initiation upon transplantation in mice. However, since CSCs represent a minor cell population, CSC-TAAs are underrepresented in the tumor bulk, where non CSC-TAAs prevail. DCs can be loaded with lysates from either sorted non-CSCs, unsorted cancer cells or sorted CSCs, generating vaccines specific for non CSC-TAAs, mixed non-CSC- and CSC-TAAs, or CSC-TAAs, respectively. Only vaccination with DCs loaded with CSC lysates prevents tumor growth in mice challenged with unsorted cancer cells, while DCs loaded with non-CSCs or unsorted cancer cells induce only a moderate protection. The immune responses activated by CSC-loaded DCs are CSC specific and thus effectively target the cells that are able to generate tumors. The immune responses elicited by DCs loaded with non-CSCs or unselected cells are shifted toward the non-CSC population due to the prevalence of non-CSC-TAAs in unsorted cancer cells, and thus eliminate non-tumor-initiating cells while leaving CSC unaffected.

namely plasmid DNA-based vaccine[206] or a Bovine Herpesvirus-4 (BoHV-4) vector encoding the full length xCT protein,[209] or a virus-like particle (VLP) displaying the sixth extracellular domain of xCT.[210] Vaccination was able to overcome immune tolerance toward xCT and to induce a humoral immune response that was able to delay primary tumor growth and to impair the formation of lung metastases in mice challenged with tumorsphere-derived cells. Moreover, vaccination against xCT increased the efficacy of doxorubicin treatment, suggesting that it may act in synergy with chemotherapy. The vaccine elicited a similar immune response across different formulations, with anti-xCT antibodies playing a pivotal role in hampering tumor growth and metastatization by directly targeting CSC and decreasing their content within the bulk cell population.[211] Another CSC-associated oncoantigen that we successfully targeted by vaccination was Cripto-1, an oncofetal antigen expressed in the majority of tumors, such as colorectal, breast, gastric, pancreatic, ovarian and lung carcinomas, and associated with undifferentiated phenotype and poor prognosis.[212] As in the case of xCT, vaccination with DNA plasmid coding the full length Cripto-1 protein induced a humoral response against Cripto-1 that protected against metastasis in the aggressive orthotopic 4T1 and the spontaneous BALB-neuT breast cancer mouse models through the clearance of breast CSCs.[213]

3.3.2 Antibody-based targeting of cancer stem cells

During the last 2 decades, several mAbs have emerged as effective drugs for the treatment of human cancer, being able to increase patients' survival and quality of life. Among these, antibodies against HER2, CD20, and EGFR for the treatment of HER2-positive breast cancer, lymphoma, and epithelial cancer, respectively, and against VEGF, to avoid pathological angiogenesis in solid tumors, have been efficacious.[214] Being highly specific for their target antigen and generally not being significantly toxic, novel mAbs are emerging as potent and specific agents for CSCs elimination. Thus, given that the combination of positive and negative expression of surface markers generally allows the identification of the CSC population of a given tumor type, attention is now turning to the possibility of using these markers as therapeutic antibody targets.

CD44 is a transmembrane glycoprotein overexpressed in many tumors and is associated with aggressive tumor growth, invasion and metastasis.[215] CD44 has been described as a marker of mammary CSCs, moreover it has been shown to be expressed in CSCs of other tumor types, such as bladder, gastric, prostate, pancreatic, ovarian, colorectal and hepatocellular cancer, head and neck squamous cell carcinomas.[215] The mAb H90 is the first that has been demonstrated to successfully target CD44$^+$ CSCs, inducing terminal differentiation and inhibition of homing, proliferation and repopulation of human AML CSCs.[216] Moreover, in a breast cancer xenograft mouse model, the antibody-mediated CD44 targeting has shown to significantly impair tumor growth and prevent tumor relapse.[217] The overexpression of CD47, a transmembrane protein acting both as the receptor for thrombospondin family members or as the ligand for SIRPα, has been linked to patients' worse clinical outcome and poor survival, thus making it a reasonable therapeutic target.[218] Two anti-CD47 mAbs, B6H12 and B6H12.2, enable the phagocytosis of CD47$^+$ cancer cell in both hematological and solid tumors. B6H12.2 has been shown to be effective in inducing phagocytosis of colorectal-derived CSCs and inhibiting tumor growth in a xenotransplantation model.[219]

Several antibody-based preclinical studies and clinical trials targeting EpCAM, a CSC overexpressed marker, have been done. Solitomab (MT110), a bispecific T cell engager antibody construct (BiTE), was able to successfully induce both CD8$^+$ and CD4$^+$ T cell-specific lysis and apoptosis of EpCAM-expressing cells in xenograft models of ovarian carcinoma and colon cancer.[220,221] Moreover, MT110 effectively eliminated CSCs isolated from patient-derived colon and pancreatic carcinomas in a xenograft mouse model.[222,223] Currently, several ongoing clinical trials aim at targeting CSCs of epithelial cancers. The BiTe Catumaxomab has shown promising results in eliminating EpCAM$^+$ CSCs from malignant ascites of patients affected by ovarian, gastric and pancreatic cancer.[224] In various cancers, including gastric, pancreatic, breast, colon and ovarian carcinomas, the expression of CD9, known as a putative CSC marker member of the tetraspanin superfamily, has been linked to the aggressiveness of the tumor.[225] In gastric cancers, the effectiveness of CD9 targeting with a mAb (ALB6) has been observed in an *in vivo* mouse model showing a significant impact on the reduction of tumor volume.[226] Finally, several mAbs targeting interleukins, which play a crucial role in helping CSCs to evade conventional therapies and maintaining their activity, are in clinical trials. Among them, the anti-IL-6 mAb is an example.

Overall, the use of antibodies directed against CSCs has produced encouraging results. Data obtained on tumor xenograft models have revealed that the combination of mAbs with conventional cytostatic drugs is more effective in inhibiting CSC-dependent tumorigenic processes, opening the possibility of developing new combinatorial approaches.[225] A step forward is represented by the approval of Phase I/Ib clinical studies of OMP21M18, a recombinant IgG mAb binding the Notch ligand Delta-like ligand 4 (DLL4) combined with conventional drugs in colon cancer-bearing patients (NCT01189929, NCT01189942, and NCT01189968). Of note, OMP21M18 in combination with irinotecan or paclitaxel is able to exert anti-tumor activity in both colon and breast cancer. Positive results were in fact obtained previously both *in vivo* and *in vitro* on human-derived colon and breast xenograft models, showing a consistent reduction in CSC frequency and in the inhibition of tumor growth.[227]

Another type of combinatorial approach is based on the use of bispecific antibodies targeting two different CSC antigens. Compared to targeting a single molecule using a specific mAb, which could trigger a compensatory increase of the activity of an alternative pathway, bispecific antibodies are able to simultaneously bind to two molecular targets and have indeed demonstrated increased efficacy.[225] As such, a bispecific antibody against an EGF receptor variant form and CD133 has demonstrated its effectiveness in reducing tumorigenicity of GBM CSCs in preclinical models better than any reagent directed against a single epitope.[228]

3.3.3 T cell-based therapy

An interesting and promising immunological approach to target CSCs is the CAR-T cell therapy. Indeed, the FDA, and more recently the European Medicines Agency (EMA), approved two CAR-T cell therapies targeting the CD19 antigen for the treatment of hematological malignancies, in particular for refractory or relapse B cell precursor acute lymphoblastic leukemia (ALL) and large B cell lymphoma.[229]

CAR-T cell therapy presents certain advantages over adoptive T cell therapy or *ex vivo* activation of autologous, non-modified, T cells. The most important is that the recognition of tumor cells is independent of the MHC complex. This is particularly important for CSC targeting, since, as discussed in Section 3.2.1, these cells often have poor antigen presentation due to the downmodulation of MHC and/or APM molecules.[184]

Several different CAR-T cell therapies specific for GBM CSCs have been developed. For example, Zhu et al. demonstrated that CAR-T cells directed against CD133 are able to kill $CD133^+$ CSCs derived from GBM patients, both *in vitro* and when injected *in vivo* in orthotopic tumor models. However, therapy was not able to eradicate the tumors. This could be due to the induction of CAR-T cell terminal differentiation or senescence by tumor cells, as suggested by the expression of the T cell aging marker CD57 on CAR-T cells upon cell-to-cell contact with CD57$^+$ GBM cells.[230] Similarly, CAR-T specific for epidermal growth factor receptor variant III (EGFRvIII), a TSA expressed on 30% of GBM and on GBM CSCs, were able to efficiently kill these cells *in vitro*.[231] However, when tested in a Phase I clinical trial in GBM patients, EGFRvIII CAR-T cells induced antigen downregulation on tumor cells and displayed elevated upregulation of inhibitory molecules.[232] These results suggest that more efforts are needed in order to improve CAR-T cell efficacy in patients, such as the combination of CAR-T cells with therapies able to hinder TME immunosuppressive properties, strategies aiming at bypassing the development of antigen loss tumor clones and/or generation of more efficient CAR-T cells.

A possible strategy could be the administration of cocktails of CAR-T cells directed against different antigens in order to prevent the development of resistance to the therapy. This possibility has recently been explored by Feng et al., who treated a patient affected by advanced cholangiocarcinoma with two different CSC-directed CAR-T cells, one directed against CD133 and the other against the EGFR. Although this therapy produced some toxicity, a partial response was obtained,[233] suggesting that combinatorial therapies could improve the efficacy of CAR-T cell-mediated CSC targeting. Additional CAR-T cell therapies specific for CSC-TAAs have been developed in the last years, such as CAR-T cells specific for the CSC marker EpCAM, which *in vitro* eliminate prostate, ovarian and colon tumor cells, while *in vivo* inhibit prostate and ovarian cancer in mouse models. These results pave the way for the development of new therapeutic protocols.[234,235]

3.3.4 Immune checkpoint inhibitors

As discussed in Section 3.2.2, evidence exists linking high expression of immune checkpoint ligands to the CSC phenotype. These molecules also can have a direct role in CSC biology in some cases. Thus, the idea of using inhibitors of these checkpoint molecules is particularly appealing in the context of CSC treatment. Immunomodulatory mAb targeting immune checkpoints, commonly referred to as "checkpoint inhibitors," have been shown to induce potent and long-lasting responses in cohorts of patients with a wide range of solid tumors.[236] Checkpoint inhibitors currently approved in the clinics include the anti-CTLA4 mAb Ipilimumab, the anti-PD-1 mAb Pembrolizumab and Nivolumab, and the anti-PD-L1 mAb Atezolizumab. Preclinical data showed that anti-CSC vaccination plus anti-PD-L1 administration in the context of resected squamous cell carcinoma significantly impaired tumor relapse and prolonged survival as compared to either treatment alone.[195] In a second study, the same authors showed that dual blockade of PD-L1 and CTLA-4 and CSC vaccine was more effective than the CSC vaccine alone in conferring tumor regression in melanoma-bearing mice.[237] These anti-tumor effects were caused by dramatic depletion of $ALDH^{+}$ CSCs following the combination treatment and were associated with significantly improved T cell expansion, suppression of TGF-β secretion, increased IFN-γ secretion, and significantly augmented host-specific $CD8^{+}$ T cell responses against CSCs. Interestingly, Shi et al. reported that a therapeutic bladder CSC vaccine not only enhanced the infiltration of IFN-γ-producing, $PD\text{-}1^{+}$ CTLs, but also increased the expression of PD-L1 in the TME, and eventually failed to control tumor progression. However, combining anti-CSC vaccination with PD-1 blockade enhanced the function of tumor-specific CTLs and protected mice against a second challenge of CSCs, thus representing a rationale for a combinatorial approach in the treatment of cancer.[238] Collectively, these data show that immune targeting CSCs in combination with checkpoint-mediated immune suppression blockade could significantly improve the efficacy of cancer immunotherapies.

3.3.5 Innate and innate-like immune response to cancer stem cells

3.3.5.1 NK cells

Natural killer (NK) cells are key components of the innate immune system, with a critical role in early host defense against virus and cancer cells. They are large granular lymphocytes that represent 5–10% of lymphocytes in blood and secondary lymphoid organs.[239] Two main subsets of NK cells are present in humans, distinguished according to the levels of expression of their surface molecules CD56 and CD16. Most human NK cells present in the circulation express low levels of CD56 ($CD56^{dim}$) and high levels of CD16 ($CD16^{high}$) and are endowed with cytotoxic activity, while $CD56^{high}$ and $CD16^{dim}$ cells mainly reside in tissues, and act mainly as potent producers of inflammatory cytokines.[240] In contrast to T lymphocytes, NK cells do not express CD3, and the balancing of signals transduced by several activating and inhibitory receptors regulates their activation. In humans, the triggering receptors include the NK-specific natural cytotoxicity receptors (NCRs) NKp30, NKp44, and NKp46, as well as the DNAX-activating molecule-I (DNAM-I) and natural killer group 2D (NKG2D), which are also expressed on T lymphocytes. The inhibitory receptors include killer

Ig-like receptors (KIRs), CD94/NKG2A, and some immune checkpoint molecules such as CD96, TIGIT, lymphocyte activation gene-3 (Lag-3) and T cell immunoglobulin-3 (Tim-3).[241] Although not all the ligands of activating receptors have been identified so far, it has been demonstrated that NCRs can recognize some viral or bacterial proteins, while NKG2D and DNAM-I are able to bind proteins expressed by infected and neoplastic cells. In particular, NKG2D ligands include cellular stress inducible molecules such as highly polymorphic MHC-I-related proteins (MIC)A and MICB and some UL16-binding proteins (ULBP), while DNAM-I recognizes PVR and Nectin-2 that are overexpressed in different tumors. On the contrary, inhibitory receptors recognize self MHC class I molecules.[242] Therefore, most normal cells that express MHC-I molecules are not killed by NK cells, while many tumor cells that downregulate MHC-I molecules and upregulate their activating ligands become a target of NK cell-mediated killing.[239]

NK cells are considered efficient anti-tumor effectors, and NK cell-based adoptive immunotherapy represents a promising strategy for both hematological and solid tumors.[243] Of note, interest in NK cells has increased after the discovery that they can target CSCs.[244] Indeed, we and other research groups have demonstrated that CSCs from different cancer types express lower levels of MHC class I molecules and higher levels of NK activating ligands than their differentiated counterpart.[127,242] Many signaling pathways are involved in the regulation of NK ligands in CSCs.[242] For example, it has been shown that MICA, MICB and ULBPs are upregulated during the EMT process, while activation of the Ras-PI3K axis induces the expression of NKG2D ligands.[245] In contrast, as discussed in Section 3.2.1, the downregulation of MHC-I molecules in cancer cells, and particularly in CSCs, can be due to genetic or regulatory mechanisms altering the expression of either MHC-I genes, genes that participate in peptide generation, such as immunoproteasomal components and endoplasmic reticulum aminopeptidases, or genes responsible for peptide transport or loading of MHC-I molecules.[246] In particular, a decrease in proteasome activity has been observed in CSCs from breast cancer, osteosarcoma and glioblastoma, resulting in a low MHC-I expression.[73,247]

Imbalance in the expression of MHC-I and NK activating ligands on CSCs results in an increased susceptibility to NK killing.[244] The first evidence of CSC sensitivity to NK cell killing came from the analysis of CSCs expanded *in vitro* from tumor specimens from nine glioblastoma patients. These CSCs expressed NKp46 and DNAM-I ligands and insufficient amounts of MHC-I to inhibit NK cell activation, thus being efficiently killed by IL-2-activated NK cells of both autologous and heterologous origin.[248] Similarly, lower expression of MHC-I and higher expression of NCR ligands were observed in CSCs from colorectal cancer patients when compared to the corresponding tumor bulk cell population.[122] This altered expression of NK activating and inhibitory ligands and rendered CSCs from colorectal cancer more sensitive to autologous and allogenic NK cytotoxicity than differentiated cells.[122,249] Many papers have reported an analogous pattern of expression of NK ligands and susceptibility to their cytotoxic effect in other tumors, including melanoma,[250] pancreatic cancer,[251] oral squamous carcinoma,[252] breast cancer,[127,253] and Ewing's sarcoma.[251]

Besides directly killing CSCs, NK cells can induce their differentiation. Indeed, Tseng and colleagues have demonstrated that CD16 triggering or activation of NK cells in the presence of normal or neoplastic stem cells and IL-2 can inhibit their cytotoxic activity and increase their cytokine production, a state named "split anergy."[252] These split anergized NK cells secrete high amounts of IFN-γ, which induce the expression of MHC-I, differentiation receptors and PD-L1 and decrease the levels of CD44 on CSCs. This induction of CSC differentiation then results in decreased growth and decreased metastatic spreading of tumors.[254] Therefore, NK cells seem to counteract tumor progression through a two-step mechanism, initially killing a subset of CSCs and then, after undergoing split anergy, inducing cell differentiation of the remaining CSC population.[255]

Treatment with radiation therapy or chemotherapy is known to increase the frequency of CSCs in different tumor types and has been reported to induce increased expression of the NKG2D ligands MICA and MICB on CSCs.[256] This suggest that NK cells could potentially be exploited to eradicate CSCs in combination with cytoreductive therapies.

Although data support a role for NK cells in counteracting CSCs, most of these studies were based on *in vitro* experiments, which do not consider the effects that tumor microenvironment exerts on NK cell activation and function.[242] Indeed, tumor cells and soluble factors present in the tumor microenvironment can induce NK immunosuppression through different mechanisms. Many tumors produce soluble factors that inhibit NK cell function, such as TGF-β1, IL-10, PGE_2, and adenosine indoleamine 2,3-dioxygenase.[257] Moreover, tumor-infiltrating NK cells often display an altered phenotype when compared to circulating NK cells, with decreased expression of the stimulatory receptors DNAM-I, CD16, NCRs and NKG2D,[258,259] and increased expression of KIRs, which is linked to poor prognosis in non-small cell lung cancer patients.[260] Recent data have revealed an additional mechanism of NK suppression exerted by tumor cells—inhibition of the terminal maturation of NK cells—as demonstrated by the increase in poorly differentiated and non-cytotoxic NK cell subsets in breast cancer patients.[261] These immature NK cells not only do not kill cancer cells but also they promote tumor neovascularization through the production of VEGF.[262] Despite the fact that the immunosuppressive activity of tumor cells must be taken in consideration in the development of immunotherapies, some hints on the possibility of harnessing NK cells to eliminate CSCs has recently been provided by *in vivo* studies in tumor-bearing animal models. For example, we recently demonstrated that autologous NK cells are able to exert a protective effect toward cancer metastatic spreading to the lungs in a mouse model of Her2/neu$^+$ breast cancer. Indeed, administration of an activator of NK cells, tilorone, to BALB/c mice strongly impaired the generation of lung metastases induced by the injection of syngeneic CSCs but not of differentiated cancer cells.[127] Moreover, Ames and colleagues demonstrated that infusion of IL-2-activated NK cells is able to target the CSC population and impair tumor growth in different human cancer models developed in NSG mice.[251] Together, data obtained in mouse models highlight the possibility of exploiting NK immunotherapy for the development of combined therapies for the eradication of CSCs in refractory solid tumors in patients.

3.3.5.2 γδ T cells

An innate-like immune cell population that is garnering great interest for its possible usage in CSC immunotargeting is constituted by gamma delta (γδ) T cells.[79] γδ T cells represent the prototype of unconventional T cells and are characterized by expression of heterodimeric T cell receptors composed of γ and δ chains, which recognize antigen in an MHC-unrestricted manner.[263] Two major subtypes of γδ T cells have been identified in humans, based on the presence or absence of the Vδ2 chain. Among the Vδ2$^+$ cells, Vγ9/Vδ2$^+$ represents the most abundant and studied population.[79] Vγ9Vδ2 T cells recognize phosphoantigens, small non-peptidic molecules containing pyrophosphate groups produced as intermediate products of the mevalonate pathway of isoprenoid biosynthesis, which are upregulated in several cancers such as leukemia, lymphoma, multiple myeloma and solid tumors such as breast, prostate, hepatic, esophageal and pancreatic cancers.[264] Indeed, γδ T lymphocytes have been identified in the immune infiltrates in different human cancers, suggesting that they can play a role in the anticancer immune response. Indeed, although the intratumor γδ1 T cell population has been shown to exert immunosuppressive activity in breast cancer by inhibiting DC maturation and naive and effector T cell function,[265] several studies have demonstrated that Vγ9Vδ2 T cells are able to kill tumor cells of different origin both *in vitro* and *in vivo*, independently of their expression of MHC molecules.[263] Vγ9Vδ2 T cells exert their anti-tumor activity through different mechanisms, since they can either directly induce cytotoxicity with mechanisms similar to those used by $CD8^+$ T lymphocytes, or indirectly stimulate the activation of other killer immune cells, such as NK cells and conventional cytotoxic T cells.[266,267] Of note, Vγ9Vδ2 T cell activation can be induced by the administration of phosphoantigens or by treating target cells with aminobisphosphonates such as zoledronate (Zometa) or pamidronate (Aredia), drugs currently used in the treatment of osteoporosis and bone metastases. Indeed, aminobisphosphonates, by acting on the mevalonate pathway, increase the presence of phosphoantigens in cancer cells, rendering them more sensitive to Vγ9Vδ2 T cell-mediated killing.[268] Moreover, Vγ9Vδ2 T lymphocytes express some activating receptors typical of NK cells, such as NKG2D and CD16, suggesting that their activation can be promoted by NK activating ligands and by immunoglobulins.[263]

The ability of Vγ9Vδ2 T cells to target cancer cells in an MHC-unrestricted manner makes them potential tools to eradicate CSCs. Different groups demonstrated that CSCs from colon, breast and neuroblastoma are relatively insensitive to Vγ9Vδ2 T cell cytotoxicity.[269–271] However, when CSCs are treated with zoledronate, they stimulate Vγ9Vδ2 T cells to secrete IFN-γ and other cytokines, express TRAIL and other pro-apoptotic molecules and release the content of their cytotoxic granules, finally inducing CSC death through a TCR-dependent manner.[269–271] Of note, zoledronate was also shown to directly induce apoptosis and inhibit motility in CSCs from different solid tumor, thus further contributing to their elimination.[272,273] Moreover, treatment of colon CSCs with the chemotherapeutic drugs doxorubicin and 5-fluorouracil induces the expression of the activating ligands of TRAIL and NKG2D on CSCs. This makes them sensitive to Vγ9Vδ2 T cell-mediated killing.[274] Intriguingly, chemotherapy renders CSCs sensitive to γδ T cells, and Vγ9Vδ2 T cells increase ovarian CSC chemosensitivity by decreasing the expression of the multidrug resistance component ABCG2, topoisomerases 2a and 2b. Thus, chemotherapy and Vγ9Vδ2 T cell activators could synergize to induce CSC eradication.[275]

Immunotherapy with γδ T cells may be accomplished using different strategies, such as adoptive transfer of autologous or allogenic Vγ9Vδ2 T cells expanded and activated *in vitro*, or with *in vivo* stimulation with aminobisphosphonates and cytokines. Until now, clinical trials with Vγ9Vδ2 T cells have been performed in many hematological and solid tumors, such as breast, prostate, lung and head and neck cancer.[263] Although these therapies induced reduction in the tumor burden, only a small augmentation in long-term disease survival was observed, suggesting that the mechanisms that regulate CSC sensitivity to γδ T cells deserve further studies and must be taken into account when designing new clinical trials.

3.3.5.3 CIK cells

Another potential player in MHC-unrestricted immunotherapy is constituted by cytokine-induced killer (CIK) cells, a population of cells that can be generated *in vitro* by culturing human PBMCs with CD3-activating monoclonal antibodies and cytokines such as IFN-γ, IL-1 and IL-2.[276] CIK cells co-express markers typical of T and NK cells, although they comprise a heterogeneous population that can express either CD3, CD56 or both.[277] Although they express αβ TCR, their activation is prompted by an MHC-independent mechanism mediated by NKG2D and lymphocyte function-associated antigen (LFA)-1.[79] CIKs can be potently activated by a vast range of tumor cells of both hematological and solid tumor origin. Once activated, CIK cells kill their targets mainly through perforin/granzymes secretion and activation of the pro-apoptotic Fas signaling pathway in the target cells. Moreover, they secrete large amounts of cytokines such as IL-2, IFN-γ, and TNF-α.[277] When injected *in vivo*, CIKs can localize in the spleen, tumors and tumor-draining lymph nodes, where they can exert their effector functions.[277]

In light of their MHC-unrestricted activation pattern, the ability of CIK cells to kill CSCs was analyzed in preclinical models of melanoma, sarcoma and liver cancer. In all these contexts, CIKs efficiently killed CSCs both *in vitro* and *in vivo*, although through different mechanisms.[278–280] Indeed, CIK cells induced caspase-3-dependent apoptosis and cell cycle arrest in G2/M in hepatic CSCs,[280] while exerted a direct cytotoxic effect on CSCs from melanoma and sarcoma patients.[278,279]

Besides the possibility of generating high amounts of CIKs in a cost-effective manner, one of the characteristics that make CIKs promising candidates for cancer immunotherapy is their relatively high resistance to chemotherapy and the possibility of deriving CIKs from patients that already underwent these treatments.[277] Indeed, chemotherapy and molecular-targeted therapies often exert positive selection pressure for CSCs, which can then be killed by CIKs, as demonstrated by the group of Prof. Sangiolo in sarcomas.[281] Therefore, the combination of CIK adoptive transfer with chemotherapy or targeted therapies to simultaneously target differentiated cells and CSCs is a hopeful strategy for cancer immunotherapy.

4. Conclusions and future perspectives

4.1 Cancer stem cell immunology and immunotherapy in short

A growing body of evidence suggests that complete eradication of a tumors relies on the effective destruction of CSCs. This task is difficult because of the therapy-resistant phenotype of CSCs, which are able to survive the action of radiotherapy and chemotherapy. In this regard, immunotherapy represents an alternative option, since its action relies on mechanisms that are different from those of standard therapies.

Although immunotherapeutic approaches to treat cancer have created great enthusiasm due to recent success in the clinic, CSCs may have immunological properties that are different compared to those of more differentiated cancer cells. This could imply that CSCs would behave differently when challenged by immunotherapy. Indeed, review of the literature suggests that, compared to differentiated cancer cells, CSCs have a different pattern of expression of some common TAAs (Section 3.1). This potentially implies that the immune response that is spontaneously mounted by the immune system in response to tumor development may be shifted toward TAAs that are expressed by differentiated cancer cells, which represent the majority of tumor cells. On the contrary, CSC-specific TAAs, which are expressed only by a minority of cells, may not be able to induce a strong immune response. Therefore, the spontaneous immune response would be shifted toward non-CSC antigens, sparing CSCs from elimination. Moreover, CSCs may escape T cell recognition by downregulating MHC-I and APM molecules or by directly suppressing anti-tumor immune responses through the expression of immune checkpoint ligands or other immunosuppressive molecules. These mechanisms are common in cancer, but appear to be particularly active in CSCs as compared to differentiated cancer cells (Section 3.2.1). This could represent a strength on one side, but an Achille's heel to be exploited on the other. Downregulation of MHC-I may indeed render CSCs more susceptible to

NK-mediated recognition and killing, while upregulation of checkpoint ligands may render them a good target for checkpoint inhibition therapy.[79]

Despite being able to evade recognition and killing from a physiologically mounted immune response, immunoevasion can be overcome by providing the right stimuli and directing the immune system against the right targets, as is the case for anti-tumor vaccination. The studies described in Section 3.3.1 clearly demonstrate that anticancer vaccination, either prophylactic or therapeutic, is more effective when CSCs are targeted. Indeed, vaccines based on DCs loaded with lysates from non-CSCs or from a mixed population (in which CSCs represent a very small minority) are not able to halt tumor growth, likely because the immune response is shifted toward antigens that are expressed by non-CSCs, which are not able to generate tumors, but not by CSCs, which are those responsible for tumor initiation. In contrast, vaccines based on DCs loaded with lysates from CSCs are able to impede tumor development and progression.

This is clearly a strong proof-of-concept that CSCs are the right target and are sensitive to vaccine-induced immune responses. However, caution should be used when transferring this approach to the clinic. Indeed, CSCs may share some, if not many, of their antigens with normal stem cells. Nowadays a clear antigen profile of CSCs is lacking, thus more effort should be directed to better characterization of the antigens expressed by them. Also, the identification of antigens derived from proteins with a non-dispensable role in CSC biology would help to avoid escape mechanisms. When these CSC-specific antigens are identified, other targeted immunotherapies are possible, such as mAb therapy and CAR-T therapies. Thus, targeting the mechanisms that are responsible for the biological features of CSCs would render their escape from targeted therapies more difficult. However, it should be noted that interactions between CSCs and their niche may contribute to a lack of treatment response *in vivo* even when sensitivity is observed *in vitro*.[19]

4.2 Integration of cancer immunology in the CSC model

In general, all the data discussed here demonstrate that there is complex crosstalk between CSCs and the immune system, which oncologists and immunologists should take into account since it has strong practical and theoretical implications. Indeed, CSCs have been defined as cells that can self-renew and give rise to all the heterogeneous cell types that compose a tumor when transplanted in an animal model (Section 2.1.3). However, since

CSCs and differentiated cancer cells display different immunogenicity, the percentage of cancer cells possessing these characteristics strongly depends on the degree of immune competence of the mouse model used.[282] Therefore, the interplay between cancer cells and immune system raises questions about the CSC model. For instance, a debate regarding melanoma CSCs arose when Quintana et al. demonstrated that in NSG mice, which lack not only functional T and B lymphocytes but also NK cells, about 25% of melanoma cells from patients were able to give rise to tumors.[283] These data were in contrast with previous research conducted in NOD/scid mice that possess functional NK cells, where the frequency of melanoma CSCs was estimated to be less than 1 in 10^6 cancer cells.[284] These different results could be due to different sensitivity of melanoma CSCs and differentiated cancer cells to NK-mediated killing, suggesting that in NOD/scid and immune competent mice NK cells can kill the majority of injected cells but not CSCs, which can thus proliferate and differentiate, giving rise to the tumor.[282] However, these data contradict results reported by Pietra et al., who showed that activated NK cells efficiently kill human melanoma CSCs,[250] similarly to what observed by us and other groups on CSCs deriving from different solid tumors.

On the one hand, the observed difference in the frequency of tumor-initiating cells would represent a proof that tumor initiation is not a prerogative of a rare population of CSCs, but is rather a property shared by many cancer cells, thus questioning the clinical relevance of the CSC model. On the other hand, it highlights the fact that different degrees of immune competence affect the percentage of cells with tumor-initiating ability within a cancer. In general, we think that the current definition of CSCs is simply not complete. Indeed, many of the studies discussed in this chapter have highlighted the immunoevasive properties of CSCs. Therefore, while both CSCs and some non-CSCs may share the tumor-initiating ability in a completely immunosuppressed host, only CSCs are effective in doing so in immunocompetent models. In this view, a better definition of CSCs would be "a cell within a tumor that possesses the capacity to self-renew and to cause the heterogeneous lineages of cancer cells that comprise the tumor in an immunocompetent host." Clearly, more studies are needed to determine whether immunoevasive properties of CSCs are an intrinsic feature of the stem-like state or are a result of immunoediting of clones with better fitness. This would better align our view of immunoediting with the one proposed by Bruttel et al., which integrates the classical model of tumor elimination, equilibrium and evasion with the CSC concept.[282]

In this new model, normal cells (either stem cells or more differentiated cells) acquire the first oncogenic hits that transform them into tumor cells. If the stem cell-like phenotype is maintained or acquired by means of de-differentiation, then the cancer cell is a CSC and is able to avoid elimination due to its intrinsic immunoevasive properties. As time goes by, slow-cycling CSCs maintain themselves and generate differentiated cancer cells by means of asymmetrical cell division. The formers survive immunosurveillance, while the latter are destroyed by the immune system, thus creating an equilibrium. Accumulation of further mutations, extrinsic stimuli from the microenvironment and progressive deterioration of the immune system can render the TME more permissive. Subsequently, new, long-lasting CSCs arise with more aggressive phenotypes and then the generation of highly proliferating clones of differentiated cells which have undergone immunoediting and a non-functioning immune system finally allow the evasion of the tumor and the appearance of an overt disease (Fig. 6). In summary, the effect of the immune system on cancer in general and, in particular, on CSCs, should always be taken in consideration, and therefore experimental treatments should be tested also in immune competent preclinical models, which better resemble the situation present in tumor-bearing immune competent patients.

4.3 Caveats associated with CSC immunology and possible solutions

Studying the immunology of CSCs, as well as their properties in general, is associated with many caveats. The first is the current lack of unequivocal CSC markers.[55] Many studies use different methods and markers to isolate CSCs from the same type of tumor. Therefore, besides the difficulties related to inter-tumor heterogeneity, studies performed on the same tumor type can provide contrasting results based on the identification method. Moreover, CSC populations within the same tumor appear to be heterogeneous and this may be reflected in heterogeneous sensitivity to therapies, including immunotherapies. Even if some CSCs are responsive, the target may not be expressed in other CSCs. This would allow the malignant cell population to escape a particular targeted therapy and produce a relapse.

Furthermore, a major hindrance to achieve efficacious anti-CSC treatments relies on the plasticity of the CSC phenotype. This concept has started to emerge relatively recently, with studies suggesting that the differentiation of CSCs into non-tumorigenic cells may be reversible,[285] and at least in some cases appears to be induced by chemotherapy and radiotherapy.[74,75,77]

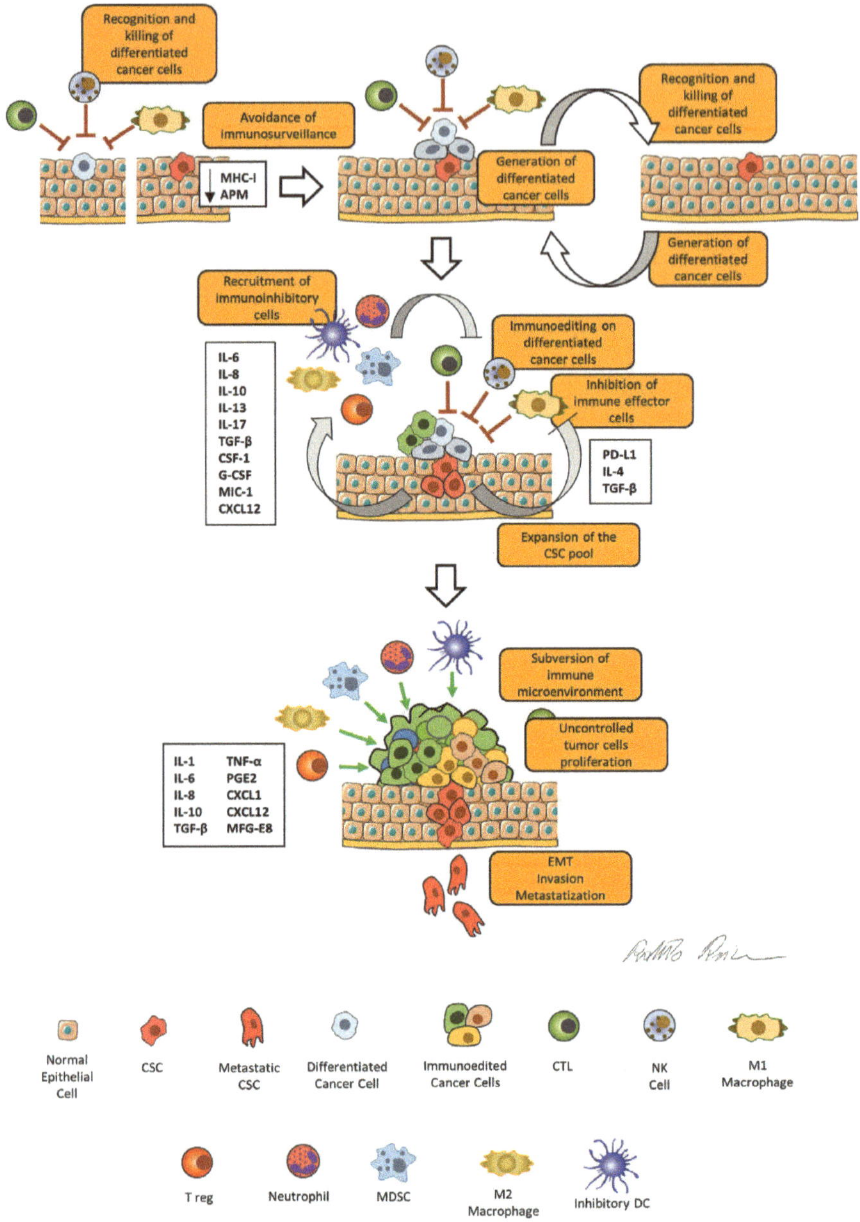

Fig. 6 Role of CSCs in tumor immunoediting. Oncogenic transformation of differentiated cells renders them targets of the immune system, leading to their elimination. On the contrary, newly generated CSCs avoid elimination thanks to their intrinsic immunoevasive properties, and can thus self-renew and generate differentiated cancer cells. CSCs survive immunosurveillance, while differentiated cells are destroyed by the immune system, thus creating an equilibrium. The accumulation of further mutations in differentiated cells, the recruitment of immunosuppressive cells and the consequent inhibition of effector immune cells generates a permissive TME and leads to uncontrolled tumor expansion (escape), with the generation of less immunogenic clones and the expansion of CSCs. CSCs can then undergo EMT and invasion, giving rise to metastasis.

It has been further suggested that the chemoresistant stem-like state is transient, occurring as a reversible adaptation to treatment. In this case, finding the driver of the acquired CSC phenotype would allow the targeting of *de novo* induced CSC following chemotherapy, leading to their depletion.[286] The plasticity and instability of the CSC state thus suggests that, compared to anti-CSC treatment alone, combinatorial treatments could be efficacious. In conclusion, the crosstalk between CSCs and immune cells deserves further study to elucidate all the complex mechanisms and interactions implied, because a deeper knowledge of CSC immunology may eventually lead to more benefit for cancer patients.

Acknowledgments

This chapter was supported by grants from the Italian Association for Cancer Research (IG 11675 and IG 21468) to F.C. and by a grant from Italian Ministry of Health (Bando 2013 Progetti Ordinari Ricerca Finalizzata, RF-2013-02354892) to L.C.; R.R. and G.B. were supported by a FIRC-AIRC fellowship for Italy.

References

1. Hanahan D, Weinberg RA. Hallmarks of cancer: the next generation. *Cell.* 2011;144:646–674.
2. Marusyk A, Almendro V, Polyak K. Intra-tumour heterogeneity: a looking glass for cancer? *Nat Rev Cancer.* 2012;12:323–334.
3. Shackleton M, Quintana E, Fearon ER, Morrison SJ. Heterogeneity in cancer: cancer stem cells versus clonal evolution. *Cell.* 2009;138:822–829.
4. Schatton T, Frank MH. Antitumor immunity and cancer stem cells. *Ann N Y Acad Sci.* 2009;1176:154–169.
5. Nowell PC. The clonal evolution of tumor cell populations. *Science.* 1976;194:23–28.
6. Clevers H. The cancer stem cell: premises, promises and challenges. *Nat Med.* 2011; 17:313–319.
7. Pierce GB, Speers WC. Tumors as caricatures of the process of tissue renewal: prospects for therapy by directing differentiation. *Cancer Res.* 1988;48:1996–2004.
8. Visvader JE, Lindeman GJ. Cancer stem cells in solid tumours: accumulating evidence and unresolved questions. *Nat Rev Cancer.* 2008;8:755–768.
9. Pattabiraman DR, Weinberg RA. Tackling the cancer stem cells—what challenges do they pose? *Nat Rev Drug Discov.* 2014;13:497–512.
10. Kreso A, Dick JE. Evolution of the cancer stem cell model. *Cell Stem Cell.* 2014; 14:275–291.
11. Hurt EM, Farrar WL. *Section I: Characterization of Cancer Stem Cells.* Cambridge University Press; 2009:11–14.
12. Nguyen LLV, Vanner R, Dirks P, Eaves CCJ. Cancer stem cells: an evolving concept. *Nat Rev Cancer.* 2012;12:133–143.
13. Kleinsmith LJ, Pierce G. Multipotentiality of single embryonal carcinoma cells. *Cancer Res.* 1964;24:1544–1551.
14. Pierce GB, Wallace C. Differentiation of malignant to benign cells differentiation of malignant to benign cells. *Cancer Res.* 1971;31:127–134.
15. Huntly BJP, Gilliland DG. Leukaemia stem cells and the evolution of cancer-stem-cell research. *Nat Rev Cancer.* 2005;5:311–321.

16. Lapidot T, Sirard C, Vormoor J, et al. A cell initiating human acute myeloid leukaemia after transplantation into SCID mice. *Nature*. 1994;367:645–648.
17. Bonnet D, Dick JE. Human acute myeloid leukemia is organized as a hierarchy that originates from a primitive hematopoietic cell. *Nat Med*. 1997;3:730–737.
18. Al-Hajj M, Wicha MS, Benito-Hernandez A, Morrison SJ, Clarke MF. Prospective identification of tumorigenic breast cancer cells. *Proc Natl Acad Sci U S A*. 2003; 100(7):3983–3988.
19. Valent P, Bonnet D, De Maria R, et al. Cancer stem cell definitions and terminology: the devil is in the details. *Nat Rev Cancer*. 2012;12:767–775.
20. Clarke MF, Dick JE, Dirks PB, et al. Cancer stem cells—perspectives on current status and future directions: AACR workshop on cancer stem cells. *Cancer Res*. 2006;66:9339–9344.
21. Visvader JE. Cells of origin in cancer. *Nature*. 2011;469:314–322.
22. White AC, Lowry WE. Refining the role for adult stem cells as cancer cells of origin. *Trends Cell Biol*. 2015;25:11–20.
23. Karamboulas C, Ailles L. Developmental signaling pathways in cancer stem cells of solid tumors. *Biochim Biophys Acta*. 2013;1830:2481–2495.
24. Pece S, Tosoni D, Confalonieri S, et al. Biological and molecular heterogeneity of breast cancers correlates with their cancer stem cell content. *Cell*. 2010;140:62–73.
25. Reya T, Morrison SJ, Clarke MF, Weissman IL. Stem cells, cancer, and cancer stem cells. *Nature*. 2001;414(6859):105–111.
26. Reya T, Clevers H. Wnt signalling in stem cells and cancer. *Nature*. 2005;434:843–850.
27. Howe LR, Brown AMC. Wnt signaling and breast cancer. *Cancer Biol Ther*. 2004; 3:36–41.
28. Imbert A, Eelkema R, Jordan S, Feiner H, Cowin P. DeltaN89Beta-catenin induces precocious development, differentiation and neoplasia in mammary gland. *J Cell Biol*. 2001;153:555–568.
29. Zhan T, Rindtorff N, Boutros M. Wnt signaling in cancer. *Oncogene*. 2017;36: 1461–1473.
30. Liu S, Dontu G, Mantle ID, et al. Hedgehog signaling and Bmi-1 regulate self-renewal of normal and malignant human mammary stem cells. *Cancer Res*. 2006; 66:6063–6071.
31. Fiaschi M, Rozell B, Bergström Å, Toftgård R. Development of mammary tumors by conditional expression of GLI1. *Cancer Res*. 2009;69:4810–4817.
32. Dontu G, Jackson KW, McNicholas E, Kawamura MJ, Abdallah WM, Wicha MS. Role of Notch signaling in cell-fate determination of human mammary stem/progenitor cells. *Breast Cancer Res*. 2004;6:R605.
33. Mani SA, Guo W, Liao MJ, et al. The epithelial-mesenchymal transition generates cells with properties of stem cells. *Cell*. 2008;133:704–715.
34. Kang Y, Pantel K. Tumor cell dissemination: emerging biological insights from animal models and cancer patients. *Cancer Cell*. 2013;23:573–581.
35. Sato R, Semba T, Saya H, Arima Y. Concise review: stem cells and epithelial-mesenchymal transition in cancer: biological implications and therapeutic targets. *Stem Cells*. 2016;34:1997–2007.
36. Nassar D, Blanpain C. Cancer stem cells: basic concepts and therapeutic implications. *Annu Rev Pathol: Mech Dis*. 2016;11:47–76.
37. Morel A-P, Lièvre M, Thomas C, Hinkal G, Ansieau S, Puisieux A. Generation of breast cancer stem cells through epithelial-mesenchymal transition. *PLoS One*. 2008;3:e2888.
38. Onder TT, Gupta PB, Mani SA, Yang J, Lander ES, Weinberg RA. Loss of E-cadherin promotes metastasis via multiple downstream transcriptional pathways. *Cancer Res*. 2008;68:3645–3654.

39. Bartucci M, Dattilo R, Moriconi C, et al. TAZ is required for metastatic activity and chemoresistance of breast cancer stem cells. *Oncogene*. 2015;34:681–690.
40. Cordenonsi M, Zanconato F, Azzolin L, et al. The hippo transducer TAZ confers cancer stem cell-related traits on breast cancer cells. *Cell*. 2011;147:759–772.
41. Abraham BK, Fritz P, Mcclellan M, Hauptvogel P, Athelogou M, Brauch H. Prevalence of CD44 +/CD24 −/low cells in breast cancer may not be associated with clinical outcome but may favor distant metastasis. *Clin Cancer Res*. 2005;11:1154–1159.
42. Marcato P, Dean CA, Da Pan RA, et al. Aldehyde dehydrogenase activity of breast cancer stem cells is primarily due to isoform ALDH1A3 and its expression is predictive of metastasis. *Stem Cells*. 2011;29:32–45.
43. Oskarsson T, Batlle E, Massagué J. Metastatic stem cells: sources, niches, and vital pathways. *Cell Stem Cell*. 2014;14:306–321.
44. Balic M, Lin H, Young L, et al. Most early disseminated cancer cells detected in bone marrow of breast cancer patients have a putative breast cancer stem cell phenotype. *Clin Cancer Res*. 2006;12:5615–5621.
45. Liu S, Cong Y, Wang D, et al. Breast cancer stem cells transition between epithelial and mesenchymal states reflective of their normal counterparts. *Stem Cell Rep*. 2014;2:78–91.
46. Creighton CJ, Li X, Landis M, et al. Residual breast cancers after conventional therapy display mesenchymal as well as tumor-initiating features. *Proc Natl Acad Sci*. 2009;106:13820–13825.
47. Koren E, Fuchs Y. The bad seed: cancer stem cells in tumor development and resistance. *Drug Resist Updat*. 2016;28:1–12.
48. Cojoc M, Mäbert K, Muders MH, Dubrovska A. A role for cancer stem cells in therapy resistance: cellular and molecular mechanisms. *Semin Cancer Biol*. 2015;31:16–27.
49. Gasch C, Ffrench B, O'Leary JJ, Gallagher MF. Catching moving targets: cancer stem cell hierarchies, therapy-resistance & considerations for clinical intervention. *Mol Cancer*. 2017;16:43.
50. Chen W, Dong J, Haiech J, Kilhoffer MC, Zeniou M. Cancer stem cell quiescence and plasticity as major challenges in cancer therapy. *Stem Cells Int*. 2016;2016:1740936.
51. Singh S, Brocker C, Koppaka V, et al. Aldehyde dehydrogenases in cellular responses to oxidative/electrophilic stress. *Free Radic Biol Med*. 2013;56:89–101.
52. Raha D, Wilson TR, Peng J, et al. The cancer stem cell marker aldehyde dehydrogenase is required to maintain a drug-tolerant tumor cell subpopulation. *Cancer Res*. 2014;74:3579–3590.
53. Duan JJ, Qiu W, Xu SL, et al. Strategies for isolating and enriching cancer stem cells: well begun is half done. *Stem Cells Dev*. 2013;22(16):2221–2239.
54. Jaggupilli A, Elkord E. Significance of CD44 and CD24 as cancer stem cell markers: an enduring ambiguity. *Clin Dev Immunol*. 2012;2012:708036.
55. Paula AC, Lopes C. Implications of different cancer stem cell phenotypes in breast cancer. *Anticancer Res*. 2017;37:2173–2183.
56. Ricardo S, Vieira AF, Gerhard R, et al. Breast cancer stem cell markers CD44, CD24 and ALDH1: expression distribution within intrinsic molecular subtype. *J Clin Pathol*. 2011;64(11):937–946.
57. Fornari C, Beccuti M, Lanzardo S, et al. A mathematical-biological joint effort to investigate the tumor-initiating ability of cancer stem cells. *PLoS One*. 2014;9(9):e106193.
58. Yan Y, Zuo X, Wei D. Concise review: emerging role of CD44 in cancer stem cells: a promising biomarker and therapeutic target. *Stem Cells Transl Med*. 2015;4:1033–1043.
59. Singh SK, Clarke ID, Terasaki M, et al. Identification of a cancer stem cell in human brain tumors. *Cancer Res*. 2003;63:5821–5828.
60. Glumac PM, LeBeau AM. The role of CD133 in cancer: a concise review. *Clin Transl Med*. 2018;7:18.

61. Ginestier C, Hur MH, Charafe-Jauffret E, et al. ALDH1 is a marker of normal and malignant human mammary stem cells and a predictor of poor clinical outcome. *Cell Stem Cell*. 2007;1(5):555–567.
62. Jones RJ, Barber JP, Vala MS, et al. Assessment of aldehyde dehydrogenase in viable cells. *Blood*. 1995;85:2742–2746.
63. Goodell BMA, Brose K, Paradis G, Conner AS, Mulligan RC. Isolation and functional properties of murine hematopoietic stem cells that are replicating in vivo. *J Exp Med*. 1996;183:1797–1806.
64. Golebiewska A, Brons NHC, Bjerkvig R, Niclou SP. Critical appraisal of the side population assay in stem cell and cancer stem cell research. *Cell Stem Cell*. 2011;8:136–147.
65. Reynolds BA, Weiss S. Nervous system generation of neurons and astrocytes from isolated cells of the adult mammalian central nervous system. *Science*. 1992;255: 1707–1710.
66. Ponti D, Costa A, Zaffaroni N, et al. Isolation and in vitro propagation of tumorigenic breast cancer cells with stem/progenitor cell properties. *Cancer Res*. 2005;65: 5506–5512.
67. Weiswald LB, Bellet D, Dangles-Marie V. Spherical cancer models in tumor biology. *Neoplasia*. 2015;17(1):1–15.
68. Kenny PA, Lee GY, Myers CA, et al. The morphologies of breast cancer cell lines in three-dimensional assays correlate with their profiles of gene expression. *Mol Oncol*. 2007;1:84–96.
69. Smart CE, Morrison BJ, Saunus JM, et al. In vitro analysis of breast cancer cell line tumourspheres and primary human breast epithelia mammospheres demonstrates inter- and intrasphere heterogeneity. *PLoS One*. 2013;8:e64388.
70. Calvet CY, Andre FM, Mir LM. The culture of cancer cell lines as tumorspheres does not systematically result in cancer stem cell enrichment. *PLoS One*. 2014;9(2):e89644.
71. Kuch V, Schreiber C, Thiele W, Umansky V, Sleeman JP. Tumor-initiating properties of breast cancer and melanoma cells in vivo are not invariably reflected by spheroid formation in vitro, but can be increased by long-term culturing as adherent monolayers. *Int J Cancer*. 2013;132(3):E94–105.
72. Chen J, Li Y, Yu TS, et al. A restricted cell population propagates glioblastoma growth after chemotherapy. *Nature*. 2012;488(7412):522–526.
73. Vlashi E, Kim K, Lagadec C, et al. In vivo imaging, tracking, and targeting of cancer stem cells. *J Natl Cancer Inst*. 2009;101:350–359.
74. Auffinger B, Tobias AL, Han Y, et al. Conversion of differentiated cancer cells into cancer stem-like cells in a glioblastoma model after primary chemotherapy. *Cell Death Differ*. 2014;21:1119–1131.
75. Ghisolfi L, Keates AC, Hu X, Dk L, Li CJ. Ionizing radiation induces stemness in cancer cells. *PLoS One*. 2012;7:1–11.
76. Hu X, Ghisolfi L, Keates AC, et al. Induction of cancer cell stemness by chemotherapy. *Cell Cycle*. 2012;11:2691–2698.
77. Lagadec C, Vlashi E, Della Donna L, Dekmezian C, Pajonk F. Radiation-induced reprogramming of breast cancer cells. *Stem Cells*. 2012;30:833–844.
78. Wang Y, Li W, Patel SS, et al. Blocking the formation of radiation-induced breast cancer stem cells. *Oncotarget*. 2014;5:3743–3755.
79. Gammaitoni L, Leuci V, Mesiano G, et al. Immunotherapy of cancer stem cells in solid tumors: initial findings and future prospective. *Expert Opin Biol Ther*. 2014;14(9): 1259–1270.
80. Parish CR. Cancer immunotherapy: the past, the present and the future. *Immunol Cell Biol*. 2003;81(2):106–113.
81. Dunn GP, Bruce AT, Ikeda H, Old LJ, Schreiber RD. Cancer immunoediting: from immuno-surveillance to tumor escape. *Nat Immunol*. 2002;3:991–998.

82. Schreiber RD. Cancer immunoediting: integrating suppression and promotion. *Science.* 2011;331(6024):1565–1570.
83. Dunn GP, Old LJ, Schreiber RD. The three Es of cancer immunoediting. *Annu Rev Immunol.* 2004;22:329–360.
84. Zitvogel L, Tesniere A, Kroemer G. Cancer despite immunosurveillance: immunoselection and immunosubversion. *Nat Rev Immunol.* 2006;6(10):715–727.
85. Hanahan D, Coussens LM. Accessories to the crime: functions of cells recruited to the tumor microenvironment. *Cancer Cell.* 2012;21:309–322.
86. Zhang D, Tang DG, Rycaj K. Cancer stem cells: regulation programs, immunological properties and immunotherapy. *Semin Cancer Biol.* 2018;52(Pt 2):94–106.
87. Hirohashi Y, Torigoe T, Inoda S, Morita R, Kochin V, Sato N. Cytotoxic T lymphocytes: sniping cancer stem cells. *Oncoimmunology.* 2012;1(1):123–125.
88. Hirohashi Y, Torigoe T, Tsujahara T, Kanaseki T, Kochin V, Sato N. Immune responses to human cancer stem-like cells/cancer-initiating cells. *Cancer Sci.* 2016; 107(1):12–17.
89. Schumacher TN, Schreiber RD. Neoantigens in cancer immunotherapy. *Science.* 2015;348:69–74.
90. Alexandrov LB, Nik-Zainal S, Wedge DC, et al. Signatures of mutational processes in human cancer. *Nature.* 2013;500:415–421.
91. McGranahan N, Furness AJS, Rosenthal R, Quezada SA, Swanton C. Clonal neoantigens elicit T cell immunoreactivity and sensitivity to immune checkpoint blockade. *Science.* 2016;351:1463–1470.
92. Allen EMV, Miao D, Schilling B, et al. Genomic correlates of response to CTLA-4 blockade in metastatic melanoma. *Science.* 2015;350:207–212.
93. Rizvi NA, Hellmann MD, Snyder A, et al. Mutational landscape determines sensitivity to PD-1 blockade in non-small cell lung cancer. *Science.* 2016;348:124–129.
94. Yuan J, Ph D, Zaretsky JM, et al. Genetic basis for clinical response to CTLA-4 blockade in melanoma. *N Engl J Med.* 2014;371:2189–2199.
95. Mennonna D, Maccalli C, Romano MC, et al. T cell neoepitope discovery in colorectal cancer by high throughput profiling of somatic mutations in expressed genes. *Gut.* 2017;66(3):454–463.
96. Oliver SE, Unger ER, Lewis R, et al. Prevalence of human papillomavirus among females after vaccine introduction—National Health and Nutrition Examination Survey, United States, 2003-2014. *J Infect Dis.* 2017;216(5):594–603.
97. Yang A, Farmer E, Wu TC, Hung CF. Perspectives for therapeutic HPV vaccine development. *J Biomed Sci.* 2016;23(1):75.
98. Lopez J, Ruiz G, Organista-Nava J, Gariglio P, Garcia-Carranca A. Human papillomavirus infections and cancer stem cells of tumors from the uterine cervix. *Open Virol J.* 2012;6:232–240.
99. Ali N, Allam H, May R, et al. Hepatitis C virus-induced cancer stem cell-like signatures in cell culture and murine tumor xenografts. *J Virol.* 2011;85(23): 12292–12303.
100. Brichard BV, Pel AV, Wlfel T, et al. The tyrosinase gene codes for an antigen recognized by autologous cytolytic T lymphocytes on HLA-A2 melanomas. *J Exp Med.* 1993;178:489–495.
101. Gjerstorff MF, Andersen MH, Ditzel HJ. Oncogenic cancer/testis antigens: prime candidates for immunotherapy. *Oncotarget.* 2015;6(18):15772–15787.
102. Schatton T, Schutte U, Frank NY, et al. Modulation of T-cell activation by malignant melanoma initiating cells. *Cancer Res.* 2010;70(2):697–708.
103. Boiko AD, Razorenova OV, van de Rijn M, et al. Human melanoma-initiating cells express neural crest nerve growth factor receptor CD271. *Nature.* 2010;466(7302): 133–137.

104. Qin J, Liu X, Laffin B, et al. The PSA(-/lo) prostate cancer cell population harbors self-renewing long-term tumor-propagating cells that resist castration. *Cell Stem Cell.* 2012;10(5):556–569.
105. Maccalli C, Parmiani G, Ferrone S. Immunomodulating and immunoresistance properties of cancer-initiating cells: implications for the clinical success of immunotherapy. *Immunol Invest.* 2017;46(3):221–238.
106. Parmiani G, Russo V, Maccalli C, Parolini D, Rizzo N, Maio M. Peptide-based vaccines for cancer therapy. *Hum Vaccin Immunother.* 2014;10(11):3175–3178.
107. Kimura T, Finn OJ. MUC1 immunotherapy is here to stay. *Expert Opin Biol Ther.* 2012;13:35–49.
108. Mallard BW, Tiralongo J. Cancer stem cell marker glycosylation: nature, function and significance. *Glycoconj J.* 2017;34(4):441–452.
109. Pauli C, Munz M, Kieu C, et al. Tumor-specific glycosylation of the carcinoma-associated epithelial cell adhesion molecule EpCAM in head and neck carcinomas. *Cancer Lett.* 2003;193(1):25–32.
110. Hakomori S. Tumor malignancy defined by aberrant glycosylation and sphingo(glyco) lipid metabolism. *Cancer Res.* 1996;56(23):5309–5318.
111. Barkeer S, Chugh S, Batra SK, Ponnusamy MP. Glycosylation of cancer stem cells: function in stemness, tumorigenesis, and metastasis. *Neoplasia.* 2018;20(8):813–825.
112. Morita R, Nishizawa S, Torigoe T, et al. Heat shock protein DNAJB8 is a novel target for immunotherapy of colon cancer-initiating cells. *Cancer Sci.* 2014;105(4):389–395.
113. Nishizawa S, Hirohashi Y, Torigoe T, et al. HSP DNAJB8 controls tumor-initiating ability in renal cancer stem-like cells. *Cancer Res.* 2012;72(11):2844–2854.
114. Morita R, Hirohashi Y, Torigoe T, et al. Olfactory receptor family 7 subfamily C member 1 is a novel marker of colon cancer-initiating cells and is a potent target of immunotherapy. *Clin Cancer Res.* 2016;22(13):3298–3309.
115. Iezzi M, Quaglino E, Amici A, Lollini PL, Forni G, Cavallo F. DNA vaccination against oncoantigens: a promise. *Oncoimmunology.* 2012;1(3):316–325.
116. Lollini PL, Cavallo F, Nanni P, Forni G. Vaccines for tumour prevention. *Nat Rev Cancer.* 2006;6(3):204–216.
117. Cordero F, Beccuti M, Fornari C, et al. Multi-level model for the investigation of oncoantigen-driven vaccination effect. *BMC Bioinf.* 2013;14(Suppl 6):S11.
118. Maccalli C, Volonte A, Cimminiello C, Parmiani G. Immunology of cancer stem cells in solid tumours. A review. *Eur J Cancer.* 2014;50(3):649–655.
119. Khong HT, Restifo NP. Natural selection of tumor variants in the generation of "tumor escape" phenotypes. *Nat Immunol.* 2002;3:999–1005.
120. Concha-benavente F, Srivastava R, Ferrone S, Ferris RL. Immunological and clinical significance of HLA class I antigen processing machinery component defects in malignant cells. *Oral Oncol.* 2016;58:52–58.
121. Tomaso TD, Mazzoleni S, Wang E, et al. Immunobiological characterization of cancer stem cells isolated from glioblastoma patients. *Clin Cancer Res.* 2010;16(3):800–813.
122. Tallerico R, Todaro M, Franco SD, et al. Human NK cells selective targeting of colon cancer-initiating cells: a role for natural cytotoxicity receptors and MHC class I molecules. *J Immunol.* 2013;190:2381–2390.
123. Volonté A, Tomaso TD, Spinelli M, et al. Cancer-initiating cells from colorectal cancer patients escape from T cell-mediated immunosurveillance in vitro through membrane-bound IL-4. *J Immunol.* 2014;192(1):523–532.
124. Grau JJ, Mesia R, De la Iglesia-Vicente M, Domingo-Domenech J. Enrichment of cells with cancer stem cell-like markers in relapses of chemoresistant patients with locally advanced head and neck squamous. *Oncology.* 2016;90(5):267–272.
125. Domingo-Domenech J, Vidal SJ, Rodriguez-Bravo V, et al. Suppression of acquired docetaxel resistance in prostate cancer through depletion of notch- and hedgehog-dependent tumor-initiating cells. *Cancer Cell.* 2012;22(3):373–388.

126. Busse A, Letsch A, Fusi A, et al. Characterization of small spheres derived from various solid tumor cell lines: are they suitable targets for T cells? *Clin Exp Metastasis.* 2013;30(6):781–791.
127. Tallerico R, Conti L, Lanzardo S, et al. NK cells control breast cancer and related cancer stem cell hematological spread. *Oncoimmunology.* 2017;6(3):e1284718.
128. Morrison BJ, Steel JC, Morris JC. Reduction of MHC-I expression limits T-lymphocyte-mediated killing of cancer-initiating cells. *BMC Cancer.* 2018; 18(1):469.
129. Wu A, Wiesner ÆS, Xiao ÆJ, et al. Expression of MHC I and NK ligands on human CD133+ glioma cells: possible targets of immunotherapy. *J Neurooncol.* 2007;83(2): 121–131.
130. Wei J, Barr J, Kong LY, et al. Glioblastoma cancer-initiating cells inhibit T-cell proliferation and effector responses by the signal transducers and activators of transcription 3 pathway. *Mol Cancer Ther.* 2010;9(1):67–78.
131. Inoda S, Hirohashi Y, Torigoe T, et al. Cytotoxic T lymphocytes efficiently recognize human colon cancer stem-like cells. *Am J Pathol.* 2011;178(4):1805–1813.
132. Jachetti E, Mazzoleni S, Grioni M, et al. Prostate cancer stem cells are targets of both innate and adaptive immunity and elicit tumor-specific immune responses. *Oncoimmunology.* 2013;2(5):e24520.
133. Pardoll DM. The blockade of immune checkpoints in cancer immunotherapy. *Nat Rev Cancer.* 2012;12(4):252–264.
134. Okazaki T, Honjo T. The PD-1–PD-L pathway in immunological tolerance. *Trends Immunol.* 2006;27(4):195–201.
135. Lee Y, Shin JH, Longmire M, et al. CD44+ cells in head and neck squamous cell carcinoma suppress T-cell-mediated immunity by selective constitutive and inducible expression of PD-L1. *Clin Cancer Res.* 2016;22(14):3571–3581.
136. Wu Y, Chen M, Wu P, Chen C, Xu ZP, Gu W. Increased PD-L1 expression in breast and colon cancer stem cells. *Clin Exp Pharmacol Physiol.* 2017;44(5):602–604.
137. Dianzani C, Minelli R, Gigliotti CL, et al. B7h triggering inhibits the migration of tumor cell lines. *J Immunol.* 2014;192(10):4921–4931.
138. Yao Y, Wang X, Jin K, et al. B7-H4 is preferentially expressed in non-dividing brain tumor cells and in a subset of brain tumor stem-like cells. *J Neurooncol.* 2008;89(2):121–129.
139. Zhi Y, Mou Z, Chen J, et al. B7H1 expression and epithelial-to-mesenchymal transition phenotypes on colorectal cancer stem-like cells. *PLoS One.* 2015;10(8): e0135528.
140. Lee Y, Sunwoo J. PD-L1 is preferentially expressed on CD44+ tumor-initiating cells in head and neck squamous cell carcinoma. *J Immunother Cancer.* 2014;2(Suppl 3): P270.
141. Terry S, Savagner P, Ortiz-Cuaran S, et al. New insights into the role of EMT in tumor immune escape. *Mol Oncol.* 2017;11(7):824–846.
142. Chen L, Gibbons DL, Goswami S, et al. Metastasis is regulated via microRNA-200/ZEB1 axis control of tumour cell PD-L1 expression and intratumoral immunosuppression. *Nat Commun.* 2014;5:5241.
143. Almozyan S, Colak D, Mansour F, et al. PD-L1 promotes OCT4 and Nanog expression in breast cancer stem cells by sustaining PI3K/AKT pathway activation. *Int J Cancer.* 2017;141(7):1402–1412.
144. Yang Y, Wu K, Zhao E, et al. B7-H1 enhances proliferation ability of gastric cancer stem-like cells as a receptor. *Oncol Lett.* 2015;9:1833–1838.
145. Zhang B, Dang J, Ba D, Wang C, Han J, Zheng F. Potential function of CTLA-4 in the tumourigenic capacity of melanoma stem cells. *Oncol Lett.* 2018;16:6163–6170.
146. Rabinovich GA, Gabrilovich D, Sotomayor EM. Immunosuppressive strategies that are mediated by tumor cells. *Annu Rev Immunol.* 2007;25:267–296.

147. Todaro M, Alea MP, Di Stefano AB, et al. Colon cancer stem cells dictate tumor growth and resist cell death by production of interleukin-4. *Cell Stem Cell.* 2007; 1:389–402.
148. Yoshimura A, Muto G. TGF-B function in immune suppression. *Curr Top Microbiol Immunol.* 2011;350:127–147.
149. Shipitsin M, Campbell LL, Argani P, et al. Molecular definition of breast tumor heterogeneity. *Cancer Cell.* 2007;11:259–273.
150. Ricciardi M, Zanotto M, Malpeli G, et al. Epithelial-to-mesenchymal transition (EMT) induced by inflammatory priming elicits mesenchymal stromal cell-like immune-modulatory properties in cancer cells. *Br J Cancer.* 2015;112(6):1067–1075.
151. Akalay I, Janji B, Hasmim M, et al. Epithelial-to-mesenchymal transition and autophagy induction in breast carcinoma promote escape from T-cell-mediated lysis. *Cancer Res.* 2013;73(8):2418–2427.
152. Kudo-Saito C, Shirako H, Takeuchi T, Kawakami Y. Cancer metastasis is accelerated through immunosuppression during snail-induced EMT of cancer cells. *Cancer Cell.* 2009;15(3):195–206.
153. Munn DH, Bronte V. Immune suppressive mechanisms in the tumor microenvironment. *Curr Opin Immunol.* 2016;39:1–6.
154. Sultan M, Coyle KM, Vidovic D, Thomas ML, Gujar S, Marcato P. Hide-and-seek: the interplay between cancer stem cells and the immune system. *Carcinogenesis.* 2017;38(2):107–118.
155. Grivennikov SI, Greten FR, Karin M. Immunity, inflammation, and cancer. *Cell.* 2010;140(6):883–899.
156. Mantovani A, Sozzani S, Locati M, Allavena P, Sica A. Macrophage polarization: tumor-associated macrophages as a paradigm for polarized M2 mononuclear phagocytes. *Trends Immunol.* 2002;23(11):549–555.
157. Marigo I, Dolcetti L, Serafini P, Zanovello P, Bronte V. Tumor-induced tolerance and immune suppression by myeloid derived suppressor cells. *Immunol Rev.* 2008;222: 162–179.
158. Zou W. Regulatory T cells, tumour immunity and immunotherapy. *Nat Rev Immunol.* 2006;6(4):295–307.
159. Ugel S, De Sanctis F, Mandruzzato S, Bronte V. Tumor-induced myeloid deviation: when myeloid-derived suppressor cells meet tumor-associated macrophages. *J Clin Invest.* 2015;125(9):3365–3376.
160. Mitchem JB, Brennan DJ, Knolhoff BL, et al. Targeting tumor-infiltrating macrophages decreases tumor-initiating cells, relieves immunosuppression, and improves chemotherapeutic responses. *Cancer Res.* 2013;73:1128–1141.
161. Gil M, Komorowski MP, Seshadri M, et al. CXCL12/CXCR4 blockade by oncolytic virotherapy inhibits ovarian cancer growth by decreasing immunosuppression and targeting cancer-initiating cells. *J Immunol.* 2014;193(10):5327–5337.
162. Levina V, Marrangoni AM, Demarco R, Gorelik E, Lokshin AE. Drug-selected human lung cancer stem cells: cytokine network, tumorigenic and metastatic properties. *PLoS ONE.* 2008;3(8):e3077.
163. Korkaya H, Liu S, Wicha MS. Regulation of cancer stem cells by cytokine networks: attacking cancer's inflammatory roots. *Clin Cancer Res.* 2011;17(19):6125–6129.
164. Plaks V, Kong N, Werb Z. The cancer stem cell niche: how essential is the niche in regulating stemness of tumor cells? *Cell Stem Cell.* 2015;16(3):225–238.
165. Jinushi M, Chiba S, Yoshiyama H, et al. Tumor-associated macrophages regulate tumorigenicity and anticancer drug responses of cancer stem/initiating cells. *Proc Natl Acad Sci U S A.* 2011;108(30):12425–12430.
166. Regis G, Icardi L, Conti L, et al. IL-6, but not IFN-gamma, triggers apoptosis and inhibits in vivo growth of human malignant T cells on STAT3 silencing. *Leukemia.* 2009;23(11):2102–2108.

167. Korkaya H, Liu S, Wicha MS. Breast cancer stem cells, cytokine networks, and the tumor microenvironment. *J Clin Invest*. 2011;121(10):3804–3809.
168. Conti L, Ruiu R, Barutello G, et al. Microenvironment, oncoantigens, and antitumor vaccination: lessons learned from BALB-neuT mice. *Biomed Res Int*. 2014;2014: 534969.
169. Raggi C, Mousa HS, Correnti M, Sica A, Invernizzi P. Cancer stem cells and tumor-associated macrophages: a roadmap for multitargeting strategies. *Oncogene*. 2016;35(6): 671–682.
170. Lu H, Clauser KR, Tam WL, et al. A breast cancer stem cell niche supported by juxtacrine signalling from monocytes and macrophages. *Nat Cell Biol*. 2014;16(11): 1105–1117.
171. Peng D, Tanikawa T, Li W, et al. Myeloid-derived suppressor cells endow stem-like qualities to breast cancer cells through IL6/STAT3 and NO/NOTCH cross-talk signaling. *Cancer Res*. 2016;76(11):3156–3165.
172. Panni RZ, Sanford DE, Belt BA, et al. Tumor-induced STAT3 activation in monocytic myeloid-derived suppressor cells enhances stemness and mesenchymal properties in human pancreatic cancer. *Cancer Immunol Immunother*. 2014;63(5):513–528.
173. Cui TX, Kryczek I, Zhao L, et al. Myeloid-derived suppressor cells enhance stemness of cancer cells by inducing microRNA101 and suppressing the corepressor CtBP2. *Immunity*. 2013;39(3):611–621.
174. Kuroda H, Mabuchi S, Yokoi E, et al. Prostaglandin E2 produced by myeloid-derived suppressive cells induces cancer stem cells in uterine cervical cancer. *Oncotarget*. 2018;9(91):36317–36330.
175. Hsu YL, Chen YJ, Chang WA, et al. Interaction between tumor-associated dendritic cells and colon cancer cells contributes to tumor progression via CXCL1. *Int J Mol Sci*. 2018;19(8):pii: E2427.
176. Wculek SK, Malanchi I. Neutrophils support lung colonization of metastasis-initiating breast cancer cells. *Nature*. 2015;528(7582):413–417.
177. Coffelt SB, Kersten K, Doornebal CW, et al. IL-17-producing gammadelta T cells and neutrophils conspire to promote breast cancer metastasis. *Nature*. 2015; 522(7556):345–348.
178. Albrengues J, Shields MA, Ng D, et al. Neutrophil extracellular traps produced during inflammation awaken dormant cancer cells in mice. *Science*. 2018;361(6409): pii: eaao4227.
179. Galluzzi L, Vacchelli E, Pedro J-MB-S, et al. Classification of current anticancer immunotherapies. *Oncotarget*. 2014;5:12472–12508.
180. Weiner GJ. Building better monoclonal antibody-based therapeutics. *Nat Rev Cancer*. 2015;15:361–370.
181. Rosenberg SA, Restifo NP. Adoptive cell transfer as personalized immunotherapy for human cancer. *Science*. 2015;348:62–68.
182. Schmitt TM, Ragnarsson GB, Greenberg PD. T cell receptor gene therapy for cancer. *Hum Gene Ther*. 2009;1248:1240–1248.
183. Jena B, Dotti G, Cooper LJN. Redirecting T-cell specificity by introducing a tumor-specific chimeric antigen receptor. *Blood*. 2010;116:1035–1045.
184. Wilkins O, Keeler AM, Flotte TR. CAR T-cell therapy: progress and prospects. *Hum Gene Ther Methods*. 2017;28(2):61 66.
185. Guo Y, Feng K, Wang Y, Han W. Targeting cancer stem cells by using chimeric antigen receptor-modified T cells: a potential and curable approach for cancer treatment. *Protein Cell*. 2018;9(6):516–526.
186. Lee S, Margolin K. Cytokines in cancer immunotherapy. *Cancer*. 2011;3:3856–3893.
187. Calogero RA, Quaglino E, Saviozzi S, Forni G, Cavallo F. Oncoantigens as anti-tumor vaccination targets: the chance of a lucky strike? *Cancer Immunol Immunother*. 2008; 57(11):1685–1694.

188. Bol KF, Schreibelt G, Gerritsen WR, Vries IJMD, Figdor CG. Dendritic cell-based immunotherapy: state of the art and beyond. *Clin Cancer Res*. 2016;22:1897–1907.
189. Pellegatta S, Poliani PL, Corno D, et al. Neurospheres enriched in cancer stem-like cells are highly effective in eliciting a dendritic cell-mediated immune response against malignant gliomas. *Cancer Res*. 2006;66(21):10247–10252.
190. Brown CE, Starr R, Martinez C, et al. Recognition and killing of brain tumor stem-like initiating cells by CD8+ cytolytic T cells. *Cancer Res*. 2009;69:8886–8894.
191. Xu Q, Liu G, Yuan X, et al. Antigen-specific T-cell response from dendritic cell vaccination using cancer stem-like cell-associated antigens. *Stem Cells*. 2009;27: 1734–1740.
192. Li X, Zhang Z, Lin G, et al. Antigen-specific T cell response from dendritic cell vaccination using side population cell-associated antigens targets hepatocellular carcinoma. *Tumour Biol*. 2016;37(8):11267–11278.
193. Dashti A, Ebrahimi M, Hadjati J, Memarnejadian A, Moazzeni SM. Dendritic cell based immunotherapy using tumor stem cells mediates potent antitumor immune responses. *Cancer Lett*. 2016;374(1):175–185.
194. Ning N, Pan Q, Zheng F, et al. Cancer stem cell vaccination confers significant antitumor immunity. *Cancer Res*. 2012;72(7):1853–1864.
195. Hu Y, Lu L, Xia Y, et al. Therapeutic efficacy of cancer stem cell vaccines in the adjuvant setting. *Cancer Res*. 2016;76(16):4661–4672.
196. Lu L, Tao H, Chang AE, et al. Cancer stem cell vaccine inhibits metastases of primary tumors and induces humoral immune responses against cancer stem cells. *Oncoimmunology*. 2015;4(3):e990767.
197. Melief CJM, Hall TV, Arens R, Ossendorp F, Burg SHVD. Therapeutic cancer vaccines. *J Clin Invest*. 2015;125:3401–3412.
198. Burg SHVD, Arens R, Ossendorp F, Hall TV, Melief CJM. Vaccines for established cancer: overcoming the challenges posed by immune evasion. *Nat Rev Cancer*. 2016; 16:219–233.
199. Vik-Mo E, Nyakas M, Mikkelsen B, Moe M, Due-Tonnesen P, Langmoen I. Therapeutic vaccination against autologous cancer stem cells with mRNA-transfected dendritic cells in patients with glioblastoma. *Cancer Immunol Immunother*. 2013;62: 1499–1509.
200. Lin M, Li SY, Xu KC, et al. Safety and efficacy study of lung cancer stem cell vaccine. *Immunol Res*. 2015;62:16–22.
201. Lin M, Ying Y, Shu Y, et al. Prospective study of the safety and efficacy of a pancreatic cancer stem cell vaccine. *J Cancer Res Clin Oncol*. 2015;141:1827–1833.
202. Butterfield LH. Cancer vaccines. *BMJ*. 2015;350:h988.
203. Aurisicchio L, Ciliberto G. Genetic cancer vaccines: current status and perspectives. *Expert Opin Biol Ther*. 2012;12(8):1043–1058.
204. Rice J, Ottensmeier CH, Stevenson FK. DNA vaccines: precision tools for activating effective immunity against cancer. *Nat Rev Cancer*. 2008;8:108–120.
205. Conti L, Lanzardo S, Arigoni M, et al. The noninflammatory role of high mobility group box 1/Toll-like receptor 2 axis in the self-renewal of mammary cancer stem cells. *FASEB J*. 2013;27(12):4731–4744.
206. Lanzardo S, Conti L, Rooke R, et al. Immunotargeting of antigen xCT attenuates stem-like cell behavior and metastatic progression in breast cancer. *Cancer Res*. 2015; 76:62–72.
207. Conti L, Lanzardo S, Ruiu R, et al. L-Ferritin targets breast cancer stem cells and delivers therapeutic and imaging agents. *Oncotarget*. 2016;7:66713–66727.
208. Lewerenz J, Hewett SJ, Huang Y, et al. The cystine/glutamate antiporter system x c-in health and disease: from molecular mechanisms to novel therapeutic opportunities. *Antioxid Redox Signal*. 2013;18:522–555.

209. Donofrio G, Tebaldi G, Lanzardo S, et al. Bovine herpesvirus 4-based vector delivering the full length xCT DNA efficiently protects mice from mammary cancer metastases by targeting cancer stem cells. *Oncoimmunology*. 2018;7(12):e1494108.
210. Bolli E, O'Rourke JP, Conti L, et al. A Virus-Like-Particle immunotherapy targeting Epitope-Specific anti-xCT expressed on cancer stem cell inhibits the progression of metastatic cancer in vivo. *Oncoimmunology*. 2017;7:e1408746.
211. Ruiu R, Rolih V, Bolli E, et al. Fighting breast cancer stem cells through the immune-targeting of the xCT cystine-glutamate antiporter. *Cancer Immunol Immunother*. 2018;68:131–141.
212. Bianco C, Salomon DS. Targeting the embryonic gene Cripto-1 in cancer and beyond. *Expert Opin Ther Pat*. 2010;20(12):1739–1749.
213. Witt K, Ligtenberg MA, Conti L, et al. Cripto-1 plasmid DNA vaccination targets metastasis and cancer stem cells in murine mammary carcinoma. *Cancer Immunol Res*. 2018;6(11):1417–1425.
214. Naujokat C. Monoclonal antibodies against human cancer stem cells. *Immunotherapy*. 2014;6(3):290–308.
215. Negi LM, Talegaonkar S, Jaggi M, Ahmad FJ, Iqbal Z, Khar RK. Role of CD44 in tumour progression and strategies for targeting. *J Drug Target*. 2012;20(7): 561–573.
216. Jin L, Hope KJ, Zhai Q, Smadja-Joffe F, Dick JE. Targeting of CD44 eradicates human acute myeloid leukemic stem cells. *Nat Med*. 2006;12(10):1167–1174.
217. Marangoni E, Lecomte N, Durand L, et al. CD44 targeting reduces tumour growth and prevents post-chemotherapy relapse of human breast cancers xenografts. *Br J Cancer*. 2009;100(6):918–922.
218. Chao MP, Weissman IL, Majeti R. The CD47-SIRPalpha pathway in cancer immune evasion and potential therapeutic implications. *Curr Opin Immunol*. 2012;24(2): 225–232.
219. Willingham SB, Volkmer JP, Gentles AJ, et al. The CD47-signal regulatory protein alpha (SIRPa) interaction is a therapeutic target for human solid tumors. *Proc Natl Acad Sci U S A*. 2012;109(17):6662–6667.
220. Brischwein K, Schlereth B, Guller B, et al. MT110: a novel bispecific single-chain antibody construct with high efficacy in eradicating established tumors. *Mol Immunol*. 2006;43(8):1129–1143.
221. Haas C, Krinner E, Brischwein K, et al. Mode of cytotoxic action of T cell-engaging BiTE antibody MT110. *Immunobiology*. 2009;214(6):441–453.
222. Cioffi M, Dorado J, Baeuerle PA, Heeschen C. EpCAM/CD3-bispecific T-cell engaging antibody MT110 eliminates primary human pancreatic cancer stem cells. *Clin Cancer Res*. 2012;18(2):465–474.
223. Herrmann I, Baeuerle PA, Friedrich M, et al. Highly efficient elimination of colorectal tumor-initiating cells by an EpCAM/CD3-bispecific antibody engaging human T cells. *PLoS One*. 2010;5(10):e13474.
224. Jager M, Schoberth A, Ruf P, et al. Immunomonitoring results of a phase II/III study of malignant ascites patients treated with the trifunctional antibody catumaxomab (anti-EpCAM x anti-CD3). *Cancer Res*. 2012;72(1):24–32.
225. Sneha S, Nagare RP, Priya SK, Sidhanth C, Pors K, Ganesan TS. Therapeutic antibodies against cancer stem cells: a promising approach. *Cancer Immunol Immunother*. 2017;66(11):1383–1398.
226. Nakamoto T, Murayama Y, Oritani K, et al. A novel therapeutic strategy with anti-CD9 antibody in gastric cancers. *J Gastroenterol*. 2009;44(9):889–896.
227. Fischer M, Yen WC, Kapoun AM, et al. Anti-DLL4 inhibits growth and reduces tumor-initiating cell frequency in colorectal tumors with oncogenic KRAS mutations. *Cancer Res*. 2011;71(5):1520–1525.

228. Emlet DR, Gupta P, Holgado-Madruga M, et al. Targeting a glioblastoma cancer stem-cell population defined by EGF receptor variant III. *Cancer Res*. 2014;74(4):1238–1249.
229. Zheng PP, Kros JM, Li J. Approved CAR T cell therapies: ice bucket challenges on glaring safety risks and long-term impacts. *Drug Discov Today*. 2018;23(6):1175–1182.
230. Zhu X, Prasad S, Gaedicke S, Hettich M, Firat E, Niedermann G. Patient-derived glioblastoma stem cells are killed by CD133-specific CAR T cells but induce the T cell aging marker CD57. *Oncotarget*. 2015;6(1):171–184.
231. Morgan RA, Johnson LA, Davis JL, et al. Recognition of glioma stem cells by genetically modified T cells targeting EGFRvIII and development of adoptive cell therapy for glioma. *Hum Gene Ther*. 2012;23(10):1043–1053.
232. O'Rourke DM, Nasrallah MP, Desai A, et al. A single dose of peripherally infused EGFRvIII-directed CAR T cells mediates antigen loss and induces adaptive resistance in patients with recurrent glioblastoma. *Sci Transl Med*. 2017;9(399):eaaa0984.
233. Feng KC, Guo YL, Liu Y, et al. Cocktail treatment with EGFR-specific and CD133-specific chimeric antigen receptor-modified T cells in a patient with advanced cholangiocarcinoma. *J Hematol Oncol*. 2017;10(1):4.
234. Deng Z, Wu Y, Ma W, Zhang S, Zhang YQ. Adoptive T-cell therapy of prostate cancer targeting the cancer stem cell antigen EpCAM. *BMC Immunol*. 2015;16:1.
235. Ang WX, Li Z, Chi Z, et al. Intraperitoneal immunotherapy with T cells stably and transiently expressing anti-EpCAM CAR in xenograft models of peritoneal carcinomatosis. *Oncotarget*. 2017;8(8):13545–13559.
236. Postow MA, Callahan MK, Wolchok JD. Immune checkpoint blockade in cancer therapy. *J Clin Oncol*. 2017;33:1974–1982.
237. Zheng F, Dang J, Zhang H, et al. Cancer stem cell vaccination with PD-L1 and CTLA-4 blockades enhances the eradication of melanoma stem cells in a mouse tumor model. *J Immunother*. 2018;41:361–368.
238. Shi X, Zhang X, Li J, et al. PD-1 blockade enhances the antitumor efficacy of GM-CSF surface-modified bladder cancer stem cells vaccine. *Int J Cancer*. 2018;142(10):2106–2117.
239. Mandal A, Viswanathan C. Natural killer cells: in health and disease. *Hematol Oncol Stem Cell Ther*. 2015;8(2):47–55.
240. Abel AM, Yang C, Thakar MS, Malarkannan S. Natural killer cells: development, maturation, and clinical utilization. *Front Immunol*. 2018;9:1869.
241. Guia S, Fenis A, Vivier E, Narni-Mancinelli E. Activating and inhibitory receptors expressed on innate lymphoid cells. *Semin Immunopathol*. 2018;40(4):331–341.
242. Voutsadakis IA. Expression and function of immune ligand-receptor pairs in NK cells and cancer stem cells: therapeutic implications. *Cell Oncol*. 2018;41(2):107–121.
243. Ljunggren HG, Malmberg KJ. Prospects for the use of NK cells in immunotherapy of human cancer. *Nat Rev Immunol*. 2007;7(5):329–339.
244. Tallerico R, Garofalo C, Carbone E. A new biological feature of natural killer cells: the recognition of solid tumor-derived cancer stem cells. *Front Immunol*. 2016;7:179.
245. Lopez-Soto A, Huergo-Zapico L, Galvan JA, et al. Epithelial-mesenchymal transition induces an antitumor immune response mediated by NKG2D receptor. *J Immunol*. 2013;190(8):4408–4419.
246. Reeves E, James E. Antigen processing and immune regulation in the response to tumours. *Immunology*. 2017;150(1):16–24.
247. Tamari K, Hayashi K, Ishii H, et al. Identification of chemoradiation-resistant osteosarcoma stem cells using an imaging system for proteasome activity. *Int J Oncol*. 2014;45(6):2349–2354.
248. Castriconi R, Daga A, Dondero A, et al. NK cells recognize and kill human glioblastoma cells with stem cell-like properties. *J Immunol*. 2009;182(6):3530–3539.

249. Kim GR, Ha GH, Bae JH, Oh SO, Kim SH, Kang CD. Metastatic colon cancer cell populations contain more cancer stem-like cells with a higher susceptibility to natural killer cell-mediated lysis compared with primary colon cancer cells. *Oncol Lett.* 2015;9(4):1641–1646.
250. Pietra G, Manzini C, Vitale M, et al. Natural killer cells kill human melanoma cells with characteristics of cancer stem cells. *Int Immunol.* 2009;21(7):793–801.
251. Ames E, Canter RJ, Grossenbacher SK, et al. NK cells preferentially target tumor cells with a cancer stem cell phenotype. *J Immunol.* 2015;195(8):4010–4019.
252. Tseng HC, Arasteh A, Paranjpe A, et al. Increased lysis of stem cells but not their differentiated cells by natural killer cells; de-differentiation or reprogramming activates NK cells. *PLoS One.* 2010;5(7):e11590.
253. Yin T, Wang G, He S, Liu Q, Sun J, Wang Y. Human cancer cells with stem cell-like phenotype exhibit enhanced sensitivity to the cytotoxicity of IL-2 and IL-15 activated natural killer cells. *Cell Immunol.* 2016;300:41–45.
254. Jewett A, Tseng HC. Tumor induced inactivation of natural killer cell cytotoxic function; implication in growth, expansion and differentiation of cancer stem cells. *J Cancer.* 2011;2:443–457.
255. Kaur K, Nanut MP, Ko MW, Safaie T, Kos J, Jewett A. Natural killer cells target and differentiate cancer stem-like cells/undifferentiated tumors: strategies to optimize their growth and expansion for effective cancer immunotherapy. *Curr Opin Immunol.* 2018;51:170–180.
256. Ames E, Canter RJ, Grossenbacher SK, et al. Enhanced targeting of stem-like solid tumor cells with radiation and natural killer cells. *Oncoimmunology.* 2015;4(9):e1036212.
257. Dianat-Moghadam H, Rokni M, Marofi F, Panahi Y, Yousefi M. Natural killer cell-based immunotherapy: from transplantation toward targeting cancer stem cells. *J Cell Physiol.* 2018;234:259–273.
258. Carlsten M, Norell H, Bryceson YT, et al. Primary human tumor cells expressing CD155 impair tumor targeting by down-regulating DNAM-1 on NK cells. *J Immunol.* 2009;183(8):4921–4930.
259. Pietra G, Manzini C, Rivara S, et al. Melanoma cells inhibit natural killer cell function by modulating the expression of activating receptors and cytolytic activity. *Cancer Res.* 2012;72(6):1407–1415.
260. He Y, Bunn PA, Zhou C, Chan D. KIR 2D (L1, L3, L4, S4) and KIR 3DL1 protein expression in non-small cell lung cancer. *Oncotarget.* 2016;7(50):82104–82111.
261. Mamessier E, Sylvain A, Thibult ML, et al. Human breast cancer cells enhance self tolerance by promoting evasion from NK cell antitumor immunity. *J Clin Invest.* 2011;121(9):3609–3622.
262. Levi I, Amsalem H, Nissan A, et al. Characterization of tumor infiltrating natural killer cell subset. *Oncotarget.* 2015;6(15):13835–13843.
263. Pauza CD, Liou ML, Lahusen T, et al. Gamma delta T cell therapy for cancer: it is good to be local. *Front Immunol.* 2018;9:1305.
264. Harly C, Peigne CM, Scotet E. Molecules and mechanisms implicated in the peculiar antigenic activation process of human Vgamma9Vdelta2 T cells. *Front Immunol.* 2014;5:657.
265. Peng G, Wang HY, Peng W, Kiniwa Y, Seo KH, Wang RF. Tumor-infiltrating gammadelta T cells suppress T and dendritic cell function via mechanisms controlled by a unique toll-like receptor signaling pathway. *Immunity.* 2007;27(2):334–348.
266. Saitoh A, Narita M, Watanabe N, et al. Anti-tumor cytotoxicity of gammadelta T cells expanded from peripheral blood cells of patients with myeloma and lymphoma. *Med Oncol.* 2008;25(2):137–147.

267. Maniar A, Zhang X, Lin W, et al. Human gammadelta T lymphocytes induce robust NK cell-mediated antitumor cytotoxicity through CD137 engagement. *Blood*. 2010; 116(10):1726–1733.
268. Caccamo N, Meraviglia S, Cicero G, et al. Aminobisphosphonates as new weapons for gammadelta T cell-based immunotherapy of cancer. *Curr Med Chem*. 2008;15(12): 1147–1153.
269. Chen HC, Joalland N, Bridgeman JS, et al. Synergistic targeting of breast cancer stem-like cells by human gammadelta T cells and CD8(+) T cells. *Immunol Cell Biol*. 2017;95(7):620–629.
270. Todaro M, D'Asaro M, Caccamo N, et al. Efficient killing of human colon cancer stem cells by gammadelta T lymphocytes. *J Immunol*. 2009;182(11):7287–7296.
271. Nishio N, Fujita M, Tanaka Y, et al. Zoledronate sensitizes neuroblastoma-derived tumor-initiating cells to cytolysis mediated by human gammadelta T cells. *J Immunother* 2012;35(8):598–606.
272. Rouhrazi H, Turgan N, Oktem G. Zoledronic acid overcomes chemoresistance by sensitizing cancer stem cells to apoptosis. *Biotech Histochem*. 2018;93(2):77–88.
273. Buhler H, Hoberg C, Fakhrian K, Adamietz IA. Zoledronic acid inhibits the motility of cancer stem-like cells from the human breast cancer cell line MDA-MB 231. *In Vivo*. 2016;30(6):761–768.
274. Todaro M, Orlando V, Cicero G, et al. Chemotherapy sensitizes colon cancer initiating cells to Vgamma9Vdelta2 T cell-mediated cytotoxicity. *PLoS One*. 2013;8(6):e65145.
275. Lai D, Wang F, Chen Y, et al. Human ovarian cancer stem-like cells can be efficiently killed by gammadelta T lymphocytes. *Cancer Immunol Immunother*. 2012;61(7): 979–989.
276. Schmidt-Wolf IG, Lefterova P, Johnston V, Huhn D, Blume KG, Negrin RS. Propagation of large numbers of T cells with natural killer cell markers. *Br J Haematol*. 1994;87(3):453–458.
277. Guo Y, Han W. Cytokine-induced killer (CIK) cells: from basic research to clinical translation. *Chin J Cancer*. 2015;34(3):99–107.
278. Gammaitoni L, Giraudo L, Leuci V, et al. Effective activity of cytokine-induced killer cells against autologous metastatic melanoma including cells with stemness features. *Clin Cancer Res*. 2013;19(16):4347–4358.
279. Sangiolo D, Mesiano G, Gammaitoni L, et al. Cytokine-induced killer cells eradicate bone and soft-tissue sarcomas. *Cancer Res*. 2014;74(1):119–129.
280. Yang T, Zhang W, Wang L, et al. Co-culture of dendritic cells and cytokine-induced killer cells effectively suppresses liver cancer stem cell growth by inhibiting pathways in the immune system. *BMC Cancer*. 2018;18(1):984.
281. Mesiano G, Grignani G, Fiorino E, et al. Cytokine induced killer cells are effective against sarcoma cancer stem cells spared by chemotherapy and target therapy. *Oncoimmunology*. 2018;7(11):e1465161.
282. Bruttel VS, Wischhusen J. Cancer stem cell immunology: key to understanding tumorigenesis and tumor immune escape? *Front Immunol*. 2014;5:360.
283. Quintana E, Shackleton M, Sabel MS, Fullen DR, Johnson TM, Morrison SJ. Efficient tumour formation by single human melanoma cells. *Nature*. 2008;456(7222):593–598.
284. Schatton T, Murphy GF, Frank NY, et al. Identification of cells initiating human melanomas. *Nature*. 2008;451(7176):345–349.
285. Doherty MR, Smigiel JM, Junk DJ, Jackson MW. Cancer stem cell plasticity drives therapeutic resistance. *Cancer*. 2016;8:1–13.
286. Goldman A, Majumder B, Dhawan A, et al. Temporally sequenced anticancer drugs overcome adaptive resistance by targeting a vulnerable chemotherapy-induced phenotypic transition. *Nat Commun*. 2015;6:1–13.

CHAPTER SIX

Immunotherapy of pancreatic cancer

Yoshiaki Sunami*, Jörg Kleeff
Department of Visceral, Vascular and Endocrine Surgery, Martin-Luther-University Halle-Wittenberg, University Medical Center Halle, Halle, Germany
*Corresponding author: e-mail address: yoshiaki.sunami@uk-halle.de

Contents

1. Introduction 190
2. Cancer immunosurveillance and immunoediting in pancreatic cancer 190
3. Immunotherapy with natural killer cells and dendritic cells in pancreatic cancer 191
4. Classical immunotherapy with immune checkpoint inhibitors in pancreatic cancer 195
5. Immunosuppressive tumor microenvironment in pancreatic cancer 196
6. Reprogramming the microenvironment by targeting tumor-associated macrophages as a therapeutic strategy for pancreatic cancer 199
7. Reprogramming the microenvironment by targeting regulators expressed by pancreatic cancer cells or stromal cells 204
8. Adaptive T-cell therapy for pancreatic cancer 206
9. Safety considerations in CAR T-cell therapy in pancreatic cancer 208
10. Conclusion 210
References 210

Abstract

Pancreatic ductal adenocarcinoma (PDAC) is predicted to become the second most common cause of cancer-related death in the United States by 2030. So far surgery remains the only curative option for pancreatic cancer, but fewer than 20% of patients have surgically resectable disease. Furthermore, pancreatic cancer exhibits a remarkable resistance to established therapeutic options, including chemotherapy, radiotherapy, and targeted therapy, because pancreatic cancer exhibits numerous mechanisms of resistance like genetic and epigenetic alterations and a complex and dense tumor microenvironment. The tumor microenvironment is populated with different types of immune cells that play a critical role in therapy resistance, tumor progression, and carcinogenesis. Cancer immunotherapy has now been recognized as the fourth pillar of cancer care and a number of preclinical and clinical studies have been conducted for pancreatic cancer. Targeting and modulating the tumor immune microenvironment could not only switch the immune system toward anti-cancer, but also may improve sensitivity toward established chemotherapy. In this review, we discuss both preclinical and clinical studies on pancreatic cancer immunotherapy with natural killer cells,

Progress in Molecular Biology and Translational Science, Volume 164
ISSN 1877-1173
https://doi.org/10.1016/bs.pmbts.2019.03.006

dendritic cells, and chimeric antigen receptor T cells. Furthermore, we summarize strategies for reprogramming the tumor immune microenvironment by targeting macrophages and stromal cell factors in pancreatic cancer. The development of systemic therapies is essential for improving the outcomes of pancreatic cancer patients, and cancer immunotherapy would improve effectiveness of other established therapeutic options, which might together improve the prognosis of pancreatic tumors.

1. Introduction

Pancreatic cancer is predicted to become the second-leading cause of cancer-related death in the United States by 2030.[1] To date, surgery remains the only curative option for pancreatic cancer, but fewer than 20% of patients have surgically resectable disease.[2] A number of excellent studies have shown significant progress in survival of pancreatic cancer patients by combination chemotherapy such as FOLFIRINOX (folinic acid, 5-fluorouracil, irinotecan, and oxaliplatin) or nab-paclitaxel plus gemcitabine, or capecitabine plus gemcitabine.[3–7] Pancreatic cancer, however, exhibits remarkable resistance to chemotherapy, radiotherapy and targeted therapy because pancreatic cancer cells exhibit numerous mechanisms of resistance, including genetic and epigenetic alterations, a complex and dense tumor microenvironment, and other factors.[2] The tumor microenvironment includes immune cells, fibroblasts, and endothelial cells resulting in an abundant collection of inflammatory cytokines and chemokines, growth factors, and components of the extracellular matrix. Cancer immunotherapy has now been recognized as the fourth pillar of cancer care, along with surgery, radiation, and chemotherapy,[8] and a number of preclinical and clinical studies have been completed or are currently ongoing for pancreatic cancer. Development of systemic therapies is essential for improving the outcome of pancreatic cancer patients and cancer immunotherapy could improve the effectiveness of current therapeutic options.

2. Cancer immunosurveillance and immunoediting in pancreatic cancer

The immune system has complex interactions with tumors, including roles as a suppressor and promotor of tumor growth. While chronic inflammation has been widely considered to be associated with tumor pathogenesis,[9] inflammation is classically associated with anti-tumor immune responses and cancer immunosurveillance acts as an extrinsic tumor suppressor by identifying and eliminating aberrant or malignant cells.[10,11] Cancer immunosurveillance and immunoediting has been summarized in detail elsewhere.[11–13] Briefly, a growing tumor causes minor cell/tissue destruction and induces inflammatory signals leading to recruitment of cells of the innate immune system such as natural killer (NK) T cells, NK cells, γδT cells, macrophages, and dendritic cells (DCs). Transformed cell structures are recognized by infiltrating lymphocytes, which are stimulated and produce interferon γ (IFN-γ).[11] IFN-γ can induce a limited amount of tumor cell death. In this context, it has been shown that IFN-γ inhibits the growth of pancreatic carcinoma cells by inducing apoptosis.[14] Tumor cell debris, which is formed as either a direct or indirect consequence of IFN-γ, is then ingested by local DCs. Subsequently DCs are homed to lymph nodes that are located immediately adjacent to tumors (tumor-draining lymph nodes). The immigrated DCs in the draining lymph nodes induce tumor-specific $CD4^+$ T helper cells expressing IFN-γ (Th1), which support the development of tumor-specific $CD8^+$ T cells. $CD4^+$ and $CD8^+$ T cells home to the tumor sites and cytotoxic T cells recognize and destroy the remaining antigen-bearing tumor cells.[12]

3. Immunotherapy with natural killer cells and dendritic cells in pancreatic cancer

NK cells are cytotoxic lymphocytes of the innate immune system. NK cell activation is controlled by a broad range of receptors that allow NK cells to sense and to rapidly respond to changes in the environment. CD226 is involved in controlling NK cell adhesion, cytokine secretion, and cytotoxicity.[15] Decreased NK cell activity has been observed in patients with PDAC,[16] and pancreatic cancer patients have less $CD226^+$ NK cells in the blood than do healthy controls.[17] Hence, reactivation of NK cells or high activity NK cell treatment may be potential pancreatic cancer therapies. A clinical study is evaluating the killing activity of *ex vivo* expanded NK cells derived from blood of pancreatic cancer patients (NCT03665571). Several phase 1/2 studies have chosen high-activity NK cells for therapy of several cancer types, including pancreatic cancer (NCT03008304, NCT03634501). In addition, a phase 1 study is running to test the efficacy of combination therapy with NK cells and Bortezomib (NCT00720785). Bortezomib (PS-341) is a proteasome inhibitor and has been approved by the FDA for treatment of multiple myeloma and mantle-cell lymphoma.[18] Bortezomib sensitizes pancreatic cancer cells to endoplasmic reticulum (ER) stress-mediated apoptosis.[19] In another phase 1/2 study NK cells are used together with aldoxorubicin HCl, ALT-803 (IL-15 complex agonist), ETBX-011 (adenoviral vector vaccine), GI-4000 (a vaccine containing a heat-killed yeast transfected with mutant forms of Ras oncogene), avelumab (recombinant human anti-PD-L1 IgG1 monoclonal antibody (discussed in detail in Section 4)), bevacizumab (recombinant human anti VEGF IgG1 monoclonal antibody), nab-Paclitaxel, lovaza (omega-3-acid ethyl esters), Cyclophosphamide, Leucovorin (folinic acid)/Fluorouracil/Oxaliplatin (FOLFOX), and Capecitabine (NCT03387098) (Table 1). IL-15 has been shown to stimulate NK cell-mediated killing of both human pancreatic cancer cells and stellate cells.[20] Although some data have demonstrated that deletion of carcinoma-associated fibroblasts or reduction of stromal content renders pancreatic tumors more aggressive (see Section 7), and the role of pancreatic stellate cells and cancer-associated fibroblasts in cancer development and progression is still not completely understood, the reduction of stellate cells and stroma by IL-15-activated NK cells could increase chemosensitivity, which could potentiate the direct tumor-elimination effects of IL-15-activated NK cells.

Besides NK cells, DCs have also important roles in immune surveillance. DCs are antigen-presenting cells and their main function is to process and present antigens to T cells. Conventional DCs (cDCs) can be divided into two subsets, cDC1 and cDC2, which specialize in $CD8^+$ and $CD4^+$ T cell activation, respectively. In humans, cDC1s are marked by CD141, and cDC2s, which are also found in both lymphoid and peripheral tissues, are marked by CD1c.[21,22] Significantly fewer $CD141^+$ cDC1s and $CD1c^+$ cDC2s are found in the bone marrow of pancreatic cancer patients. Mechanistically, pancreatic cancer cells interrupt Interferon regulatory factor 8 (IRF8)-dependent DC development to overcome immune surveillance.[23] Hence, supporting development or reintroduction of DCs can be a promising option for recovering the immune surveillance system. To that end, vaccination strategies directed against a single antigen, as well as DCs pulsed with tumor-cell lysate as the source of antigen, have been used for pancreatic cancer immunotherapy. In an *in vitro* model, pancreatic tumor cell lysate-pulsed human DCs induced T cell proliferation and responses against pancreatic cancer cells.[24] Several clinical studies have used vaccination strategies with DCs. It has been shown that antigen-treated autologous DCs induce propagation of major histocompatibility complex (MHC) class I-"restricted" and antigen-specific $CD8^+$ T cells in cancer patients. Antigen-producing tumors are expected to process and present antigen-derived peptide fragments on their cell surface with MHC class I molecules. These

Table 1 Pancreatic cancer immunotherapy with natural killer cells and dendritic cells.

Intervention/Treatment	NCT number	Stage of clinical trial	Recruitment status	Last update
Ex vivo expansion of NK cells from blood in patients with pancreatic cancer	NCT03665571	Not mentioned	Recruiting	September 11, 2018
High activity NK cells	NCT03008304	Phase 1, 2	Recruiting	January 4, 2017
Activated NK cells	NCT03634501	Phase 1, 2	Recruiting	January 11, 2019
NK cells with Bortezomib	NCT00720785	Phase 1	Recruiting	December 18, 2018
High-affinity NK, Aldoxorubicin-HCl ALT-803 ETBX-011 GI-4000 Avelumab Bevacizumab Capecitabine Cyclophosphamide Fluorouracil Leucovorin nab-Paclitaxel Lovaza Oxaliplatin Stereotactic body radiotherapy	NCT03387098	Phase 1,2	Active but not recruiting	November 6, 2018
DC Poly-ICLC	NCT01677962	Phase 1	Completed	December 15, 2016
Peptide-pulsed DC Poly-ICLC	NCT01410968	Phase 1	Completed	December 09, 2016
DC pulsed with tumor lysate DC pulsed with MUC-1/WT-1 peptides	NCT03114631	Phase 1,2	Enrolling by invitation	January 9, 2019
mDC3/8-KRAS vaccine	NCT03592888	Phase 1	Recruiting	December 17, 2018
Peptide loaded DC vaccine	NCT03300843	Phase 2	Recruiting	January 10, 2019

MHC-"restricted" antigen-derived peptides can be recognized by T-cells. To that end, autologous DCs prepared from the peripheral blood of human leukocyte antigen (HLA)-A2 (a major type of MHC class I cell surface receptor)-positive pancreatic cancer patients were treated with HLA-A2-"restricted" peptides derived from antigens overexpressed in pancreatic cancer, including human telomerase reverse transcriptase (hTERT), a cell surface glycoprotein, carcinoembryonic antigen (CEA), or survivin, an inhibitor of apoptosis. TERT is expressed in more than 90% of the pancreatic tumors, CEA is expressed in more than 30%, and survivin is expressed in more than 80% of pancreatic tumors. These DCs were administered in combination with toll-like receptor (TLR)-3 agonist poly(IC: LC), which can activate DCs. Several phase 1 studies have suggested that vaccination with peptide-pulsed DCs in combination with poly(IC:LC) can induce tumor-specific T cell population in patients with advanced pancreatic cancer (NCT01410968, NCT01677962,[25]). Another phase 1/2 study uses DCs pulsed with tumor lysate or with MUC-1/WT-1 peptides (NCT03114631). More than 90% of pancreatic cancer patients have oncogenic KRAS mutations. In this context, a recent phase 1 study aims to establish immunotherapy where autologous DCs are pulsed with mutant KRAS peptides corresponding to the patient's specific tumor mutation (NCT03592888). A similar strategy has been chosen in a phase 2 study for treating several cancer types. Including pancreatic cancer, where DCs are loaded with mutant peptides according to tumor-specific mutations identified by exome sequencing (NCT03300843). Several combination studies, such as DCs and chemotherapy, have been completed or are currently ongoing. It has been shown that in an *in vivo* mouse model, vaccination with pancreatic tumor cell-pulsed bone marrow-derived DCs prevented orthotopic pancreatic tumor growth, and combination therapy together with gemcitabine further increased survival of mice.[26] Combination therapy with gemcitabine, zoledronate-pulsed DCs, and activated T cells demonstrated stable disease for some patients.[27] Zoledronate (zoledronic acid) is a type of bisphosphonate used for patients with osteoporosis because it inhibits bone degradation. It has been shown that zoledronate induces antiproliferative and apoptotic effects in human pancreatic cells.[28] In conclusion, reactivation of NK cells or high activity NK cell treatment for invigorating the innate immune system, as well as vaccination of DCs for induction of tumor-specific T-cells, are important strategies. There are a number of ongoing clinical studies using mono- or combination therapies for pancreatic cancer patients.

4. Classical immunotherapy with immune checkpoint inhibitors in pancreatic cancer

After activation of tumor-specific cytotoxic T cells by antigen-presenting cells (APCs), they home to the tumor and recognize and destroy tumor cells expressing distinctive tumor antigens during the elimination phase of cancer immunoediting. The system should be specific for maintaining self-tolerance as the so-called immune checkpoint system acts to avoid autoimmunity or to inhibit excessive immune reaction. T cell activation requires both T cell receptor (TCR) and CD3 binding to major histocompatibility complex (MHC) and co-stimulatory molecules, as with CD28 binding to CD80 (B7-1) or CD86 (B7-2).[29,30] Cytotoxic T-lymphocyte-associated antigen 4 (CTLA-4, also known as CD152) is homologous to CD28 and both bind to CD80 and CD86. CTLA-4 binds CD80 and CD86 with higher affinity than does CD28 and is a critical negative regulator of T-cell responses[31] because CTLA-4-deficiency results in fatal autoimmunity and the release of self-antigen-reactive T cells.[32–34] CTLA-4 is also expressed by $CD45RA^-$/$FOXP^{high}$/$CD4^+$/$CD25^{high}$ effector regulatory T (T_{reg}) cells that play an important role in autoimmune pathogenesis.[35] It has been shown that CTLA-4 is essential for T_{reg} cell function *in vivo*, and for this reason, the general strategy of blocking CTLA-4 was initially questioned since T_{reg}-specific CTLA-4 deletion has been shown to induce spontaneous development of systemic lymphoproliferation and fatal T cell-mediated autoimmune disease.[36] A study has however suggested that *in vivo* administration of antibodies to CTLA-4 resulted in the rejection of tumors, presumably *via* lowering the overall threshold of T cell activation.[37] After several clinical studies, the FDA, in 2011, approved a monoclonal antibody targeting CTLA-4, ipilimumab (Yervoy, also known as MDX-010), for treating melanoma.[38] Tremelimumab is another monoclonal antibody (human IgG2) targeting CTLA-4 and several clinical studies with tremelimumab or with ipilimumab for treating pancreatic cancer are currently ongoing.

Another key player of the immune checkpoint is a receptor called programmed death 1 (PD-1, also known as CD279), which was initially identified as a gene upregulated in a murine T-cell hybridoma and a murine hematopoietic progenitor cell line undergoing cell death.[39] Deletion of PD-1 leads to the development of autoimmune diseases and lupus-like proliferative glomerulonephritis in mice,[40] suggesting that PD-1 downregulates immune responses. PD-1 is expressed predominantly by activated $CD4^+$

and $CD8^+$ T cells, T_{reg} cells, B cells, and monocytes, but also by some DC subsets and NKT cells.[41] The ligands for PD-1 are PD-1 ligand 1 (PD-L1, also known as CD274 or B7 Homolog 1) and PD-L2 (also known as CD273 or B7-DC), and PD-1 can transmit an inhibitory signal by counteracting TCR signal transduction and CD28-co-stimulation signaling.[30,41] PD-L1 is expressed on activated B, T, myeloid cells, and DCs, but it also is expressed in non-hematopoietic cells such as microvascular endothelial cells and in non-lymphoid organs including the pancreas. PD-L2 is expressed on macrophages and DCs.[42] Furthermore, PD-L1 is expressed on a variety of tumors, including pancreatic cancer. Binding of PD-L1 to PD-1 leads to the exhaustion of effector T cells and immune escape of tumor cells. PD-L1 expression is inversely correlated with the number of T lymphocytes in pancreatic tumor microenvironment. Furthermore, PD-L1 expression is correlated with poor clinical outcome in pancreatic cancer.[43] Hence, blockage of PD-1/PD-L1 interactions has been considered a promising anti-cancer strategy. In this context, at least 500 clinical studies have been carried out with nine types of antibodies against PD-1 or PD-L1 on 23 different types of cancer.[44] An antagonistic monoclonal antibody against PD-1 (nivolumab, also known as Opdivo, BMS-936558, MDX-1106, ONO-4538) has been approved by the FDA and EMA for treatment of melanoma, lung cancer, kidney cancer, and Hodgkin lymphoma. Pembrolizumab, another antibody against PD-1, has been approved for treatment of melanoma, lung cancer, and head/neck cancer. A PD-L1 antagonistic antibody, atezolizumab has been approved for urothelial and lung cancer treatment.[44] Other approved drugs targeting PD-L1 that followed atezolizumab include avelumab and durvalumab.[45] There have been a number of clinical trials targeting CTLA-4 or PD-1/PD-L1 in pancreatic cancer patients, which are in part summarized in Table 2.

5. Immunosuppressive tumor microenvironment in pancreatic cancer

Initially ipilimumab monotherapy and monotherapy targeting PD-1/PD-L1 interactions had been considered to treat pancreatic cancer. However, a phase 2 trial of ipilimumab monotherapy was ineffective for the treatment of advanced pancreatic cancer (NCT00112580,[52]). Therapy of pancreatic cancer with an anti-PD-L1 antibody (BMS936559, MDX1105) also was ineffective (NCT00729664,[53]). A large subset of cancer patients do not respond to CTLA-4 and PD-1 immunotherapy.[54] The mechanisms of CTLA-4 or PD-1/PD-L1 resistance in pancreatic cancer are not

Table 2 Clinical trials targeting immune checkpoint in pancreatic cancer.

Intervention/Treatment	NCT number	Stage of clinical trial	Recruitment status	Last update
Ipilimumab	NCT00112580	Phase 2	Completed	October 12, 2015
Anti-PD-L1 antibody	NCT00729664	Phase 1	Completed	September 7, 2015
Anti-PD-1 antibody radiation	NCT03374293	Phase 2	Recruiting	December 15, 2017
Durvalumab Stereotactive ablative body radiotherapy	NCT03245541	Phase 1, 2	Recruiting	August 2, 2018
Anti-PD-1 antibody INT230–6	NCT03058289	Phase 1, 2	Recruiting	November 21, 2018
Pembrolizumab PEGPH20	NCT03634332	Phase 2	Not yet recruiting	August 20, 2018
Nivolumab Ipilimumab Radiation therapy	NCT03104439	Phase 2	Recruiting	October 12, 2018
Nivolumab cyclophosphamide GVAX Urelumab	NCT02451982	Phase 1, 2	Recruiting	October 26, 2018
Pembrolizumab Cyclophosphamide GVAX Stereotactic body radiation	NCT02648282	Phase 2	Recruiting	January 11, 2019
Nivolumab Ipilimumab Cyclophosphamide GVAX CRS-207	NCT03190265	Phase 2	Recruiting	April 18, 2018
Nivolumab Cyclophosphamide GVAX Stereotactic body radiation	NCT03161379	Phase 2	Recruiting	November 6, 2018
Pembrolizumab Cyclophosphamide Epacadostat CRS-207 GVAX	NCT03006302	Phase 2	Recruiting	January 30, 2018

Continued

Table 2 Clinical trials targeting immune checkpoint in pancreatic cancer.—cont'd

Intervention/Treatment	NCT number	Stage of clinical trial	Recruitment status	Last update
Nivolumab Ipilimumab Radiotherapy	NCT02866383	Phase 2	Recruiting	September 13, 2018
Ipilimumab Evofosfamide	NCT03098160	Phase 1	Recruiting	October 30, 2017
Nivolumab Ipilimumab Niraparib	NCT03404960	Phase 1, 2	Recruiting	February 21, 2018
Nivolumab Ipilimumab VX15/2503 Surgery	NCT03373188	Phase 1	Recruiting	May 23, 2018
Nivolumab Ipilimumab	NCT02834013	Phase 2	Recruiting	January 18, 2019
Pembrolizumab paricalcitol	NCT03331562	Phase 2	Recruiting	September 19, 2018

INT230-6 is a supermolecule complex of cisplatin, vinblastine, and an amphiphilic penetration enhancer,[46] PEGPH20 is PEGylated hyaluronidase,[47] Urelumab is a CD137 (4-1BB) agonist antibody (NC T02451982), CRS-207 is a live, attenuated, double deleted *Listeria monocytogenes* strain with a mesothelin expression cassette inserted,[48] epacadostat is a potent and selective indoleamine 2,3-dioxygenase 1 enzyme inhibitor,[49] evofosfamide is a 2-nitroimidazole prodrug of the DNA alkylator, bromo-isophosphoramide mustard, which is selectively activated under hypoxia,[50] niraparib is a poly(ADP-ribose) polymerase inhibitor,[51] VX15/2503 is an anti-SEMA4D monoclonal antibody (NCT03373188).

fully understood. Increasing evidence, however, suggests that the PDAC tumor microenvironment is immunosuppressive, *e.g.*, PDAC does not commonly exhibit an effective number of intratumor effector T cells. A potential explanation is that oncogenes such as *RAS* drive an inflammatory program that establishes an immune privilege in the tumor microenvironment.[55] To understand the molecular mechanisms of pancreatic cancer development, a mouse model with pancreatic expression of oncogenic $Kras^{G12D}$ and mutant *Trp53* under the control of *Pdx-1* promoter (KPC mouse) has been generated that recapitulates the human disease histologically and molecularly.[56] In KPC mice, immunosuppressive cells such as tumor-associated macrophages (TAMs), $Gr\text{-}1^+/CD11b^+$ myeloid cells, and T_{reg} cells are prominent even at the earliest stages. On the other hand, effector T cells are not commonly observed. T cell antitumor immunity is a prerequisite for therapy targeting

CTLA-4 or PD-1/PD-L1, potentially explaining why classical cancer immunotherapy did not work for pancreatic cancer. As mentioned above, under the standard concept of immunoediting, exhaustion of effector T cells and immune escape of tumor cells can be induced by CTLA-4 binding to CD80 and CD86 or PD-L1 binding to PD-1. The absence of T cell immune surveillance in the case of pancreatic cancer may indicate that the immunoediting theory is in part not applicable for this type of cancer.[57] On the other hand, PDAC may be unexpectedly sensitive to T cell killing, because the tumors may never have been exposed to selective pressure by T cells.[55] Therapeutics that can reprogram the immune microenvironment, such as targeting TAMs, may overcome immunosuppression and therefore enhance responses to immunotherapy.

6. Reprogramming the microenvironment by targeting tumor-associated macrophages as a therapeutic strategy for pancreatic cancer

Extensive immune cell infiltration and fibrosis characterize the tumor microenvironment of human PDAC. Macrophages are frequently a dominant component among the infiltrating cells and inhibiting macrophage recruitment can be a promising strategy for reprogramming the tumor microenvironment. In a mutant KRAS/p53-induced pancreatic cancer mouse model (doxycycline-inducible $Kras^{G12D}$ and $Trp53^{R172H}$ regulated under the Pancreas transcription factor 1 (Ptf1)-Cre promoter), depletion of $CD11b^{+}$ myeloid cells increased intratumor $CD8^{+}$ T cells.[58] The CC-chemokine ligand 2 (CCL2, also known as monocyte chemoattractant protein-1, MCP-1)/CC-chemokine receptor Type 2 (CCR2, or MCP-1 receptor) axis plays an important role in monocyte recruitment.[59] To that end, targeting CCL2/CCR2 could reduce infiltrating monocytes/macrophages leading to microenvironment reprograming. CCR2 deficiency leads to a 50% reduction of PDAC-infiltrating macrophages. However, despite reduced macrophage numbers, CCR2 deficiency did not affect orthotopically implanted pancreatic tumor growth derived from $Kras^{G12D}$-expressing and *Trp53*-deficient mice. Treatment with a pharmacological CCR2 inhibitor (PF-04136309) does not result in changes in pancreatic tumor burden in the $Kras^{G12D}$-expressing, *Trp53*-deficient mice.[60] It has not been clarified whether CCR2 inhibition has therapeutic effects in pancreatic cancer patients. Due to ineffectiveness of CCR2 inhibitor monotherapy in the preclinical study, a phase 1/2 clinical study is currently testing a combination therapy with pre-operative stereotactic radiation with or without CCX872-B, a CCR2 inhibitor (NCT03778879) (Table 3). Another phase 1/2 study is analyzing the effects of the combination of nivolumab and a CCR2/CCR5 dual antagonist (BMS-813160) with a cancer vaccine: GVAX (tumor cells genetically modified to secrete granulocyte-macrophage colony stimulating factor (GM-CSF) for DC cell activation) (NCT03767582). One of the main purposes of this study is to see the effect of GVAX by enhancing the infiltration of $CD8^+/CD137(4\text{-}1BB)^+$ cells in PDAC. It is also interesting whether CCR2 or CCR2/CCR5 blockade itself has therapeutic effects in pancreatic cancer by reprogramming the tumor microenvironment. A decrease in circulating monocytes may have promising therapeutic effects in pancreatic cancer, since PF-04136309 in combination with FOLFILINOX does not result in additional toxicity, but leads to a reduction in the infiltration of TAMs or T_{reg} cells, more $CD4^+$ and $CD8^+$ T cells are observed.[61] The PF-04136309/FOLFILINOX combination treatment enhances chemotherapeutic efficacy and improves survival in mice with orthotopic tumors derived from the Ptf1-Cre promoter-regulated *Kras*^{G12D}-expressing and *Trp53*-deficient mice.[62]

Substantial evidence suggests that inhibition of colony stimulating factor (CSF)-1/CSF-1 receptor (CSF-1R) signaling is one of most promising approaches to target macrophage activation.[63] Targeting TAMs with an anti-CSF-1R antibody called RG7155 efficiently depletes CSF-$1R^+/CD163^+$ macrophages in the liver and colon of monkeys.[64] Furthermore, CSF-1/CSF-1R blockade by a CSF-1-neutralizing antibody or by the CSF-1R inhibitor pexidartinib (PLX3397, Plexxikon) leads to a reduction of TAMs in orthotopically implanted PDAC tumors in mice.[65] Treatment with a CSF-1-neutralizing antibody in combination with clodronate-liposomes for depleting tissue-resident macrophages leads to a significant reduction in pancreatic tumor burden in the Ptf1-Cre promoter-regulated *Kras*^{G12D}-expressing and *Trp53*-deficient mice.[60] CSF-1R inhibition with the small molecule inhibitor AZD7507 depletes $CD11b^+/F4/80^+$ macrophages, regresses tumor growth, and prolongs survival in KPC mice.[66] In addition, macrophage depletion restores the capacity of gemcitabine and a CD40 agonist to induce T cell-dependent tumor regression in KPC mice.[67] Taken together, targeting the CSF-1/CSF-1R axis has been shown to eliminate TAMs and repress pancreatic cancer in several preclinical models. In line with this, a phase 1 clinical study to evaluate the safety and activity of pexidartinib with the anti PD-L1 antibody durvalumab in patients with metastatic, advanced pancreatic or colorectal cancers has been

Table 3 Clinical trials targeting stroma/microenvironment in pancreatic cancer.

Intervention/Treatment	NCT number	Stage of clinical trial	Recruitment status	Last update
CCX872-B Stereotactic body radiation	NCT03778879	Phase 1, 2	Not yet recruiting	December 19, 2018
BMS-813160 Nivolumab GVAX Stereotactic body radiation	NCT03767582	Phase 1, 2	Not yet recruiting	December 6, 2018
Pexidartinib durvalumab	NCT02777710	Phase 1	Active, not recruiting	January 17, 2019
Pexidartinib pembrolizumab	NCT02452424	Phase 1, 2	Completed	November 15, 2018
Emactuzumab (RG7155) RO7009789	NCT02760797	Phase 1	Completed	May 22, 2018
IMC-CS4 Pembrolizumab Cyclophosphamide GVAX	NCT03153410	Early Phase 1	Recruiting	November 6, 2018
Cabiralizumab Nab-paclitaxel Onivyde Nivolumab Fluorouracil Gemcitabine Oxaliplatin Leucovorin Irinotecan Hydrochloride	NCT03336216	Phase 2	Recruiting	December 20, 2018
Cabiralizumab nivolumab stereotactic body radiation	NCT03599362	Phase 2	Recruiting	December 19, 2018
Cabiralizumab gemcitabine nivolumab (OPDIVO)	NCT03697564	Phase 2	Not yet recruiting	January 9, 2019

Continued

Table 3 Clinical trials targeting stroma/microenvironment in pancreatic cancer.—cont'd

Intervention/Treatment	NCT number	Stage of clinical trial	Recruitment status	Last update
MCS110 spartalizumab (PDR001)	NCT02807844	Phase 1, 2	Recruiting	October 11, 2018
ARRY-382 pembrolizumab	NCT02880371	Phase 1, 2	Recruiting	October 18, 2018
GSK3145095 pembrolizumab	NCT03681951	Phase 1, 2	Recruiting	January 9, 2019
Defactinib pembrolizumab	NCT02758587	Phase 1, 2	Recruiting	March 19, 2018
Defactinib pembrolizumab gemcitabine	NCT02546531	Phase 1	Recruiting	November 1, 2018
Defactinib pembrolizumab	NCT03727880	Phase 2	Not yet recruiting	December 13, 2018
Plerixafor	NCT02179970	Phase 1	completed	December 24, 2018

initiated (NCT02777710). Treatment with ARRY-382, another CSF-1R inhibitor, in combination with the PD-1 antagonist pembrolizumab, has been considered for the treatment of patients with advanced solid tumors (Phase 1/2, NCT02880371). Another phase 1 study with RG7155 and a CD40 agonist RO7009789 in patients with advanced tumors including pancreatic cancer has been completed (NCT02760797). In a preclinical study, treatment with an agonistic CD40 antibody (FGK45, BioXcell) in combination with gemcitabine/nab-paclitaxel induces T-cell immunity and reverses the complete resistance to anti-PD-1 (RMP1–14, BioXcell) and anti-CTLA-4 (9H10, BioXcell) treatments for subcutaneously injected pancreatic tumors derived from KPC mice.[68] IMC-CS4 (LY3022855) as a CSF-1R inhibitor combined with cyclophosphamide, GVAX (see above), and pembrolizumab has been tested in patients with borderline resectable pancreatic cancer (early phase 1, NCT03153410). A CSF-1 inhibitor MCS110 and other anti-PD1 monoclonal antibody spartalizumab (PDR001) has been analyzed in a phase 1/2 study for patients with advanced malignancies including pancreatic cancer (NCT02807844). A phase 2 study is trying to determine whether therapy with the CSF-1R inhibitor cabiralizumab (FPA008) in combination with nivolumab, with or without chemotherapy, is effective for the treatment of advanced pancreatic cancer (NCT03336216). Additional studies include cabiralizumab with nivolumab and stereotactic body radiotherapy for locally advanced unresectable pancreatic cancer (Phase 2, NCT03599362) and cabiralizumab with nivolumab and gemcitabine (*versus* gemcitabine) for patients with stage IV pancreatic cancer (GemCaN trial) (NCT03697564). In summary, several inhibitors have been generated targeting CSF-1 or CSF-1R. For patients with pancreatic cancer, most of clinical studies have been designed as combination therapies with anti-PD-1 or anti-PD-L1 antibodies.

Other approaches targeting macrophages include targeting nucleotide oligomerization domain (NOD), leucine rich repeat (LRR) and pyrin domain-containing 3 (NLRP3) signaling, receptor-interacting serine/threonine protein kinase 1 (RIP1), and dectin 1. NLRP3 is best-characterized inflammasome sensor molecule. Once activated, NLRP3 induces the assembly of an inflammasome, leading to caspase 1-mediated proteolytic activation of the interleukin-1β (IL-1β) family of cytokines.[69] NLRP3 signaling is highly activated in $CD206^+$ macrophages in pancreata of Ptf1-Cre $Kras^{G12D}$ mice, and NLRP3 is highly expressed in pancreatic cancer patients. NLRP3 deletion inhibits oncogenic KRAS-induced pancreatic fibrosis and improves survival of mice. Furthermore, growth of orthotopically implanted pancreatic tumor from KPC mice is significantly reduced in pancreata of NLRP3-deficient mice.[70] A small molecule inhibitor of the NLRP3 inflammasome, MCC950, has been generated[71] and shown to attenuate colonic inflammation and IL-1β release in colitis in a mouse model.[72] It is of interest whether NLRP3 inhibition together with MCC950 attenuates pancreatic cancer development. RIP1 kinase is a key upstream regulator that controls inflammatory signaling as well as the activation of multiple cell death pathways such as apoptosis and necroptosis.[73] RIP1 is highly expressed in TAMs in PDAC, and oral treatment with a small-molecule RIP1 inhibitor called GSK547 ((*S*)-6-(4-(5-(3,5-difluorophenyl)-4,5-dihydro-1H-pyrazole-1-carbonyl)piperidin-1-yl)pyrimidine-4-carbonitrile) in $Kras^{G12D}$-expressing mice (under the control of *Ptf-1* promoter) attenuates fibrosis and tumor development and improves survival.[74] A phase 1/2 study with GSK3145095 (also a RIP-1 inhibitor) combined with the PD-1 antagonist pembrolizumab has been initiated to test whether RIP-1 inhibition in human pancreatic cancer subjects will modulate the immune infiltrate to sensitize tumors to immune checkpoint blockade (NCT03681951). Another factor that is highly expressed on macrophages in PDAC is dectin-1.[75] Dectin-1, also known as C-type lectin domain containing

7A (CLEC7A), is a non-toll-like receptor (TLR) pattern-recognition receptor (PRR) which is expressed on macrophages, DCs, and neutrophils.[76,77] Dectin-1 recognizes β-glucans that are mainly found in the cell walls of fungi and some bacteria[76] Ligation of dectin-1 by β-glucans phosphorylates spleen tyrosine kinase (Srk), recruits caspase recruitment domain-containing protein 9 (CARD9), which then forms a complex with B cell leukemia lymphoma 10 (Bcl10) and mucosa-associated lymphoid tissue 1 (Malt1) for activating immune response and ERK activation.[78] Dectin-1 is highly expressed in macrophages of *Ptf-1-Kras*^{G12D}-expressing mice and in the KPC mice as well as in human pancreatic cancer.[75] A member of the β-galactoside-binding family of lectins and a ligand for dectin-1, galectin-9, is highly expressed in pancreatic cancer-infiltrating myeloid cells and in cancer cells. Galectin-9 blockade by monoclonal antibody treatment extends survival in mice harboring orthotopic KPC tumors. Consistently, dectin-1 deletion decreases macrophage infiltration, recovers T cell differentiation, and attenuates pancreatic cancer development in *Ptf-1-Kras*^{G12D}-expressing mice.[75]

7. Reprogramming the microenvironment by targeting regulators expressed by pancreatic cancer cells or stromal cells

The immunosuppressive tumor microenvironment with poor $CD8^+$ cytotoxic T cell infiltration in pancreatic cancer is in part driven by factors from tumor cells. Focal adhesion kinase (FAK) is a cytoplasmic protein tyrosine kinase that is overexpressed and activated in several cancer types.[79] FAK is hyperactivated (phosphorylated) in human PDAC, and patients with high phosphorylated FAK levels and a low $CD8^+$ cytotoxic T lymphocyte count have shorter overall survival.[80] Administration of a FAK inhibitor (PF-562,271) leads to attenuation of macrophage infiltration and orthotopically implanted pancreatic tumor growth in mice.[81] Targeting FAK by another inhibitor, *VS*-4718, limits infiltration of $F4/80^+$ and $CD206^+$ macrophages, fibrosis, and tumor progression, resulting in a doubling of survival of *Ptf1-Cre-Kras*^{G12D}-*Trp53*$^{flox/+}$ mice. Furthermore, FAK inhibition renders PDACs more sensitive to chemo- and immunotherapy. The combination of the FAK inhibitor plus gemcitabine doubles the median survival duration of *Ptf1-Cre-Kras*^{G12D}-*Trp53*$^{flox/flox}$ mice. Treatment of the mice with gemcitabine, a PD1 antagonist (RMP1–14, BioXCell), a CTLA4 antagonist (UC10-4F10–11, BioXCell) does not alter survival compared to vehicle-treated mice. By contrast, additional FAK inhibitor treatment in combination with gemcitabine/PD-1 antagonist/CTLA antagonist results in a 2.5-fold increase in median survival.[80] Several clinical studies targeting FAK with defactinib have been initiated. Defactinib in combination with pembrolizumab has been tested in a phase 1/2 study for treating non-small-cell lung cancer, mesothelioma and pancreatic neoplasms (NCT02758587). A phase 1 study aims to treat patients with advanced solid tumors and pancreatic cancer with defactinib, pembrolizumab and gemcitabine (NCT02546531). A phase 2 study tests the anti-tumor activity and ability to increase the immune system by pembrolizumab with or without defactinib following chemotherapy as neoadjuvant or adjuvant treatment for resectable pancreatic cancer (NCT03727880).

Fibroblast activation protein (FAP) is one of the four members of the dipeptidyl peptidase (DPP) four protein family. This serine protease has dipeptidyl peptidase and endopeptidase activity. FAP expression has been observed during wound healing, at sites of inflammation, and in cirrhosis.[82] FAP is also frequently (90%) expressed, predominantly in adjacent tumor-associated myofibroblasts, in patients with pancreatic cancer.[83] Ablation of FAP-positive stromal cells reduced pancreatic tumor growth derived from the KPC mice in a transplanted murine tumor model.[84] FAP-positive cancer-associated fibroblasts are a major source of CCL2. FAP-positive fibroblasts attract Gr-1$^+$/CD11b$^+$ cells, and co-injection of tumor cells and FAP-positive fibroblasts into immunocompetent mice increases the frequency of macrophage infiltration *in vivo*.[85] As mentioned, the CCL2/CCR2 axis plays an important role in monocyte recruitment (see Section 6). To this end, anti-CCL2 antibody or CCL2 knockdown blocks migration of Gr-1$^+$/CD11b$^+$ cells toward supernatants from FAP-positive fibroblasts. Furthermore, FAP-positive fibroblasts fail to promote tumor growth in CCR2-deficient mice.[85] FAP-positive cells also highly express another chemokine, CXCL12.[86] CXCL12 is the primary chemokine ligand for CXCR4. The CXCL12/CXCR4 axis promotes pancreatic cancer development, invasion, and metastasis.[87,88] Treatment with a CXCR4 inhibitor AMD3100 (also known as Plerixafor) attenuates pancreatic cancer development in the KPC mice. Furthermore, AMD3100 treatment in combination with αPD-L1 amplifies the inhibitory effect on cancer development.[86] A phase 1 trial with Plerixafor in patients with advanced pancreatic, ovarian, or colorectal cancers (CAM-PLEX study) has been recently completed but the results have not been published yet (NCT02179970).

In summary, targeting regulators expressed by cancer cells, cancer-associated fibroblasts, or by stellate cells are promising therapeutic strategies. Deletion of cancer-associated fibroblasts however may have tumor promoting effects. Deletion of alpha-smooth muscle actin (αSMA)-positive cells in the Ptf1-Cre promoter-regulated *Kras*^{G12D}-expressing and transforming growth factor beta receptor 2 (*Tgfbr2*)-deficient pancreatic cancer mouse model leads to invasive, undifferentiated tumors with enhanced hypoxia and diminishes animal survival. Deletion of αSMA-positive cells also diminishes animal survival in KPC mice.[89] Sonic hedgehog (Shh) is known to drive formation of a fibroblast-rich desmoplastic stroma. Deletion of *Shh* reduces stromal content, but tumors are more aggressive, undifferentiated, and display increased vascularity in the *Pdx1-Cre-Kras*^{G12D}*-Trp53*$^{flox/+}$ pancreatic cancer mouse model.[90]

8. Adaptive T-cell therapy for pancreatic cancer

Strategies of immunotherapy for pancreatic cancer, such as reactivation of NK cells, development or reintroduction of DCs, vaccination strategies with DCs, reprogramming the microenvironment by targeting macrophages and regulators expressed by cancer cells or by stromal cells, can change the tumor microenvironment and support T-cell immunity. In this section, immunotherapies with chimeric antigen receptor (CAR) T cells will be summarized. The CAR is designed for specific recognition of tumor antigens. To generate the appropriate cell therapy product for adaptive transfer, T cells are collected from patients *via* leukapheresis, manipulated to target specific antigens, expanded, and reinfused. This approach is widely utilized as an individualized therapy with genetically modified autologous T cells from patients.[91] A CAR is a modulator protein consisting of an extracellular component that recognizes a cell surface protein and an intracellular component with specific signaling domains from proteins such as TCRζ, CD28 and CD137 (4-1BB) for T cell activation.[92] Several tumor antigens, including mesothelin, CD24, carcinoembryonic antigen (CEA), and human epidermal growth factor receptor 2 (HER2), have been considered as targets in pancreatic cancer.

Mesothelin is a cell surface glycoprotein highly expressed on cancers, including pancreatic cancer,[93] and overexpression of mesothelin increases cell proliferation and migration.[94] Co-expression of mesothelin and another membrane glycoprotein, CA-125, also known as mucin 16 (MUC16), correlates with significantly shorter relapse-free survival after surgical resection in pancreatic cancer patients.[95] CA-125/MUC16 expression is positively correlated with the oncogenic *KRAS*^{G12D} mutation in pancreatic cancer patients.[96] A phase 1 clinical trial using autologous mesothelin-specific

CAR T cells (CARTmeso cells) has demonstrated that patients did not develop a cytokine release syndrome or neurologic symptoms and that two out of six patients with chemotherapy-refractory metastatic PDAC showed stable disease.[97] A study of CARTmeso cells for patients with metastatic pancreatic cancer, where autologous T cells are lentivirally transduced to express anti-mesothelin scFv fused to TCRζ and CD137 (4-1BB) costimulatory domains (NCT03638193) (Table 4), and a study with lentivirally transduced CARTmeso cells in patients with histologically confirmed unresectable or metastatic pancreatic cancer (NCT03323944), have been initiated.

CD24 is a small, heavily glycosylated, mucin-like glycosylphosphatidylinositol-anchored cell surface protein that has been shown to be expressed in 72% of PDAC and to correlate with higher tumor grades.[98] Along with CD44 and CD133, CD24 is a cancer stem cell marker.[99] Administration with CAR T cells containing anti-CD24 single chain Fv and CD28 extracellular domain, in combination with irradiation, IL-2 injection, eliminates tumor in an orthotopic human pancreatic cancer xenograft model.[100]

Carcinoembryonic antigen (CEA) is a 180 kDa heavily glycosylated membrane protein involved in cell adhesion and expressed in pancreatic cancer cells.[101] In a preclinical setting, mouse CAR T-cells have been generated that harbor the murine anti-CEA scFV in the extracellular domain for targeting and the CD28 and CD3ζ signaling endodomains for T-cell activation. Administration of murine CAR T-cells reduces the size of orthotopically injected CEA-expressing pancreatic cancer.[102] In a clinical trial,

Table 4 Clinical trials with CAR T-cells in pancreatic caner.

Intervention/ Treatment	NCT number	Stage of clinical trial	Recruitment status	Last update
CART-meso (mesothelin) cells	NCT03638193	Not applicable	Recruiting	December 27, 2018
huCART-meso (mesothelin) cells	NCT03323944	Phase 1	Recruiting	November 29, 2018
Anti-CEA-CAR T	NCT02349724	Phase 1	Recruiting	April 25, 2017
CAR T cell	NCT03267173	Early phase 1	Recruiting	August 30, 2017
Anti-HER2 CAR-T	NCT02713984	Phase 1, 2	Recruiting	April 25, 2017

treatment with CEA (CEACAM5)-specific CAR T cells containing anti-CEA single chain Fv fused to the CD3ζ chain of the T-cell receptor/CD3 complex were short-lived and treatment induced acute respiratory toxicity with elevated levels of IL-6.[103] Currently another clinical study is ongoing to verify the safety of CEA targeted CAR T-cells for targeting several cancer types including pancreatic cancer (NCT02349724).

Human epidermal growth factor receptor 2 (HER2), also known as CD340, proto-oncogene Neu, or ERBB2, has been linked to aggressiveness in breast, lung, and gastric carcinomas. Furthermore, HER2 overexpression correlates with survival after curative resection of pancreatic cancer.[104] In a murine preclinical study, CARs were constructed in different configurations. The scFv was linked through the extracellular Fc domain to the transmembrane and intracellular CD28 domain followed by the CD3ζ chain in one configuration (αHER2-743). In the other configuration, the CAR is composed of the scFv of trastuzumab (Herceptin, monoclonal antibody against HER2), the CD8 extracellular hinge, and the intracellular CD28, CD137 (4-1BB), and CD3ζ signaling domains (αHER2-4D5). Administration of αHER2-743 CAR T-cells showed an antitumor activity in subcutaneously transplanted tumors. Administration of αHER2-4D5 dramatically reduced the size of the tumors transplanted in the murine pancreas.[100] A phase 1 CAR T-cell study is evaluating the efficacy of CAR-T targeting several factors, including HER2, in the treatment of pancreatic cancer (NCT03267173) and a phase 1/2 study target sHER2-positive cancers, including pancreatic cancer (NCT02713984).

9. Safety considerations in CAR T-cell therapy in pancreatic cancer

The use of CAR T-cells can trigger a range of side effects, particularly severe complications like cytokine release syndrome, where the infused CAR T-cells release cytokines causing a constellation of symptoms including fever and hypotension. CAR T-cells can also elicit neurologic toxicity. The ideal target antigen is restricted to the tumor cells, but most targets of CAR T-cells are also expressed on normal tissues, inducing so-called on-target/off-tumor toxicity on nonpathogenic tissues.[105,106] As a case report example of "on-target/off-tumor" recognition, a patient with colon cancer metastatic to the lungs and liver received HER2-specific CAR T-cells. Prior lymphodepleting conditioning was chosen to improve the efficacy of the CAR T-cell treatment. These CAR T-cells contained single chain Fv from trastuzumab (Herceptin) linked the CD8α-chain hinge and transmembrane regions with CD28, CD137 (4-1BB), and CD3ζ intracellular signaling domains (αHER2-4D5), which are inserted into a retroviral vector (MSGV1). Following treatment with the CAR T-cells, the patient developed rapid respiratory distress, displayed a pulmonary infiltrate, decreased blood oxygen saturation, severe hypotension, multi-organ dysfunction, bradycardia as well as gastrointestinal bleeding resulting in cardiac arrest.[107] However, another clinical study in which patients with refractory HER2-positive sarcoma received escalating doses of CAR T-cells concluded that HER2 CAR T-cells are well tolerated with no dose-limiting toxicity. Prior lymphodepletion was not performed. In this study, HER2-specific CAR T-cells containing an ectodomain derived from the HER2-specific monoclonal antibody (FRP5) and a CD28ζ endodomain in patients with refractory HER2-positive sarcoma were used.[108] A reason for the contradictory outcomes between the two HER2 CAR T-cell studies could be due to lymphodepletion and/or different configuration of the HER2 CAR.[108] A clinical challenge is the degree of immunosuppression that limits both the ability of CAR T-cells to effectively traffic to the tumor site and the persistence of CAR T cells. To that end, lymphodepletion has been shown to enhance expansion and persistence of the CAR T cells.[91] Lymphodepletion has so far been included in most CAR T-cell therapy protocols, and variable levels of persistence have been observed.[109] Nonetheless, the safety of lymphodepletion for conventional CAR T-cell therapies has to be considered. A recent breakthrough, *viz.* "switchable" CAR T-cells, may improve safety of CAR T-cell therapy and may overcome the "on-target/off-tumor" problem. To generate a "switchable" system, HER2-specific antigen binding fragment (Fab) fused with peptide neoepitope (PNE) tag and CAR T-cells targeting PNE has been generated. The activity of the CAR T-cell can therefore be controlled by dosage of a HER2-specific recombinant Fab-mediated "switch." It has been shown that treatment with switchable HER2 CAR T-cells eradiated pancreatic cancer cells derived from patients with stage IV PDAC in an orthotopic model. The primary patient cancer cells used in the study expressed HER2 at relatively modest levels and established extremely aggressive primary tumors with rapid-onset metastasis to the liver and lung, suggesting that both local and disseminated pancreatic cancer can be targeted by the switchable CAT T-system even when expressing relatively modest levels of HER2.[110] As mentioned in Section 7, ablation of FAP-positive stromal cells reduced PDAC growth derived from the KPC mice in a transplanted murine tumor model, hence FAP can be an attractive target for CAR T-cell therapy. The adoptive transfer of FAP-targeted CAR-transduced T cells into lymphodepleted mice (Rag1-deficient mice[111]) bearing a stroma-rich, intensely FAP-positive pancreatic cancer xenograft leads to cachexia and significant morbidity.[112] These CAR T-cells contain single chain Fv from FAP-specific antibody FAP5 linked the CD28, CD137 (4-1BB), and CD3ζ intracellular signaling domains, which are inserted into a retroviral vector (MSGV1). Osteogenic cells of non-hematopoietic origin are enriched with FAP^+ cells,[112] an "on-target/off-tumor" problem. Selecting proper target and vector and improving "on-target/off-tumor" effects may achieve effective treatment outcomes in pancreatic cancer.

10. Conclusion

Recent advances in the understanding of the tumor immune microenvironment have created great opportunities for the development of more effective immunotherapy approaches and combination strategies with chemotherapy for pancreatic cancer patients. Furthermore, from preclinical cancer models to clinical trials, it has now become evident that therapies with dendritic cells, natural killer cells, and chimeric antigen receptor T cells are highly promising for pancreatic cancer patients. Further reducing off target effects of chimeric antigen receptor T cell therapies and establishing combination therapy may result in better and more effective treatment outcomes in pancreatic cancer in the future.

References

1. Rahib L, Smith BD, Aizenberg R, Rosenzweig AB, Fleshman JM, Matrisian LM. Projecting cancer incidence and deaths to 2030: the unexpected burden of thyroid, liver, and pancreas cancers in the United States. *Cancer Res.* 2014;74:2913–2921.
2. Kleeff J, Korc M, Apte M, et al. Pancreatic cancer. *Nat Rev Dis Primers.* 2016;2:16022.
3. Cunningham D, Chau I, Stocken DD, et al. Phase III randomized comparison of gemcitabine versus gemcitabine plus capecitabine in patients with advanced pancreatic cancer. *J Clin Oncol.* 2009;27:5513–5518.
4. Conroy T, Desseigne F, Ychou M, et al. FOLFIRINOX versus gemcitabine for metastatic pancreatic cancer. *N Engl J Med.* 2011;364(19):1817–1825.
5. Von Hoff DD, Ervin T, Arena FP, et al. Increased survival in pancreatic cancer with nab-paclitaxel plus gemcitabine. *N Engl J Med.* 2013;369(18):1691–1703.
6. Neoptolemos JP, Palmer DH, Ghaneh P, et al. Comparison of adjuvant gemcitabine and capecitabine with gemcitabine monotherapy in patients with resected pancreatic cancer (ESPAC-4): a multicentre, open-label, randomised, phase 3 trial. *Lancet.* 2017;389(10073):1011–1024.
7. Conroy T, Hammel P, Hebbar M, et al. FOLFIRINOX or gemcitabine as adjuvant therapy for pancreatic cancer. *N Engl J Med.* 2018;379(25):2395–2406.

8. Emens LA, Ascierto PA, Darcy PK, et al. Cancer immunotherapy: opportunities and challenges in the rapidly evolving clinical landscape. *Eur J Cancer*. 2017;81:116–129.
9. Hanahan D, Weinberg RA. Hallmarks of cancer: the next generation. *Cell*. 2011;144:646–674.
10. Burnet M. Cancer: a biological approach. III. Viruses associated with neoplastic conditions. IV. Practical applications. *Br Med J*. 1957;1(5023):841–847.
11. Dunn GP, Old LJ, Schreiber RD. The three Es of cancer immunoediting. *Annu Rev Immunol*. 2004;22:329–360.
12. Dunn GP, Bruce AT, Ikeda H, Old LJ, Schreiber RD. Cancer immunoediting: from immunosurveillance to tumor escape. *Nat Immunol*. 2002;3:991–998.
13. Schreiber RD, Old LJ, Smyth MJ. Cancer immunoediting: integrating immunity's roles in cancer suppression and promotion. *Science*. 2011;331:1565–1570.
14. Detjen KM, Farwig K, Welzel M, Wiedenmann B, Rosewicz S. Interferon gamma inhibits growth of human pancreatic carcinoma cells via caspase-1 dependent induction of apoptosis. *Gut*. 2001;49:251–262.
15. Martinet L, Smyth MJ. Balancing natural killer cell activation through paired receptors. *Nat Rev Immunol*. 2015;15:243–254.
16. Funa K, Nilsson B, Jacobsson G, Alm GV. Decreased natural killer cell activity and interferon production by leucocytes in patients with adenocarcinoma of the pancreas. *Br J Cancer*. 1984;50:231–233.
17. Peng YP, Xi CH, Zhu Y, et al. Altered expression of CD226 and CD96 on natural killer cells in patients with pancreatic cancer. *Oncotarget*. 2016;7:66586–66594.
18. Raedler L. Velcade (Bortezomib) receives 2 new FDA indications: for retreatment of patients with multiple myeloma and for first-line treatment of patients with mantle-cell lymphoma. *Am Health Drug Benefits*. 2015;8:135–140.
19. Nawrocki ST, Carew JS, Pino MS, et al. Bortezomib sensitizes pancreatic cancer cells to endoplasmic reticulum stress-mediated apoptosis. *Cancer Res*. 2005;65:11658–11666.
20. Van Audenaerde JRM, De Waele J, Marcq E, et al. Interleukin-15 stimulates natural killer cell-mediated killing of both human pancreatic cancer and stellate cells. *Oncotarget*. 2017;8:56968–56979.
21. Palucka K, Banchereau J. Dendritic-cell-based therapeutic cancer vaccines. *Immunity*. 2013;39(1):38–48.
22. Worbs T, Hammerschmidt SI, Förster R. Dendritic cell migration in health and disease. *Nat Rev Immunol*. 2017;17:30–48.
23. Meyer MA, Baer JM, Knolhoff BL, et al. Breast and pancreatic cancer interrupt IRF8-dependent dendritic cell development to overcome immune surveillance. *Nat Commun*. 2018;9:1250.
24. Schnurr M, Galambos P, Scholz C, et al. Tumor cell lysate-pulsed human dendritic cells induce a T-cell response against pancreatic carcinoma cells: an in vitro model for the assessment of tumor vaccines. *Cancer Res*. 2001;61:6445–6450.
25. Mehrotra S, Britten CD, Chin S, et al. Vaccination with poly(IC:LC) and peptide-pulsed autologous dendritic cells in patients with pancreatic cancer. *J Hematol Oncol*. 2017;10:82.
26. Bauer C, Bauernfeind F, Sterzik A, et al. Dendritic cell-based vaccination combined with gemcitabine increases survival in a murine pancreatic carcinoma model. *Gut*. 2007;56:1275–1282.
27. Hirooka Y, Kawashima H, Ohno E, et al. Comprehensive immunotherapy combined with intratumoral injection of zoledronate-pulsed dendritic cells, intravenous adoptive activated T lymphocyte and gemcitabine in unresectable locally advanced pancreatic carcinoma: a phase I/II trial. *Oncotarget*. 2017;9:2838–2847.
28. Tassone P, Tagliaferri P, Viscomi C, et al. Zoledronic acid induces antiproliferative and apoptotic effects in human pancreatic cancer cells in vitro. *Br J Cancer*. 2003;88:1971–1978.

29. Smith-Garvin JE, Koretzky GA, Jordan MS. T cell activation. *Annu Rev Immunol.* 2009;27:591–619.
30. Pardoll DM. The blockade of immune checkpoints in cancer immunotherapy. *Nat Rev Cancer.* 2012;12:252–264.
31. Rowshanravan B, Halliday N, Sansom DM. CTLA-4: a moving target in immunotherapy. *Blood.* 2018;131:58–67.
32. Tivol EA, Borriello F, Schweitzer AN, Lynch WP, Bluestone JA, Sharpe AH. Loss of CTLA-4 leads to massive lymphoproliferation and fatal multiorgan tissue destruction, revealing a critical negative regulatory role of CTLA-4. *Immunity.* 1995;3:541–547.
33. Waterhouse P, Penninger JM, Timms E, et al. Lymphoproliferative disorders with early lethality in mice deficient in Ctla-4. *Science.* 1995;270:985–988.
34. Ise W, Kohyama M, Nutsch KM, et al. CTLA-4 suppresses the pathogenicity of self antigen-specific T cells by cell-intrinsic and cell-extrinsic mechanisms. *Nat Immunol.* 2010;11:129–135.
35. Shitara K, Nishikawa H. Regulatory T cells: a potential target in cancer immunotherapy. *Ann N Y Acad Sci.* 2018;1417:104–115.
36. Wing K, Onishi Y, Prieto-Martin P, et al. CTLA-4 control over Foxp3+ regulatory T cell function. *Science.* 2008;322:271–275.
37. Leach DR, Krummel MF, Allison JP. Enhancement of antitumor immunity by CTLA-4 blockade. *Science.* 1996;271:1734–1736.
38. Ledford H. Melanoma drug wins US approval. *Nature.* 2011;471:561.
39. Ishida Y, Agata Y, Shibahara K, Honjo T. Induced expression of PD-1, a novel member of the immunoglobulin gene superfamily, upon programmed cell death. *EMBO J.* 1992;11:3887–3895.
40. Nishimura H, Nose M, Hiai H, Minato N, Honjo T. Development of lupus-like autoimmune diseases by disruption of the PD-1 gene encoding an ITIM motif-carrying immunoreceptor. *Immunity.* 1999;11:141–151.
41. Francisco LM, Sage PT, Sharpe AH. The PD-1 pathway in tolerance and autoimmunity. *Immunol Rev.* 2010;236:219–242.
42. Greenwald RJ, Freeman GJ, Sharpe AH. The B7 family revisited. *Annu Rev Immunol.* 2005;23:515–548.
43. Wang X, Teng F, Kong L, Yu J. PD-L1 expression in human cancers and its association with clinical outcomes. *Onco Targets Ther.* 2016;9:5023–5039.
44. Iwai Y, Hamanishi J, Chamoto K, Honjo T. Cancer immunotherapies targeting the PD-1 signaling pathway. *J Biomed Sci.* 2017;24:26.
45. Hsu FS, Su CH, Huang KH. A comprehensive review of US FDA-approved immune checkpoint inhibitors in urothelial carcinoma. *J Immunol Res.* 2017;2017:6940546.
46. Bloom AC, Bender LH, Tarabe M, Berzofsky JA, Walters IB. INT230-6 shows strong synergy with anti-PD-1 and can induce high complete response rates with T cell memory response in a colon cancer mouse model. *J Immunother Cancer.* 2015;3(Suppl 2): P352.
47. Hingorani SR, Zheng L, Bullock AJ, et al. HALO 202: randomized phase II study of PEGPH20 plus nab-paclitaxel/gemcitabine versus nab-paclitaxel/gemcitabine in patients with untreated, metastatic pancreatic ductal adenocarcinoma. *J Clin Oncol.* 2018;36:359–366.
48. Le DT, Wang-Gillam A, Picozzi V, et al. Safety and survival with GVAX pancreas prime and listeria monocytogenes-expressing mesothelin (CRS-207) boost vaccines for metastatic pancreatic cancer. *J Clin Oncol.* 2015;33:1325–1333.
49. Mitchell TC, Hamid O, Smith DC, et al. Epacadostat plus pembrolizumab in patients with advanced solid tumors: phase I results from a multicenter, open-label phase I/II trial (ECHO-202/KEYNOTE-037). *J Clin Oncol.* 2018;36:3223–3230, JCO2018789602.

50. Laubach JP, Liu CJ, Raje NS, et al. A phase I/II study of evofosfamide, a hypoxia-activated prodrug with or without bortezomib in subjects with relapsed/refractory multiple myeloma. *Clin Cancer Res.* 2019;25:478–486.
51. Berek JS, Matulonis UA, Peen U, et al. Safety and dose modification for patients receiving niraparib. *Ann Oncol.* 2018;29:1784–1792.
52. Royal RE, Levy C, Turner K, et al. Phase 2 trial of single agent Ipilimumab (anti-CTLA-4) for locally advanced or metastatic pancreatic adenocarcinoma. *J Immunother.* 2010;33:828–833.
53. Brahmer JR, Tykodi SS, Chow LQ, et al. Safety and activity of anti-PD-L1 antibody in patients with advanced cancer. *N Engl J Med.* 2012;366:2455–2465.
54. Jenkins RW, Barbie DA, Flaherty KT. Mechanisms of resistance to immune checkpoint inhibitors. *Br J Cancer.* 2018;118:9–16.
55. Vonderheide RH, Bayne LJ. Inflammatory networks and immune surveillance of pancreatic carcinoma. *Curr Opin Immunol.* 2013;25:200–205.
56. Hingorani SR, Wang L, Multani AS, et al. Trp53R172H and KrasG12D cooperate to promote chromosomal instability and widely metastatic pancreatic ductal adenocarcinoma in mice. *Cancer Cell.* 2005;7:469–483.
57. Clark CE, Beatty GL, Vonderheide RH. Immunosurveillance of pancreatic adenocarcinoma: insights from genetically engineered mouse models of cancer. *Cancer Lett.* 2009;279:1–7.
58. Zhang Y, Velez-Delgado A, Mathew E, et al. Myeloid cells are required for PD-1/PD-L1 checkpoint activation and the establishment of an immunosuppressive environment in pancreatic cancer. *Gut.* 2017;66:124–136.
59. Tacke F. Targeting hepatic macrophages to treat liver diseases. *J Hepatol.* 2017;66:1300–1312.
60. Zhu Y, Herndon JM, Sojka DK, et al. Tissue-resident macrophages in pancreatic ductal adenocarcinoma originate from embryonic hematopoiesis and promote tumor progression. *Immunity.* 2017;47:323–338.e6.
61. Nywening TM, Wang-Gillam A, Sanford DE, et al. Targeting tumour-associated macrophages with CCR2 inhibition in combination with FOLFIRINOX in patients with borderline resectable and locally advanced pancreatic cancer: a single-centre, open-label, dose-finding, non-randomised, phase 1b trial. *Lancet Oncol.* 2016;17:651–662.
62. Nywening TM, Belt BA, Cullinan DR, et al. Targeting both tumour-associated $CXCR2^+$ neutrophils and $CCR2^+$ macrophages disrupts myeloid recruitment and improves chemotherapeutic responses in pancreatic ductal adenocarcinoma. *Gut.* 2018;67(6):1112–1123.
63. Hume DA, MacDonald KP. Therapeutic applications of macrophage colony-stimulating factor-1 (CSF-1) and antagonists of CSF-1 receptor (CSF-1R) signaling. *Blood.* 2012;119:1810–1820.
64. Ries CH, Cannarile MA, Hoves S, et al. Targeting tumor-associated macrophages with anti-CSF-1R antibody reveals a strategy for cancer therapy. *Cancer Cell.* 2014;25:846–859.
65. Zhu Y, Knolhoff BL, Meyer MA, et al. CSF1/CSF1R blockade reprograms tumor-infiltrating macrophages and improves response to T-cell checkpoint immunotherapy in pancreatic cancer models. *Cancer Res.* 2014;74:5057–5069.
66. Candido JB, Morton JP, Bailey P, et al. $CSF1R^+$ macrophages sustain pancreatic tumor growth through T cell suppression and maintenance of key gene programs that define the squamous subtype. *Cell Rep.* 2018;23:1448–1460.
67. Beatty GL, Winograd R, Evans RA, et al. Exclusion of T cells from pancreatic carcinomas in mice is regulated by Ly6C(low) F4/80(+) extratumoral macrophages. *Gastroenterology.* 2015;149:201–210.

68. Winograd R, Byrne KT, Evans RA, et al. Induction of T-cell immunity overcomes complete resistance to PD-1 and CTLA-4 blockade and improves survival in pancreatic carcinoma. *Cancer Immunol Res*. 2015;3:399–411.
69. Mangan MSJ, Olhava EJ, Roush WR, Seidel HM, Glick GD, Latz E. Targeting the NLRP3 inflammasome in inflammatory diseases. *Nat Rev Drug Discov*. 2018;17:688.
70. Daley D, Mani VR, Mohan N, et al. NLRP3 signaling drives macrophage-induced adaptive immune suppression in pancreatic carcinoma. *J Exp Med*. 2017;214:1711–1724.
71. Coll RC, Robertson AA, Chae JJ, et al. A small-molecule inhibitor of the NLRP3 inflammasome for the treatment of inflammatory diseases. *Nat Med*. 2015;21:248–255.
72. Perera AP, Fernando R, Shinde T, et al. MCC950, a specific small molecule inhibitor of NLRP3 inflammasome attenuates colonic inflammation in spontaneous colitis mice. *Sci Rep*. 2018;8:8618.
73. Ofengeim D, Yuan J. Regulation of RIP1 kinase signalling at the crossroads of inflammation and cell death. *Nat Rev Mol Cell Biol*. 2013;14:727–736.
74. Wang W, Marinis JM, Beal AM, et al. RIP1 kinase drives macrophage-mediated adaptive immune tolerance in pancreatic cancer. *Cancer Cell*. 2018;34:757–774.e7.
75. Daley D, Mani VR, Mohan N, et al. Dectin 1 activation on macrophages by galectin 9 promotes pancreatic carcinoma and peritumoral immune tolerance. *Nat Med*. 2017;23:556–567.
76. Brown GD. Dectin-1: a signalling non-TLR pattern-recognition receptor. *Nat Rev Immunol*. 2006;6:33–43.
77. Goodridge HS, Reyes CN, Becker CA, et al. Activation of the innate immune receptor Dectin-1 upon formation of a 'phagocytic synapse'. *Nature*. 2011;472:471–475.
78. Jia XM, Tang B, Zhu LL, et al. CARD9 mediates Dectin-1-induced ERK activation by linking Ras-GRF1 to H-Ras for antifungal immunity. *J Exp Med*. 2014;211:2307–2321.
79. Sulzmaier FJ, Jean C, Schlaepfer DD. FAK in cancer: mechanistic findings and clinical applications. *Nat Rev Cancer*. 2014;14:598–610.
80. Jiang H, Hegde S, Knolhoff BL, et al. Targeting focal adhesion kinase renders pancreatic cancers responsive to checkpoint immunotherapy. *Nat Med*. 2016;22:851–860.
81. Stokes JB, Adair SJ, Slack-Davis JK, et al. Inhibition of focal adhesion kinase by PF-562,271 inhibits the growth and metastasis of pancreatic cancer concomitant with altering the tumor microenvironment. *Mol Cancer Ther*. 2011;10:2135–2145.
82. Hamson EJ, Keane FM, Tholen S, Schilling O, Gorrell MD. Understanding fibroblast activation protein (FAP): substrates, activities, expression and targeting for cancer therapy. *Proteomics Clin Appl*. 2014;8:454–463.
83. Cohen SJ, Alpaugh RK, Palazzo I, et al. Fibroblast activation protein and its relationship to clinical outcome in pancreatic adenocarcinoma. *Pancreas*. 2008;37:154–158.
84. Kraman M, Bambrough PJ, Arnold JN, et al. Suppression of antitumor immunity by stromal cells expressing fibroblast activation protein-alpha. *Science*. 2010;330:827–830.
85. Yang X, Lin Y, Shi Y, et al. FAP promotes immunosuppression by cancer-associated fibroblasts in the tumor microenvironment via STAT3-CCL2 signaling. *Cancer Res*. 2016;76:4124–4135.
86. Feig C, Jones JO, Kraman M, et al. Targeting CXCL12 from FAP-expressing carcinoma-associated fibroblasts synergizes with anti-PD-L1 immunotherapy in pancreatic cancer. *Proc Natl Acad Sci U S A*. 2013;110:20212–20217.
87. Sleightholm RL, Neilsen BK, Li J, et al. Emerging roles of the CXCL12/CXCR4 axis in pancreatic cancer progression and therapy. *Pharmacol Ther*. 2017;179:158–170.
88. Nagarsheth N, Wicha MS, Zou W. Chemokines in the cancer microenvironment and their relevance in cancer immunotherapy. *Nat Rev Immunol*. 2017;17:559–572.

89. Özdemir BC, Pentcheva-Hoang T, Carstens JL, et al. Depletion of carcinoma-associated fibroblasts and fibrosis induces immunosuppression and accelerates pancreas cancer with reduced survival. *Cancer Cell.* 2014;25:719–734.
90. Rhim AD, Oberstein PE, Thomas DH, et al. Stromal elements act to restrain, rather than support, pancreatic ductal adenocarcinoma. *Cancer Cell.* 2014;25:735–747.
91. Akce M, Zaidi MY, Waller EK, El-Rayes BF, Lesinski GB. The potential of CAR T cell therapy in pancreatic cancer. *Front Immunol.* 2018;9:2166.
92. Beatty GL, O'Hara M. Chimeric antigen receptor-modified T cells for the treatment of solid tumors: defining the challenges and next steps. *Pharmacol Ther.* 2016;166:30–39.
93. Pastan I, Hassan R. Discovery of mesothelin and exploiting it as a target for immunotherapy. *Cancer Res.* 2014;74:2907–2912.
94. Li M, Bharadwaj U, Zhang R, et al. Mesothelin is a malignant factor and therapeutic vaccine target for pancreatic cancer. *Mol Cancer Ther.* 2008;7:286–296.
95. Einama T, Kamachi H, Nishihara H, et al. Co-expression of mesothelin and CA125 correlates with unfavorable patient outcome in pancreatic ductal adenocarcinoma. *Pancreas.* 2011;40:1276–1282.
96. Liang C, Qin Y, Zhang B, et al. Oncogenic KRAS targets MUC16/CA125 in pancreatic ductal adenocarcinoma. *Mol Cancer Res.* 2017;15:201–212.
97. Beatty GL, O'Hara MH, Lacey SF, et al. Activity of Mesothelin-specific chimeric antigen receptor T cells against pancreatic carcinoma metastases in a phase 1 trial. *Gastroenterology.* 2018;155:29–32.
98. Jacob J, Bellach J, Grützmann R, et al. Expression of CD24 in adenocarcinomas of the pancreas correlates with higher tumor grades. *Pancreatology.* 2004;4:454–460.
99. Li C, Heidt DG, Dalerba P, et al. Identification of pancreatic cancer stem cells. *Cancer Res.* 2007;67:1030–1037.
100. Maliar A, Servais C, Waks T, et al. Redirected T cells that target pancreatic adenocarcinoma antigens eliminate tumors and metastases in mice. *Gastroenterology.* 2012;143:1375–1384.e5.
101. Sarobe P, Huarte E, Lasarte JJ, Borrás-Cuesta F. Carcinoembryonic antigen as a target to induce anti-tumor immune responses. *Curr Cancer Drug Targets.* 2004;4: 443–454.
102. Chmielewski M, Hahn O, Rappl G, et al. T cells that target carcinoembryonic antigen eradicate orthotopic pancreatic carcinomas without inducing autoimmune colitis in mice. *Gastroenterology.* 2012;143:1095–1107.e2.
103. Thistlethwaite FC, Gilham DE, Guest RD, et al. The clinical efficacy of first-generation carcinoembryonic antigen (CEACAM5)-specific CAR T cells is limited by poor persistence and transient pre-conditioning-dependent respiratory toxicity. *Cancer Immunol Immunother.* 2017;66:1425–1436.
104. Komoto M, Nakata B, Amano R, et al. HER2 overexpression correlates with survival after curative resection of pancreatic cancer. *Cancer Sci.* 2009;100:1243–1247.
105. Brudno JN, Kochenderfer JN. Toxicities of chimeric antigen receptor T cells: recognition and management. *Blood.* 2016;127:3321–3330.
106. Bonifant CL, Jackson HJ, Brentjens RJ, Curran KJ. Toxicity and management in CAR T-cell therapy. *Mol Ther Oncolytics.* 2016;3:16011.
107. Morgan RA, Yang JC, Kitano M, Dudley ME, Laurencot CM, Rosenberg SA. Case report of a serious adverse event following the administration of T cells transduced with a chimeric antigen receptor recognizing ERBB2. *Mol Ther.* 2010;18:843–851.
108. Ahmed N, Brawley VS, Hegde M, et al. Human epidermal growth factor receptor 2 (HER2) -specific chimeric antigen receptor-modified T cells for the immunotherapy of HER2-positive sarcoma. *J Clin Oncol.* 2015;33:1688–1696.
109. Maus MV, June CH. Making better chimeric antigen receptors for adoptive T-cell therapy. *Clin Cancer Res.* 2016;22:1875–1884.

110. Raj D, Yang MH, Rodgers D, et al. Switchable CAR-T cells mediate remission in metastatic pancreatic ductal adenocarcinoma. *Gut*. 2018; https://doi.org/10.1136/gutjnl-2018-316595. [Epub ahead of print].
111. Mombaerts P, Iacomini J, Johnson RS, Herrup K, Tonegawa S, Papaioannou VE. RAG-1-deficient mice have no mature B and T lymphocytes. *Cell*. 1992;68:869–877.
112. Tran E, Chinnasamy D, Yu Z, et al. Immune targeting of fibroblast activation protein triggers recognition of multipotent bone marrow stromal cells and cachexia. *J Exp Med*. 2013;210:1125–1135.

CHAPTER SEVEN

Synergistic combination of oncolytic virotherapy with CAR T-cell therapy

Adam Ajina[a,*], John Maher[a,b,c]

[a]King's College London, Division of Cancer Studies, Guy's Hospital, London, United Kingdom
[b]Department of Clinical Immunology and Allergy, King's College Hospital NHS Foundation Trust, London, United Kingdom
[c]Department of Immunology, Eastbourne Hospital, East Sussex, United Kingdom
*Corresponding author: e-mail address: adam.ajina@nhs.net

Contents

1. Background 218
2. Oncolytic virotherapy—An overview 219
   2.1 Oncolytic virus diversity 220
   2.2 Oncolytic virus tumor tropism 221
   2.3 Recombinant OVs 224
   2.4 Options for OV administration 225
   2.5 Issues of OV clearance 226
3. The development of chimeric antigen receptor T-cell therapy 227
   3.1 Chimeric antigen receptor design 229
   3.2 Clinical data 230
   3.3 Barriers to CAR efficacy in solid tumors 231
   3.4 Safety concerns 232
4. Synergies between OVs and CAR T-cell therapy 234
   4.1 OV-mediated induction of local inflammation 234
   4.2 OV-induced immunogenic cell death 235
   4.3 OV vaccination strategies 238
5. Engineered OVs with payloads conducive to CAR function 239
   5.1 OV-mediated delivery of cytokines supportive of professional DCs 239
   5.2 OV-mediated delivery of cytokines favorable to CAR T-cell function 240
   5.3 OV-mediated delivery of CAR T-cell attracting chemokines 243
   5.4 OV expression of bispecific T-cell engagers 249
   5.5 CAR carriage of OV 251
   5.6 Chemokine matching 252
   5.7 Delivery of pro-T-cell cytokines and chemokines in combination 255
   5.8 Delivery of immune co-stimulatory molecules 255
   5.9 Delivery of molecules targeting immune checkpoints 256
   5.10 Delivery of molecules targeting immunosuppressive cells 261

Progress in Molecular Biology and Translational Science, Volume 164
ISSN 1877-1173
https://doi.org/10.1016/bs.pmbts.2019.06.015

5.11 Delivery of molecules targeting immunosuppressive metabolic pathways within the tumor microenvironment 265
5.12 Delivery of pro-apoptotic or cytotoxic molecules 268
5.13 Delivery of molecules capable of structurally altering the tumor microenvironment 269
6. Conclusions 273
Acknowledgments 273
Ethical approval and consent to participate 274
Consent for publication 274
Availability of supporting data 274
Competing interests 274
Funding 274
Authors' contributions 274
References 274

Abstract

For patients with advanced hematological malignancies the therapeutic landscape has been transformed by the emergence of adoptive cell transfer utilizing autologous chimeric antigen receptor (CAR)-redirected T-cells. However, solid tumors have proved far more resistant to this approach. Here, we summarize the numerous challenges faced by CAR T-cells designed to target solid tumors, highlighting, in particular, issues related to impaired trafficking, expansion, and persistence. In parallel, we draw attention to exciting developments in the burgeoning field of oncolytic virotherapy and posit strategies for the synergistic combination of oncolytic viruses with CAR T-cells to improve outcomes for patients with advanced solid tumors.

1. Background

Major achievements in our understanding of anti-tumoral immunity have led to a plethora of novel immunotherapeutic agents gaining regulatory approval for patients with advanced solid tumors and hematological malignancies over the past decade.[1] Foremost, we have witnessed striking clinical efficacy across a wide range of tumors with the use of monoclonal antibodies targeting the PD-1/PD-L1 axis and CTLA-4.[2] These immune checkpoint inhibitors have already had a transformational impact on the lives of a diverse group of patients with advanced malignant melanoma, non-small cell lung cancer, renal cell carcinoma, urothelial cancer, head and neck squamous cell carcinoma, Merkel cell carcinoma, microsatellite instability-high/mismatch repair deficient tumors, PD-L1$^+$ gastric or gastro-esophageal junction and classical Hodgkin lymphoma.[3] More recent approvals in 2018 have further extended the use of these agents for patients with small cell lung cancer, primary mediastinal large B-cell lymphoma, hepatocellular carcinoma and metastatic PD-L1$^+$ cervical cancer. On October 1, 2018 the Nobel Prize in Physiology or Medicine was jointly awarded to James P. Allison and Tasuku Honjo in recognition of their pioneering preclinical research culminating in the discovery of immune checkpoint inhibition as a potent new form of anti-cancer therapy.[4] Alongside these achievements and in concert with rapid developments seen more broadly in the field of biotechnology and genetic engineering, we have witnessed the landmark approvals of the first recombinant oncolytic virotherapy (T-VEC) for patients with locally advanced inoperable malignant melanoma based upon the phase III OPTiM trial[5] and the first chimeric antigen receptor (tisagenlecleucel) for children and young adults with relapsed or refractory B-cell ALL based upon the phase II ELIANA trial.[6] CAR T-cell therapy has subsequently demonstrated efficacy in patients with advanced diffuse large B-cell NHL and shown promise in patients with CLL, mantle cell lymphoma, multiple myeloma and Hodgkin's lymphoma.[7]

Despite these successes, ultimately only a minority of cancer patients will respond to immune checkpoint blockade,[8] and both oncolytic virotherapy and CAR T-cell therapy remain in a nascent stage of development. Indeed, the latter has thus far proven relatively ineffective in the management of solid tumors.[9] Considerable focus in the field has therefore led to the development of a huge number of therapeutic agents designed to augment patient responses and/or render a greater proportion of patients amenable to immunotherapy. Such agents have been designed to target the adaptive and/or innate immune system either by providing a direct boost to endogenous anti-tumoral immunity or by inhibiting negative regulators of response, potentially in synergistic combination with other approved or experimental agents. The list of such agents is long but includes inhibitors of novel immune checkpoints, costimulatory agonists, cytokines, antagonists of TME-related immunosuppression, neoantigen vaccines, oncolytic viruses and adoptive cell transfer using CAR or recombinant TCR-engineered T-cells.[1]

2. Oncolytic virotherapy—An overview

Historically, attempts to deploy oncolytic viruses have relied upon their capacity to directly kill tumor cells during their replication cycle, complemented by innate or engineered tumor-tropism.[10] Over recent years, however, greater potential has been attributed to their capacity to augment host anti-tumoral immunity by resculpting the tumor microenvironment as an in-situ vaccine.[11] This process can occur in a multitude of ways but at its core lies an ability to recruit and activate professional antigen presenting cells via the local release of tumor antigens and the triggering of danger signals including include small molecules (such as uric acid[12] and adenosine triphosphate (ATP)), and protein mediators such as high-mobility group box 1 (HMGB1).[13] Ultimately, viral-induced multimodal immunogenic cell death elicits an inflammatory cascade characterized by the release of type I interferons and other cytokines such as TNFα, and chemokines able to attract effector T-cells and other players involved in an adaptive immune response.[14] Antigen cross-presentation in the context of epitope spreading is thus capable of eliciting clonal T-cell expansion and ultimately tumor rejection. Epitopes shared at distant metastatic sites may also be targeted by the systemic recirculation of clonally expanded and activated effector T-cells irrespective of local viral infection. The intrinsic capacity of OVs to modulate the TME may be further enhanced by using recombinant engineering to insert transgenes encoding various pro-inflammatory cytokines, chemokines, immune checkpoint inhibitors and/or co-stimulatory molecules into viral virulence genes.[15] Taking into account their intrinsic qualities and the almost limitless potential for customization, OV's lend themselves particularly well to combination immunotherapy, with considerable scope for combination with cellular therapies designed to target solid tumors.

2.1 Oncolytic virus diversity

Currently, we are witnessing a renaissance in oncolytic virotherapy with a panoply of different OV vectors currently undergoing preclinical and clinical evaluation.[16,17] The search for efficacious clinical candidates spans numerous viral orders and families, including both DNA and RNA viruses, either in attenuated or unattenuated form. Of the DNA viruses, the most commonly studied include members of the families: adenoviridae (such as ONYX-015[18]); herpesviridae (including talimogene laherparepvec, NV1020 and G207[19]); poxviridae (including vaccinia strains such as pexastimogene devacirepvec[20] and myxoma viruses[21]); and parvoviridae (e.g., the rat parvovirus H-1PV).[22,23] Examples of RNA OV's include the picornaviridae (such as the coxsackievirus CVA21[24] or the polio/rhinovirus recombinant PVSRIPO[25] and Seneca Valley virus[26]); reoviridae (e.g., Reolysin, an unmodified type 3 Dearing strain reovirus[27]); togaviridae (particularly using certain alphavirus strains such as M1 and Sindbis AR339[28]); paramyxoviridae (including measles viruses[29] and Newcastle disease virus (NDV)[30]); and rhabdoviruses (e.g., vesicular stomatitis virus (VSV)[31] and Maraba virus[32]).[14] Viruses that display synergy with existing clinically approved anti-cancer agents have been prioritized for development and currently a vast array of early and late phase clinical trials are recruiting patients globally. Thus far, two OV's (namely talimogene laherparepvec (T-VEC)[5] and H101[33]) have gained regulatory approval worldwide. T-VEC is a first in class modified herpes simplex virus, type 1 (HSV-1) with two copies of the human cytokine granulocyte macrophage-colony stimulating factor (GM-CSF) gene[34] that won approval by the U.S. Food and Drug Administration (FDA) in October 2015 for the treatment of patients with local unresectable malignant melanoma based mainly upon significant overall survival data gleaned from the pivotal phase III OPTiM trial. H101, on the other hand, is a genetically modified oncolytic adenovirus that received approval in China in November 2005 in combination with cytotoxic chemotherapy for the treatment of patients with advanced nasopharyngeal carcinoma.[35] The safety and tolerability of both these agents has been determined over many years as have the safety of many other OVs due to their use in national and international vaccination programs. Indeed, certain strains of vaccinia, a large enveloped group I DNA virus from the order poxviridae, have the most extensive pedigree of testing in man of any known virus due to their deployment in the global vaccination against smallpox in the 1960s and 1970s.[36] The wide variety of available natural and recombinant OVs, each with their own individual characteristics with regard to tumor tropism, oncolysis, immunogenicity and their potential to resculpt the TME permits a highly selective and rational strategy when attempting to combine their use with adoptive cell transfer and indeed CAR T-cell therapy in particular.

2.2 Oncolytic virus tumor tropism

One of the hallmarks of effective oncolytic virotherapy relates to the natural capacity of these agents to preferentially target and replicate in cancer cells while sparing normal tissue.[37] Although tumor tropism is incompletely understood for many OV strains, certain features predominate. In general rapidly dividing tumors provide a nucleotide-rich environment replete with the cellular machinery required for viral genomic replication. This feature can be exploited further by editing OVs in particular ways.[38] For example, the genomic DNA of HSV and vaccinia viruses incorporate a viral thymidine kinase (TK) utilized to increase the availability of deoxynucleotide triphosphates for enhanced viral replication. Recombinant HSV or VV strains with deletion of the TK gene are thus reliant on the elevated endogenous TK levels found in tumor cells compared to normal tissue.[39] Similarly, the adenoviral strains KD1 and KD3 have been engineered with mutations to the E1A virulence gene, preventing the encoded E1A proteins from binding to host p300 and pRB, effectively preventing the virus from driving cells from G_0 to S phase, blocking efficient viral replication in non-dividing cells.[40] These vectors have also been engineered to overexpress Ad death protein (ADP), enhancing cell lysis and the release of viral particles from infected cells. Other OVs are essentially reliant on host cell replication, such as the S-phase-dependent C-type retroviruses, of which Toca-511, a replication competent C-type retrovirus encoding an enzymatic activator of 5-FU (namely cytosine deaminase) is one such example.[41] An alternative target of engineered OV tumor tropism is the interplay between tumor telomere length and telomerase activity, which has been shown to have a critical role in tumor initiation and progression.[42] Kawashima et al. have utilized this target to engineer a recombinant adenovirus 5 vector (termed tumor- or telomerase-specific replication-competent adenovirus (TRAD)), whereby the human telomerase reverse transcriptase (hTERT) promoter is able to drive expression of the viral E1A and E1B genes.[43] In nude mice harboring human non-small cell lung cancer xenografts, delivery of TRAD was able to elicit tumor regression without histopathological evidence of infected normal tissues. Further improvements can be elicited with the co-expression of a small non-coding inhibitory RNA sequence targeting the host endoribonuclease Dicer.[44]

The tumor tropism of other OVs may reflect preferential viral entry into tumor cells due to a differential in surface expression of certain adhesion molecules. A classic example is the ability of coxsackievirus strains to bind intercellular adhesion molecule 1 (ICAM-1), a cell adhesion molecule that is upregulated in many tumors.[45] Off-target viral tropism may be prevented by the insertion of tissue-specific microRNAs.[46] For example, the insertion of a muscle-specific miRNA into the genome of a coxsackievirus A21 OV was able to ensure the rapid destruction of viral RNA in myocytes, preventing fatal myositis in murine models of melanoma and myeloma.[47] An additional innovative strategy for identifying OVs with enhanced tumor tropism utilizes in vitro directed evolution in an ovarian cancer 3D-Matrigel culture-based expansion system.[48] Here, the Matrigel-derived virus OvAd1 was found to suppress tumor growth twice as efficiently as a corresponding non-Matrigel-derived virus. Bioengineering techniques may also be deployed to enhance tumor specificity. Examples include the development of a pH-sensitive and bio-reducible polymer (PPCBA)-coated oncolytic adenovirus, which is able to infect cells via macropinocytosis and exhibit greater efficacy in a typically low pH hypoxic TME[49]; and a pancreatic cancer-specific adenovirus with neurotensin peptide (NT)-conjugated polyethylene glycol (PEG) chemically crosslinked to its surface.[50]

Host cell apoptosis is often able to curtail viral infection and OVs may therefore take advantage of a tumor's defective apoptotic machinery.[51] Furthermore OVs with deletions or modifications to anti-apoptotic virulence genes may be rendered more amenable to clearance by normal tissue but remain relatively protected in tumor cells due to their intrinsic resistance to apoptosis. Thus NDV strains may exploit the upregulation of B-cell lymphoma (BCL) pro-survival proteins such as BCL-xL by tumor cells to proliferate and form syncytia,[52] while p53-deficient tumor cells may be more amenable to infection by adenoviral strains that have been engineered with a mutated E1B protein.[53] In the unedited virus the E1B gene encodes proteins capable of preventing p53 from inhibiting cell cycling and thus blocks it from inducing apoptosis. Alternatively, OVs may co-opt various aberrant cell signaling pathways found in tumor cells but not in normal tissue. For example, in a malignant peripheral nerve sheath tumor (MPNST)-derived cell line model, overactivation of the Ras pathway was associated with lower levels of apoptosis in host tumor cells following exposure to the oncolytic herpes simplex virus 1 (G207), conferring greater susceptibility to viral infection and proliferation.[54] Similarly, untransformed Schwann cells exhibiting low Ras activation may be rendered more permissive to infection following their conditional immortalization with the SV40 large T antigen, a process known to induce Ras-Raf signaling. Similar findings have been demonstrated using oncolytic reovirus.[55] In addition, the proliferation of vaccinia may be enhanced in cells with overactive epidermal growth factor receptor (EGFR)-RAS signaling,[56] commonly encountered in many solid tumors. Vaccinia may also exhibit tumor-tropism by dint of its tendency to be sequestered by a disrupted and tortuous tumor vasculature.[57]

Finally, and perhaps most relevant to this class of agents' potential synergistic combination with CAR T-cell therapy, many OVs are capable of actively or passively exploiting the impaired antiviral state often encountered in the TME. Specifically, tumor cells often display an intrinsically impaired or indeed absent response to type I interferon signaling, rendering them more permissive to viral infection.[58–60] For example, tissue microarray data obtained from patient-derived invasive bladder tumors and normal surrounding tissue has revealed a divergence in type I IFN receptor expression, with lower levels of IFNAR found in tumor specimens.[58] The divergence was particularly marked in cases of advanced malignancy. Low expression of IFNAR conferred susceptibility to VSV, which could be recapitulated in resistant tumor specimens following siRNA knockdown of IFNAR or with neutralizing antibodies. Tumor tropism may be further engendered by deleting or modifying virulence genes acting on the type I IFN pathway. For example, the matrix protein of VSV has been shown to suppress IFNβ gene expression,[61] while the HSV γ-34.5 protein is able to circumvent the innate immune system by shutting down the production of type I IFNs and aiding neurovirulence.[62]

2.3 Recombinant OVs

Several families of DNA viruses such as adenoviruses and poxviruses have genomes of a sufficient size to allow the incorporation of specific transgenes. The natural capacity of these OVs to sculpt and modulate the TME may therefore be further enhanced by delivering genes encoding a vast array of immunomodulatory agents.[14] These may include (but are not limited to) immunostimulatory cytokines, co-stimulatory molecules, chemokines, stromal disruptors, bispecific T-cell engagers, immune checkpoint inhibitors, chemoradiation sensitizers, microRNAs with a plethora of targets and apoptosis-inducers. Additionally, their targeted incorporation into viral virulence genes can accomplish a dual strategy of further inflaming the TME while improving tumor tropism and reducing off-target infection. Engineered OVs may be designed to bypass many of the intrinsic mechanisms of immune evasion exhibited by tumors, which are also often shared to a greater or lesser extent with mechanisms of resistance to CAR T-cell therapy. Great synergy with the latter may be elicited by the former's ability to actively remediate or reverse many of these shared mechanisms of resistance. For example, recombinant OVs have been shown capable of enhancing local type I IFN signaling leading to an upregulation of interferon-responsive chemokines such as CXCL-9 and CXCL-10[63] able to attract CXCR3-expressing CAR T-cells to the TME. MHC class I may also be upregulated locally,[64] and while this would not be expected to synergize directly with CAR T-cell therapy (which is MHC-unrestricted), the shared ability of OV infection and CAR T-cell therapy to induce epitope spreading and a bystander effect would certainly be enhanced by this feature, resulting in broader anti-tumor immunity. Furthermore, OVs may be designed to disrupt enhanced transforming growth factor beta (TGF-β)/Wnt/β-catenin signaling,[50] which has been shown to negatively regulate antigen presentation.[65] Alternatively, OVs may target immunosuppressive pathways in the TME, e.g., by blocking the function of prostaglandin E2 (PGE2)[66] or adenosine A2a receptors (A2ARs). The secretion of "naked" or chimeric death receptor ligands and/or pro-apoptotic warheads (such as apoptin)[67] may further enhance the oncolytic potential of these viruses by allowing them to target uninfected cells in the locality.

2.4 Options for OV administration

Oncolytic viruses have tended to be administered either by direct intratumoral injection or via parenteral delivery directly into the circulation.[68] Each approach is associated with certain advantages and disadvantages dependent upon contextual factors such as the anatomical position of the tumor, patient-specific characteristics and the nature of the particular OV being used. Successful systemic deployment of OVs, while preferential to elicit widespread infection of micro- and macrometastatic disease may be compromised by a patient's capacity to elicit an innate and/or adaptive anti-viral immune response. For example, individual viral particles may be targeted by neutralizing antibodies and/or form complexes with complement proteins.[69] Alternatively they may be sequestered by non-target tissue microvasculature such as that found in pulmonary, hepatic or splenic tissues. They may be blocked from entering the TME by the presence of dense tumor-associated stroma, creating a physical barrier, or may be cleared by a local infiltration of myeloid cells (such as neutrophils or macrophages) recruited to the TME via various mechanisms.[38] Aberrant microvasculature, while conducive to the infection of certain vaccinia strains,[57] may impede OV entry as may central tumor necrosis and/or a hostile microenvironment characterized by hypoxia, low pH, electrolyte disturbances and high hydrostatic pressure.[70] Naturally, all of these factors also pose a potent barrier to CAR T-cell entry as will be discussed in greater depth later in this chapter. Local delivery of the OV directly into the TME may bypass several of these natural barriers. However, this approach becomes problematic in the face of a poorly accessible deep-seated tumor or indeed one intimately involving or closely apposed to critical structures such as nerves and blood vessels. Although clinical trials involving T-VEC have revealed a capacity of OVs to elicit an abscopal regression of metastatic deposits located distant from the injection site,[71] improved efficacy has been seen in patients without measurable dissemination. Thus, the ability to deliver OVs systemically is likely to be seized upon in patients with demonstrable metastatic disease and is currently being evaluated in clinical trials involving coxsackievirus A21 (CAVATAK)[72] and the vaccinia virus GL-ONC1.[73] A number of strategies have been deployed with a view to aiding the systemic delivery of OVs. These have included the use of plasmapheresis,[74] complement pathway antagonists,[69] B-cell depleting agents (such as Rituximab) or preconditioning with cytokine cocktails and/or immunomodulatory cytotoxic agents such as cyclophosphamide.[75] Intriguing pre-clinical data has been obtained showing promise with the delivery of OV particles or genomic DNA using cytotoxic T-cell "Trojan horses,"[76,77] which would also be expected to elicit an anti-tumor immunity in the form of adoptive cell transfer. Such an approach may be particularly attractive for combination with CAR T-cell therapy.[77]

2.5 Issues of OV clearance

An issue that has to some extent plagued the field of oncolytic virotherapy relates to pre-existing immunity and the rapid emergence of anti-viral antibodies[78] in patients that may have been vaccinated against related vectors. And even in those who have not been vaccinated, studies suggest that following the administration of many OVs patients often develop rapid immunity associated with the clearance of viral particles by antibody and complement complexes. Furthermore, potential safety concerns may come into play due to the relative immunosuppression of cancer patients who may have undergone multiple rounds of cytotoxic chemotherapy. Specifically, an impaired host antiviral response may subvert the exquisite tumor tropism of these viruses leading to unforeseen off-target effects.[79] Likewise, safety concerns exist with regard to the potential for horizontal transmission.[80] Rapid evaluation in combination studies (e.g., with CAR T-cell therapy) is also likely to be hindered by the paucity of agency-approved OVs in the clinic; the limited number of patients with tumors accessible to direct injection of OV (and/or CAR T-cells) and the low number OVs suitable for systemic delivery. Furthermore, the use of CAR T-cell ACT would be expected to limit the evaluation of combination therapy to large academic centers with the necessary experience and expertise to manage potential toxicities and risks.

3. The development of chimeric antigen receptor T-cell therapy

The development of CAR T-cell therapy has proven to be a truly ground-breaking innovation in the field of cancer immunotherapy, and has already produced durable remissions in patients with advanced hematological malignancies refractory to all standard treatments.[81] At their most fundamental, CARs are structured as cell surface fusion proteins linking an extracellular binding domain (typically designed to recognize tumor-specific or tumor-associated cell surface antigens (TSAs or TAAs)) with an intracellular activating domain capable of eliciting (in the case of cancer immunotherapy) target cell cytotoxicity.[82] Further customization of CARs and/or ancillary proteins expressed by engineered T-cells are able to recapitulate signals 1, 2 and 3 required for robust T-cell activation and effector function. Numerous CARs have already demonstrated promise in the field of hemato-oncology with the anti-CD19 CARs tisagenlecleucel and axicabtagene ciloleucel already a standard therapeutic option for patients with relapsed ALL or NHL. Other clinical candidates include CARs designed to target multiple myeloma (most notably targeting BCMA, but also CD19, kappa light chain, CD138, CD38, CD44v6, SLAMF7 or CD229)[83]; relapsed B-cell leukemias and lymphomas (particularly targeting CD22 but also CD20, ROR1, TSLPR or CD133)[84]; Hodgkin's lymphoma (CD30)[85]; AML (CD33, CD123, CD44v6 and Flt3)[86]; and T-cell leukemias and lymphomas (e.g., CD5, CD7 or TRBC1).[87] Despite slower progress in the deployment of CARs targeting solid tumors, numerous targets are currently being evaluated in early phase trials. A heavily selected list of notable targets include: CEA, the ErbB family (including EGFR, EGFRvIII, HER2, etc.), B7-H3, CD133, Claudin 18.2, CD70, EpCAM, GD2, GP3, IL13Ra2, L1CAM, mesothelin, cMET, MUC1 (including non-glycosylated and/or truncated variants), PSCA and PSMA.[17,88,89] CARs may also be designed to target immunosuppressive cells in the TME such as CAFs, e.g., by targeting FAP.[90]

The development of CAR T-cell therapy is occurring in concert with other forms of ACT including T-cell receptor (TCR)-engineered antigen-specific T-cells and TIL therapy, whereby endogenous T-cells are harvested from a tumor biopsy and expanded ex vivo with or without positive or negative selection before being reinfused into the patient as an autologous cellular product.[91] CAR T-cells are associated with several advantages over TIL or TCR-engineered therapy. Most notably, CAR T-cell therapy is by design MHC-unrestricted, meaning that CARs can function independently of host antigen processing and presentation. This attribute is particularly attractive when targeting solid tumors that have downregulated MHC class I (for example via HLA loss of heterozygosity (LOH)[92] or mutations involving β2-microglobulin[93]) or exhibit dysregulation of the antigen processing machinery (e.g., via defects in TAP or various proteosomal subunits[94]), both of which are frequently encountered mechanisms of immune escape by tumors. In this common scenario CARs may be able to outperform both immune checkpoint inhibition using monoclonal antibodies targeting the PD-1/PD-L1 axis and/or CTLA-4 as well as other forms of ACT. Similarly, second generation CARs able to provide their own costimulation (signal 2) are able to bypass defects in antigen cross-priming and presentation caused by the suppression or absence of professional DCs in the TME or surrounding tissues, a phenomenon common to many tumors exhibiting a so-called "immune desert" immunophenotype.[95] Furthermore, this feature also means that CARs may lend themselves particularly well to both autologous and allogeneic ACT given that CARs are capable of recognizing antigen on any HLA background, whereas engineered TCRs must be matched to a patient's own specific HLA haplotype.[96]

An additional advantage of CARs is that they may be designed to target non-protein-based TAAs such as glycolipid structures or carbohydrates.[97] And while classical CARs are limited to targeting cell surface rather than cytosolic or nuclear proteins, significantly reducing the repertoire of potential targets, recent designs have been able to address this issue, for example, by designing scFvs exquisitely selective and specific for cytoplasmic protein-derived peptides in complex with MHC. A notable example presented by Liu et al. describes a CAR designed to target the $AFP_{158-166}$ peptide complexed with HLA-A*02:01 in hepatocellular carcinoma.[98] In parallel, an evolving understanding of the human "surfaceome" of both neoplastic and normal tissue is aiding in the design of CARs targeting solid tumors.[99] Furthermore, due to the modular structure of CARs, they lend themselves to customization and adaptation in a manner that would be far more difficult to replicate in engineered TCRs. Thus, numerous systems have been developed to control CAR functionality both temporally[100] and spatially,[101,102] with the ultimate goal of creating complex algorithmic circuits composed of both stimulatory and inhibitory inputs able to elicit a form of pattern recognition and a bespoke anti-tumoral immune response.[103]

3.1 Chimeric antigen receptor design

Having been first described by Eshhar and colleagues in 1989,[104,105] CAR design has undergone a number of incremental changes over the intervening decades. However, today's designs remain in many ways wedded to the first generation "T-bodies" that fused an antigen-binding ectodomain, typically derived from an antibody-derived scFv or endogenous receptor ligand, with a transmembrane domain derived from CD8a, CD4, CD25 or CD16 and an activating endodomain containing immunoreceptor tyrosine-based activation motifs (ITAMs) derived from the CD3ζ chain or Fc receptor γ.[82] Engagement of the CAR ectodomain with its cognate binding partner is able to elicit the formation of an immune synapse with many structural features common to the endogenous MHC/TCR-mediated synapse (as well as a number of important differences). CAR engagement is subsequently able to co-opt proximal TCR signaling via adapter proteins including Lck and ZAP70[106,107] to elicit T-cell activation and, ultimately, induce target cell cytolysis via the transfer and local release of cytotoxic molecules such as perforin and granzyme B. CAR activation is also able to induce the release of a number of immunostimulatory cytokines capable of fostering T-cell proliferation and expansion both in an autocrine and paracrine manner. First generation CAR constructs lacking a co-stimulatory endodomain upstream of the activating domain are unable to generate a sufficiently strong activation signal, particularly following repeated exposure to antigen and are susceptible to anergy and AICD.[108] Second generation CARs, however, are designed to incorporate a discrete endodomain derived from one of a number of co-stimulatory receptors (most commonly CD28 or 4-1BB but others may include OX40, CD27, ICOS and DAP10) positioned in *cis* with a downstream activating domain.[109] Activation of these second generation receptors is able to recapitulate both signal 1 and 2, mimicking antigen cross-presentation between effector T-cells and professional DCs in the tumor draining lymph node. Third generation constructs incorporate more than one co-stimulatory endodomain in *cis* with a downstream activating domain (e.g., CD28 followed by 4-1BB and CD3ζ or CD28 followed by OX40 and CD3ζ).[110] However, these are yet to demonstrate any consistent benefits over second generation constructs with a number of pre-clinical studies suggesting that this format may be disadvantageous in certain circumstances.[111] Fourth generation constructs are now legion and may be split into at least four broad camps. The first involves additional engineering to the CAR itself, such as the introduction of a truncated IL-2Rβ-derived cytoplasmic domain with a STAT3-binding motif[112] conferring signal 3; the separation of costimulatory endodomains from the activating domain by expressing a chimeric costimulatory receptor in trans[113]; or modifications to CD3ζ chain itself.[114] The second involves co-expression, alongside a second or third generation CAR, of an auxiliary surface-bound or secreted immunomodulatory molecule such as a cytokine (e.g., IL-18[115] or membrane-tethered IL-15[116]), chemokine (e.g., CCR2b[117] or CXCR4[118] to aid trafficking into the TME), costimulatory ligand (e.g., 4-1BBL[119] or CD40L[120]), chimeric switch receptor (e.g., PD-1/CD28),[121] dominant negative receptor (e.g., truncated Fas,[122] PD-1[123] or TGFβR[124]), stromal disruptor,[125] checkpoint inhibitor (e.g., an scFv targeting PD-1[126] or CD47[127]) or BiTE.[128] For example, the combination of auto-co-stimulation utilizing 4-1BBL and a second generation CD19-redirected CAR (1928z_41BBL) is associated with profound CAR T-cell expansion and in vivo persistence when targeting the aggressive NALM6 human ALL cell line in NSG mice.[129] Furthermore, the improved ability of this construct to elicit leukemic control was found to be dependent on activation of the IFNβ pathway and sustained expression of the transcription factor IRF7, a type I IFN regulator.[119] 1928z_41BBL-mediated secretion of IFNβ may therefore provide signal 3 in an autocrine fashion, further highlighting the potential for the synergistic combination of CAR T-cell therapy with oncolytic virotherapy, while at the same time clearly demonstrating the capacity of CAR T-cell therapy itself to modulate endogenous innate and adaptive anti-tumor immunity. The third innovation relates to the control of CAR expression, most notably featured in the technique of targeting the CAR to the TRAC locus either using CRISPR/Cas9[130] or with site-specific engineered homing endonucleases.[131] A fourth category encompasses technologies developed to elicit the controlled functionality of CAR T-cells such as that afforded by the UniCAR,[132] ZipCAR[102] or GoCAR[133] systems as well as Boolean logic-based systems such as Syn-Notch.[101] Additional strategies have been developed that either move across these categories or function independently, e.g., those related to the development of allogeneic CAR T-cell, gamma delta T-cell or NK cell products.[134]

3.2 Clinical data

A number of second generation CARs targeting CD19 have demonstrated particular potency against leukemia and lymphoma in both pre-clinical models and clinical trials. Tisagenlecleucel (Kymriah or CTL019), a second generation CD19-directed construct incorporating a 4-1BB-derived costimulatory endodomain and CD3ζ tail, was the first autologous CAR T-cell therapy approved by the United States Food and Drug Administration (FDA) in August 2017.[135] Approval was granted, based upon data derived from the phase II ELIANA trial, for the treatment of children or young adults aged 25 or below with B-cell precursor ALL refractory to standard treatments or in second or third relapse.[6] In this context tisagenlecleucel was associated with a complete remission rate of over 80%, historically unprecedented in this target population. The agent was subsequently approved in May 2018 for the management of a broader group of patients with relapsed or refractory large B-cell non-Hodgkin lymphoma after two or more lines of systemic therapy.[136] Additional trials are ongoing in patients with relapsed mantle cell lymphoma, CLL and other B-cell malignancies. A second agent, axicabtagene ciloleucel (Yescarta or KTE-C19), which contains a CD28-derived endodomain, was approved by the FDA in October 2017 for adult patients with diffuse large B-cell lymphoma (DLBCL) not otherwise specified, primary mediastinal large B-cell lymphoma (PMBCL), high grade B-cell lymphoma, or DLBCL transformed from follicular lymphoma following two prior therapies.[137] Approval was based upon data from the phase II single arm multicenter ZUMA-1 trial, wherein the agent was associated with an ORR assessed by independent central review of 72% and CR of 51%.[138] Very limited head to head data comparing the CD28 versus 4-1BB endodomains are currently available aside from a small early phase Chinese study (NCT03173417) with initial data presented at ASCO in 2018.[139] Here, 28 patients with relapsed or refractory B-cell ALL were treated with a 4-1BB-containing CD19-directed second generation CAR and 19 others were treated with a corresponding CD28-containing CAR. Peak CAR T-cell expansion was found to be significantly increased in the 4-1BB group, which also exhibited lower levels of CRS and neurotoxicity. In general it is recognized that CAR T-cells exhibiting a central memory or stem cell memory phenotype can continue to function for prolonged periods of time as evidenced by the phenomenon of prolonged B-cell aplasia[140] in patients that have achieved durable remissions at a number of leading institutions including Memorial Sloan-Kettering Cancer Centre, the University of Pennsylvania and the National Cancer Institute.

3.3 Barriers to CAR efficacy in solid tumors

Despite the progress of CAR T-cell therapies targeting B-cell malignancies, similar achievements have not yet been realized in patients with advanced solid tumors.[89] Here, a number of potential issues will need to be addressed before significant progress can be made. Specifically, while CD19 can be regarded as a "dispensable" antigen due to the ability of patients to survive long-term with chronic hypogammaglobulinemia caused by B-cell aplasia, no obvious dispensable TAAs have been identified in solid tumors. Furthermore, solid tumor TAA's currently being targeted by CARs in the clinic are all expressed (albeit at a lower level) on normal tissue,[141] substantially increasing the risk of unacceptable toxicity due to "on target off tumor" activity. Physical barriers present in and around solid tumors (such as tumor-associated stroma, raised hydrostatic pressure) are also liable to preclude CAR T-cell entry, while effector function may be compromised by a hostile TME characterized by hypoxia, acidosis, electrolyte or metabolic imbalances and a plethora of immunosuppressive molecules.[142] Mismatched chemokine gradients and/or disrupted endothelium may impair CAR T-cell trafficking, while the T-cells may themselves be sequestered by off-target tissues.[143] Another issue relates to the intrinsic heterogeneity of solid tumors, which through a process of clonal selection dictated in part by their metastatic niche are liable to exhibit variability in TAA expression both temporally and spatially, increasing the likelihood of incomplete tumor targeting, antigen escape and immune evasion.[144] Indeed, the sheer diversity of solid tumors and the variability of TAA-expression may necessitate the bespoke design of multiple pattern-recognizing CAR products enhancing the complexity and cost of the therapy even if the component parts are developed in a tumor-agnostic fashion. Finally, certain aspects of treating patients with hematological malignancies do not necessarily feature in the solid tumor arena. Most obviously, leukemic cells are immediately accessible to CAR T-cells within the circulation, bone marrow and splenic tissue. In addition, it has been posited that often these cells express co-stimulatory ligands capable of amplifying signal 2 and aiding CAR T-cell expansion.[145] Finally, patients with relapsed or refractory hematological malignancies will almost certainly have received multiple cycles of lymphodepleting chemotherapy, providing a more favorable niche for CAR T-cell expansion.[146]

3.4 Safety concerns

CAR T-cell safety remains an ongoing concern, particularly in patients with solid tumors. Several clinical trials using CD19-directed CARs have featured reports of grade IV or grade V toxicity due to neurotoxicity, cytokine release syndrome or the related macrophage activation syndrome.[147,148] The former, which has also been encountered in patients receiving anti-CD22 and anti-BCMA CAR products,[149] is poorly understood and while usually entirely reversible has occasionally proved resistant to prevention or supportive management. Separately, a recently reported case whereby a patient's CD19-directed CAR product was contaminated with a single transduced leukemic cell, leading to the CD19 epitope being masked by co-expression of the CAR and ultimately disease relapse raises further questions regarding the technical challenges of safely delivering this complex therapy.[150] The issue of insertional mutagenesis due to the use of integrating viral vectors is also theoretically possible, potentially leading to cases of T-cell leukemia, although none have yet been reported in the literature.[151]

The risks of inducing "on target off tumor" toxicity using CARs designed to target solid tumor TAAs is borne out in one particular case report wherein a patient enrolled in a first in man phase I trial of a third generation HER2-directed autologous CAR T-cell product unfortunately succumbed to severe CRS, pulmonary toxicity and multiple organ failure.[152] The purported mechanism involved the targeting of physiological low level HER2 expression in pulmonary tissues, exacerbated by first-pass passage and/or sequestration of CAR T-cells by the pulmonary vasculature. Many trials currently recruiting patients now adopt more stringent dose escalation protocols and incorporate inducible safety systems or suicide switches built-in to the CAR product. Common examples include an inducible caspase 9 gene cassette (CaspaCIDe), whereby CAR T-cells may be induced to undergo apoptosis following the systemic delivery of a lipid-permeable tacrolimus analogue (rimiducid)[153]; suicide systems utilizing herpes simplex virus-thymidine kinase (HSV-TK), rendering T-cells susceptible to ganciclovir-induced cytotoxicity[154]; or potentially reversible tet-on/tet-off systems whereby CAR T-cells may only be activated in the presence or absence of a tetracycline-analogue such as doxycycline.[155] The co-expression of a truncated surface protein (e.g., truncated human EGFR[156]) or mimotopes (such as RQR8, derived from CD34 and CD20[157]) potentially targetable via ADCC using clinical grade monoclonal antibodies (cetuximab or rituximab, respectively) is an alternative strategy, albeit one associated with potential concerns regarding the rapidity of CAR clearance in the face of severe toxicity.

Separately, the CAR structure itself may be redesigned to function only in the presence of a systemically delivered exogenous small molecule or adapter protein. For example, optimal activation of the GoCAR system requires both engagement of the CAR with its cognate antigen (e.g., PSCA) and the presence of rimiducid, which permits a co-expressed chimeric costimulatory molecule comprised of MyD88 and a CD40-derived endodomain to homodimerize and provide signal 2.[133] Similarly, various switchable "universal" CARs have been designed that rely on adapter proteins to bridge the CAR ectodomain and a desired TAA, facilitating synapse formation. Examples include CARs with scFvs targeting FITC[158] or PNE[159] and so-called ZipCARs which rely upon homo- or hetero-dimerization with paired leucine zipper-based adapter proteins.[102] Such an approach is designed to tune CAR activation, while providing a means to address antigen escape via the simultaneous or sequential delivery of multiple adapter proteins targeting different TAAs. Such CARs would be expected to be controllable remotely in a precise fashion over a period of time.

4. Synergies between OVs and CAR T-cell therapy

4.1 OV-mediated induction of local inflammation

Preclinical studies involving multiple OVs have demonstrated a shared capacity to induce an enhanced type I IFN signature in the TME.[14] Alongside the ability of type I interferons to negatively regulate cellular proliferation, this feature, which is central to ability of this class of agents to resculpt and modulate the TME, thereby rendering it more conducive to host innate and adaptive immunity, is also ripe for exploitation when considering combination strategies with ACT, including CAR T-cell therapy. The ability of an effector T-cell to respond robustly following target cell engagement requires the presence of three signals: signal 1 provided by TCR activation; signal 2 provided by the engagement of costimulatory receptors with cognate ligands (typically expressed by professional antigen presenting cells); and signal 3 usually provided by local proinflammatory cytokines. The engagement of second or third generation CARs will recapitulate signals 1 and 2. Signal 3 is provided (at least initially) by the exposure of these cells to supportive cytokines during ex vivo expansion and may be further aided either by the contemporaneous delivery of cytokines at the time of ACT or by rendering the CAR T-cells capable of secreting their own cytokines or by tuning their responsiveness to them in fourth generation constructs.[160] Type I interferons may also provide signal 3, and if delivered by OVs in a tumor tropic fashion, would be expected to exert their maximum effect in the TME, potentially heightening efficacy and improving the safety of CAR T-cell therapy. More specifically, type I interferons have been demonstrated to enhance the clonal proliferation of effector T-cells, improve their cytolytic function and encourage memory formation.[161,162] Alongside these effects, IFNβ has been shown to exert a positive effect on antigen cross-presentation mediated by professional DCs as well as positively disrupt tumor-associated microvasculature and inhibit the local proliferation of MDSCs and Tregs.[119] The ultimate impact of OV-derived type I interferons is therefore to convert a tumor from being immunologically "cold" to one that is "hot" and permissive CAR T-cell entry, activation and proliferation.[51]

The importance of type I interferons in CAR T-cell therapy is further exemplified by preclinical data gleaned from second generation CARs containing a 4-1BB endodomain, whereby downstream activation of TNF receptor-associated factor 2 (TRAF2) is associated with the induction of IFNβ gene expression and the development of autocrine signaling via interferon receptors located on the CAR T-cells themselves.[119]

4.2 OV-induced immunogenic cell death

The release of endogenous danger molecules, so-called damage-associated molecular patterns (DAMPs) by OV-infected dying tumor cells is a characteristic hallmark of immunogenic cell death (ICD)[163] and a crucial component in a cascade of events that can ultimately result in effective adaptive anti-tumor immunity. Non-programmed (accidental) cell death in the form of necrosis, characterized by an uncoordinated leakage of cell contents through a ruptured plasma membrane, is particularly pro-inflammatory and immunogenic due to the associated release of multiple DAMPs such as HMGB1, ATP, heat shock proteins (hsp70 & hsp90),[164] endoplasmic reticulum (ER) sessile proteins,[44] IL-1a, IL-33, mRNA and genomic DNA.[165] While programmed cell death (PCD) has historically been viewed as tolerogenic, certain morphological forms of PCD (particularly pyroptosis and to a lesser extent necroptosis) can also induce the release of DAMPs and the secretion of pro-phagocytic "eat-me" signals such as lysophosphatidylcholine (LPC), phosphatidylserine (PS) and cell membrane-associated calreticulin (ecto-CRT).[166] The latter, in particular, facilitates the engulfment of released tumor associated antigens (TAAs) by resident and migratory scavenging macrophages and Batf3$^+$ DCs, activated by the engagement of pattern recognition receptors (PRRs) such as the receptor for advanced glycation end-products (RAGE) and Toll-like receptors by specific viral pathogen-associated molecular patterns (PAMPs) and DAMPs.[167] This cascade induces MyD88-NFκB/JAK-STAT signaling, the induction of viral clearance genes, IFNα and β release and the local generation of interferon-inducible chemokine gradients capable of drawing in effector T-cells as well as enhanced DC cell maturation and cross-priming capabilities.[34] Indeed, the infection of tumor cells by OVs often leads to large quantities of viral proteins being produced, triggering ER stress which is further exacerbated by the production of reactive oxygen species (ROS),[44] culminating in RNA-dependent protein kinase (PKR)-like ER kinase (PERK)-dependent mitochondrial apoptosis[43] and ICD, further promoting viral replication and virulence. In the absence of defective HLA expression (e.g., due to acquired mutations in β2-microglobulin[93]) or an impaired IFN-sensing pathway (e.g., due to JAK1/2 mutations[168]), the presence of type I IFN in the TME may be associated with an upregulation of class I MHC on infected and uninfected tumor cells,[169] and would be expected to be particularly synergistic with non-CAR based ACT. CAR T-cell therapy, on the other hand, might be expected to fulfill a gap in patients whose tumors have lost MHC or, to a lesser extent, tumors that are interferon-unresponsive. Furthermore, T-cells themselves are able to respond to PAMPs and DAMPs. Most notably, HMGB1 can engage RAGE present on effector T-cells, leading to the latter's activation and clonal proliferation.[170,171]

This pro-inflammatory cascade is exemplified in many preclinical studies of OV infection and ICD. For example, the infection of human non-small cell lung cancer (NSCLC) cell lines by the enterovirus coxsackievirus B3 (CVB3) is associated with the abundant surface expression of ecto-CRT and the active secretion of ATP and HMGB1.[172] Intra-tumoral delivery of CVB3 into human NSCLC xenografts in BALB/c nude mice was associated with a marked local recruitment of NK cells and CD107a$^+$ granulocytes, which, through in vivo depletion assays, were shown to contribute significantly to the CVB3's anti-tumor effect. Separately, an oncolytic reovirus has been shown to confer human myeloid DC activation and phenotypic maturation, stimulating the production of several proinflammatory cytokines including IFNa TNFα, IL-12p70, and IL-6. Co-culture of reovirus-activated DCs with autologous NK or T-cells led to their enhanced cytolytic function and elevated production of cytokines such as IFNγ.[173] In a further example, the intratumoral delivery of a replication-restricted mutant herpes simplex virus (HSV-1716) in a syngeneic murine model of ovarian carcinoma has been shown to activate monocytes and DCs, leading to enhanced local secretion of type I and II interferons as well as the IFN-inducible C-X-C chemokines CXCL-9 and CXCL-10. The latter was associated with a significant influx of CXCR3$^+$ NK and effector CD8$^+$ T-cells mediating an anti-tumoral effect.[174]

The cGAS-cGAMP-STING (Stimulator of Interferon Genes) cytosolic DNA sensing pathway has also been implicated in the ability of certain OVs to engender innate an adaptive immune responses.[175] Engagement of STING by viral genomic DNA is known to induce type I IFN gene expression, with concomitant benefits seen in terms of DC activation, maturation and cross-presentation. While STING hypofunction (potentially caused by acquired mutations affecting the pathway) may disrupt type I interferon signaling in tumors[176,177] and stoke resistance to many forms of immunotherapy (including immune checkpoint blockade and many forms of ACT), this state is likely to be conducive to OV infection and replication, which via mechanisms independent of STING may to some extent warm up these immunologically cold tumors. Indeed, STING and/or cyclic GMP-AMP synthase gene expression was observed to be absent in approximately 40% of human melanoma cells examined in one recent study, with pathway suppression most frequently caused by epigenetic silencing of either STING or cGAS.[177] STING hypofunction was found to render melanoma cells unable to produce type I IFN and other immune cytokines following exposure to OV-derived cytosolic DNA species. These cells were highly susceptible to infection with DNA OVs including HSV-1, potentially revealing a new biomarker with utility for selecting patients most suitable for treatment with T-VEC or other DNA OVs. In addition, a number of STING "agonists" (including a heat shocked OV) have entered clinical practice[178] and may prove to be synergistic with CAR T-cell therapy.

Thus far, several preclinical studies have explored the synergistic combination of OVs with ACT, including CAR T-cell therapy targeting solid tumors. Many involve arming the OV with transgenes encoding various immunomodulatory agents and these are described later in relevant sections of the review. Nevertheless, studies with unarmed OVs suggest the potential for direct synergy with CAR T-cell therapy. For example, use of attenuated herpes virus (HSV1716) has been shown to enhance GD2-redirected CAR T-cell expansion in co-cultures with GD2$^+$ tumor cells (including melanoma and rhabdomyosarcoma cell lines).[179] In particular, GD2$^+$ melanoma cells, that demonstrated an intrinsic resistance to CAR-mediated cytolysis, could yet induce T-cell expansion in the presence of HSV1716. Similarly, albeit utilizing CAR-transduced NK-92 cells, an oncolytic HSV (oHSV) strain was able to augment the in vitro cytolytic function of EGFR-redirected CAR NK cells against human breast cancer cell lines.[180] While in an intracranial breast cancer metastasis model, dual treatment with CAR NK cells and oHSV via intra-tumoral injection outperformed either treatment delivered as monotherapies, leading to prolonged animal survival.

4.3 OV vaccination strategies

A good example of the ability of OVs to synergize with ACT and potentially break tumor-mediated peripheral tolerance is illustrated by a study involving a syngeneic B16 melanoma model in immunocompetent mice.[181] Here, melanoma cells were engineered to express chicken ovalbumin (OVA) as a target antigen. Alone, the intraperitoneal delivery of polyclonally activated OVA-specific effector T-cells failed to elicit durable disease control. However, their delivery in combination with an adenoviral OV was able to successfully target the tumor, and was associated with the release of inflammatory cytokines, the local recruitment of macrophages, an upregulation of co-stimulatory ligands on $CD11c^+$ APCs, the recruitment and activation of OVA-specific $CD8^+$ effector T-cells as well as evidence of a bystander effect with a concomitant increase in $CD8^+$ T-cells specific for endogenous TAAs including TRP-2 and gp100.

More nuanced approaches to optimizing interferon-mediated adaptive immune responses may involve the sequential delivery of multiple serologically distinct OVs using a so called prime-boost strategy.[182] Here, an adaptive immune infiltrate is enhanced by the synergistic delivery of multiple OVs that are selected to minimize immunodominance and aid epitope spreading. The coordinated delivery of such a strategy, potentially in concert with viral-specific CAR T-cells may prove to be a particularly attractive when targeting solid tumors. An example of the latter includes the use of Epstein Barr virus (EBV)-specific T-cells transduced with a first generation anti-GD2 CAR in a phase I study targeting neuroblastoma.[183] Here, virus-specific CAR T-cells were found to outperform conventional polyclonal redirected CAR T-cells in terms of initial expansion and short-term persistence. A more involved strategy may be to initiate vaccination with an OV at a suitable time point prior to obtaining virus-specific PBMCs to be transduced with a relevant CAR. This approach has been utilized in patients following the intratumoral delivery of the recombinant vaccinia strain JX-594.[184] Here, researchers were able to stimulate harvested PBMCs with overlapping peptide libraries spanning the sequences of six vaccinia antigens. Activated PBMCs were then transduced with a HER2-redirected CAR and expanded. These cells were capable of being activated either via engagement of their CAR with HER2 or via their TCR with vaccinia-derived peptides.

5. Engineered OVs with payloads conducive to CAR function

The almost limitless ability to re-engineer OVs with therapeutic payloads designed to modulate the TME has led to a vast array of potential clinical candidates, many of which will be eminently suitable for synergistic combination with ACT and CAR T-cell therapy more specifically. Perhaps most notably, the infection of tumors by these recombinant OVs has the potential to repurpose the cancer cells themselves into cytokine and/or chemokine factories that can remodel the TME into one that is at once less hostile and more permissive to CAR T-cell entry, activation, proliferation, persistence and memory. In parallel, these OVs are likely to augment endogenous anti-tumor immunity, epitope spreading and ultimately an effective combined anti-cancer response. However, these two arms (delivered by endogenous versus adoptively transferred redirected T-cells) are non-distinct, with considerable cross-over between the two and the daunting task of eradicating an advanced solid tumor will likely rely upon the optimization of both arms.

5.1 OV-mediated delivery of cytokines supportive of professional DCs

The impaired function or absence of DCs in the TME has been highlighted in many models of tumor immune evasion. The development of T-VEC is a prototypical example of an attempt to address this issue using an oncolytic viral vector. Here, an HSV-1 viral virulence gene (ICP34.5) has been replaced with two copies of the human GM-CSF gene, with the aim of aiding the recruitment and expansion of professional antigen presenting cells in the TME.[34] The deletion of ICP34.5 is able to restore viral-induced type I IFN production by preventing its interaction with TANK-binding kinase 1 (TBK-1), which has been shown to block autocrine type I IFN signaling. The removal of ICP34.5 also restores functionality of the stress response protein kinase R (PKR) pathway, further enhancing the local production of type I IFNs. MHC class I antigen presentation is also augmented by the deletion of a second viral virulence gene (ICP47) that acts by inhibiting the host transporter associated with antigen processing and presentation (TAP). Preclinical testing of T-VEC has been able to demonstrate a clear abscopal effect in an immune competent murine melanoma model, whereby regression of both injected and non-injected contralateral flank tumors was associated with a significant overall survival benefit.[185] Pharmacodynamic analysis has revealed improvements in antigen cross presentation and the recruitment of an activated effector T-cell population. This approach would be expected to be synergistic with CAR T-cell therapy by fostering the generation of T-cell attracting chemokine gradients and the local release of pro-inflammatory cytokines favoring an adaptive immune response.

Furthermore, OVs may be rationally designed to optimize DC-mediated anti-tumor responses. For example, an attenuated measles virus[186] and an adenovirus[187] have both been engineered to secrete *Helicobacter pylori* Neutrophil Activating Protein (HP-NAP), a potent TLR-2 agonist, leading to the local activation of DCs with a Th1-skewed cytokine profile. Interestingly, CAR T-cells expressing HP-NAP under the control of CAR-activation inducible promoter (IL-2/NFAT) have been developed and shown enhanced cytotoxicity of target cells in vitro when co-cultured with immature DCs (imDCs).[188,189]

5.2 OV-mediated delivery of cytokines favorable to CAR T-cell function

Cytokines have recently re-emerged as a viable therapeutic modality for the treatment of patients with advanced cancer.[190] Currently, clinical trials are being conducted using a variety of pro-inflammatory cytokines either as monotherapies or in combination with immune checkpoint inhibitors or chemotherapy. While IL-2 and both type I and type II interferons have been utilized for many decades, recent examples include second generation IL-2-based agents (such as Bempegaldesleukin or NKTR-214, a pegylated CD122 agonist[191]), potentially conjugated to tumor-targeting monoclonal antibodies (e.g., cergutuzumab amunaleukin, comprising an engineered variant of IL-2 conjugated to an antibody targeting CEA[192]); recombinant human IL-15 delivered subcutaneously[193] or IL-15/IL-15Rα sushi domain-based fusion proteins[194]; a pegylated form of IL-10 (pegilodecakin), which although historically regarded as immunosuppressive has gained favor through its context-dependent pleiotropic ability to protect activated T-cells from AICD and enhance persistence[195]; IL-21[196] (which may be particularly efficacious with CAR T-cell therapy); and IL-12, which may be more safely delivered by in vivo electroporation of an encoding plasmid.[197]

Likewise, the carriage of immunostimulatory cytokines by OVs may enhance the safety profile of the former, while augmenting the pro-inflammatory effects of the latter, tipping the balance in the TME in favor of an enhanced innate and adaptive immune response. The local delivery of cytokines afforded by OVs would also be expected to have advantages when combining their use with CAR T-cell therapy and may prove to be safer than utilizing so called "armored CARs,"[198] which in the absence of a control mechanism may be liable to induce systemic cytokine-related toxicity. One caveat, however, is that the injudicious carriage of certain cytokines by OVs may compromise the virus' ability to proliferate within the TME, and a balance may need to be achieved to prevent the premature clearance of virus before it is able to elicit an optimal inflammatory response.

To date, a large number of cytokine-armed OVs have been engineered, including agents designed to produce type I interferons,[199] TNFα,[200] IL-2,[201] IL-4,[202] IL-12,[203] IL-15[204] and IL-18.[205] Aside from being pro-inflammatory, several of these cytokines (most notably TNFα and type I IFNs) may exhibit direct cytotoxicity on bystander tumor cells depending upon their intrinsic susceptibility.[206] In several preclinical murine glioblastoma models, an oncolytic HSV armed with a number of cytokines (including IL-4 and IL-12) was able to demonstrate more potent anti-tumor efficacy.[207] Likewise, an oncolytic adenovirus dually armed with both IL-12 and IL-18 yielded improved efficacy in a B16-F10 murine melanoma model with enhanced differentiation and cytotoxicity exhibited by tumor-specific immune cells.[208] Similarly, arming an oncolytic NDV strain with human IL-2 in combination with TRAIL led to a synergistic improvement in virally-mediated tumor cell apoptosis as well as robust proliferation of both $CD4^+$ and $CD8^+$ T-cells in murine hepatocellular carcinoma and melanoma models.[209] And in a syngeneic lung cancer model, the intratumoral injection of an oncolytic VSV strain armed with IFNβ was associated with improved disease control, tumor regression and cures in up to 30% of mice studied.[210] Analysis of treated tumors revealed an associated reduction of Tregs and an increase in $CD8^+$ effector T-cells. Similar studies have been conducted using oncolytic vaccinia strains, armed with a variety of cytokines including IFNβ, IL-1β, IL-2, IL-4, IL-6,[211] IL-10[212] and IL-12. Indeed, further evidence supportive of IL-10's pleiotropic effects in the TME is provided by preclinical data from a study evaluating a recombinant thymidine kinase (TK)-deleted Lister strain vaccinia OV armed with murine IL-10, which was able to elicit durable disease control in two murine pancreatic ductal adenocarcinoma (PDAC) models. Here, the presence of IL-10 led to impaired viral clearance and prolonged viral oncolysis with the emergence of adaptive immunity, evidenced by the ability of treated mice to reject tumor re-challenge.[212]

An approach that might be expected to exert indirect synergy with CAR T-cell therapy include arming OVs with agents supportive of precursor DC mobilization and expansion in the TME. A notable example is an oncolytic HSV strain engineered to express soluble FMS-like Tyrosine Kinase 3 Ligand (Flt3L).[213] In a syngeneic murine GBM model, the local deployment of this OV (termed G47Δ-hFLT3-L) led to elevated systemic plasma levels of Flt3L and significantly enhanced survival compared to an empty vector control.

Alongside these developments, CAR T-cells have themselves been armed with a number of immunostimulatory cytokines. Prior attempts to deliver IL-12 systemically[214] or via engineered TIL[215] have met with problematic toxicity. However, CD19-directed CAR T-cells armed with IL-12 were found to be safe in an immune competent conditioning-dependent murine model utilizing human CD19 modified EL4 thymoma cells.[216] In addition, IL-12-armed CAR T-cells could elicit tumor eradication even in the absence of prior cyclophosphamide conditioning and demonstrated modified IFNγ secretion with intrinsic resistance to Treg-mediated immunosuppression. Similarly, CD19-redirected CAR T-cells engineered to constitutively express the γc-cytokines IL-2, IL-7, IL-15 or IL-21 all demonstrated enhanced eradication of systemic $CD19^+$ B-cell malignancies in NSG mice.[217] Despite IL-2 and IL-15-transduced CAR T-cells exhibiting greater effector function, IL-7 and IL-21-transduced cells proved most efficacious in vivo with prolonged persistence most apparent in the IL-15 and IL-21-transduced constructs. This type of approach may prove synergistic with OVs armed with complementary cytokines. Thus far, few preclinical studies have been published exploring the combination of cytokine-armed OVs with CAR T-cell therapy. One example has utilized a mesothelin-redirected 4-1BB-containing second generation CAR with an adenoviral OV armed with either IL-2 or TNFα or both in a highly immunosuppressive human PDAC xenograft model.[218] Here, combination therapy was shown to outperform CAR T-cell monotherapy with the carriage of IL-2 and TNFα by the OV preventing the emergence of metastatic disease. In an immune competent model, the treatment was associated with the polarization of macrophages toward an M1 phenotype with evidence of enhanced dendritic cell maturation, the generation of TNFα-inducible chemokines (including CCL-2, CCL-5 and CXCL-10) and the local infiltration of both non-CAR host T cells in addition to adoptively transferred donor CAR T cells.

The IL-1 family member IL-33 is also emerging as a potentially important modulator of anti-tumoral immune responses. This cytokine has typically been characterized as an "alarmin," capable of alerting the immune system to tissue damage, necrosis and infections.[219] Various isoforms have been identified with context-dependent effects on a plethora of immune cells depending upon the tumor type, site of expression and local concentration. CEA-redirected CARs armed with IL-33 are being evaluated in pre-clinical studies.[220]

Similarly, CD19-redirected CAR T-cells armed with a membrane-bound IL-15/IL-15Rα fusion molecule (delivered via electroporation utilizing a Sleeping Beauty DNA transposon plasmid) are able to demonstrate both *trans*- and *cis*-mediated IL-15-induced activation, leading to superior in vitro expansion and in vivo persistence.[116] These CAR T-cells also exhibited a significantly enhanced $CD45RO^-$ $CCR7^+$ $CD95^+$ T_{SCM} memory phenotype despite multiple rounds of ex vivo activation.

In a similar approach to G47Δ-hFLT3-L, HER2-redirected CAR T-cells have been engineered to constitutively secrete Flt3L, leading to an expansion of mature $CD103^+$ DCs with enhanced antigen presenting capabilities.[221] Adoptive transfer of Flt3L-secreting CAR T-cells was associated with significantly enhanced tumor growth inhibition and improved mouse survival compared to control CARs.

5.3 OV-mediated delivery of CAR T-cell attracting chemokines

CAR T-cells are designed to target tumor cells in a manner that bypasses DC-mediated antigen cross-presentation. This is both a strength and a weakness of the CAR paradigm. On the one hand, CAR T-cells are able to function in an MHC-unrestricted manner and do not require direct interactions with DCs to exert their cytolytic function. Yet on the opposing side, the chemokine gradients, cell adhesion molecules and cytokines that foster the trafficking of endogenous activated CTLs into the TME may be unavailable to adoptively transferred CAR T-cells. Thus, these cells commonly encounter significant hurdles in homing to tumors and, as evidenced by bio-distribution data, are prone to mis-localization/sequestration in non-target tissues. For example, shortly following parenteral administration, CAR T-cells are commonly detected in the lungs, having exited the right ventricle and become trapped in alveolar microvasculature.[222] Low level expression of target antigen by the pulmonary tissues can therefore induce significant local and systemic toxicity, notwithstanding a potential impact on onward CAR T-cell trafficking. Should CAR T-cells reach the peripheral margins of tumors and/or their draining lymph nodes, they continue to encounter specific issues with adhesion, rolling and extravasation from the circulation due to: mismatched chemokine/chemokine receptor expression; the downregulation of relevant cell adhesion molecules by endothelial cells; aberrant microvasculature; dense tumor-associated stroma; and an immunosuppressive milieu within the TME.[223]

In an attempt to overcome this issue, CAR T-cells may be delivered directly into the tumor by intra-tumoral or regional injection.[224] Tumors predominantly localized to the pleural or peritoneal cavity may be particularly amenable to direct instillation. Several clinical trials are currently exploring this approach, including the phase I T4 trial based at Guy's and St Thomas' NHS Foundation Trust which involves the direct intra-tumoral delivery of an autologous CAR T-cell product incorporating a second generation CAR targeting ErbB homo- and heterodimers (T1E.28ζ) with a chimeric cytokine receptor fusion protein (4αβ) in patients with locally advanced or metastatic squamous cell carcinoma of the head and neck.[225] Promising preliminary data for this approach has recently been reported at AACR 2019. In a phase I clinical trial based at MSKCC patients with mesothelin positive malignant pleural disease were treated with a mesothelin-redirected CD28-containing second generation CAR in conjunction with an inducible caspase 9 safety switch (icasM28z).[226] Transduced autologous T-cells were injected directly into the pleural cavity. A total of 21 patients were treated over a period of 38 weeks. In 13 patients CAR T-cells were found to persist in the peripheral blood with a concomitant fall in serum free mesothelin. The majority of patients went on to receive an anti-PD-1 immune checkpoint inhibitor with several prolonged responses on cross-sectional imaging. However, while this approach may prove effective for locally advanced disease, more widespread disseminated metastatic disease would be expected to benefit more from systemic delivery. In addition, due to their location and size many solid tumors may not be ideally suited to targeting in this fashion.

The molecular basis of CAR T-cell exclusion from the TME relates in part to the aberrant secretion of chemokines by both tumor cells and tumor-resident host cells that are often subject to reprogramming. Cell adhesion molecules may be downregulated or aberrantly expressed on the luminal surfaces of tumor-associated vascular and lymphatic endothelium.[227]

The CAR T-cells themselves may lack coordinated expression of chemokine receptors and/or cell adhesion molecules that would assist otherwise assist tumor entry. The latter may stem in part from the inability of CAR-mediated activation to recapitulate physiological TCR signaling but can also reflect the phenotypic maturation of CAR T-cells, which may not be favorable to migration and tumor entry. A number of important preclinical studies in mice have shed light on the ways in which tumors can evade immune surveillance by modulating the trafficking of key players involved in T-cell priming. It has become abundantly clear that the role of professional $Batf3^+$ DCs (in humans the $CD103^+$ $CD141^+$ subset in particular) is central to mounting an effective anti-tumor immune response.[228,229] In the afferent limb of the host anti-tumoral immune response (which is bypassed in the context of CAR T-cell therapy), these specialized cells, which are actively recruited into the TME by tumor-derived chemokines such as CCL-4,[230] are able to scavenge TAAs and subsequently activate $CD8^+$ effector T-cells in draining lymph nodes using antigen cross-presentation (or cross-dressing[231]). In the efferent limb, these DCs also mediate effective $CD8^+$ T-cell trafficking and tumor entry by responding to danger signals within the TME and secreting type I interferons, which in turn switch on interferon-inducible genes including the T-cell chemoattractant chemokines CXCL-9 and CXCL-10. Concurrently, primed T-cells upregulate cognate chemokine receptors, particularly CXCR3, and home to the inflamed TME.

In experiments utilizing transgenic $Batf3^+$ knock out mice, the vital role of $Batf3^+$ $CD8\alpha^+$ professional DCs in the immunogenic rejection of tumors has been elucidated. In a melanoma model, $Batf3^+$ DC function was found to be critically dependent upon Wnt/β-catenin signaling within the tumor cells themselves. Overactivation of this pathway was found to mediate transcriptional repression of CCL-4, which in turn impaired the trafficking and function of $Batf3^+$ DCs within the TME.[230] The enforced absence of these cells leads in turn to a non-inflamed TME unable to support an effective adaptive immune response. CAR T-cells commonly express high levels of CXCR3 at the time of ACT, but despite benefiting from signal 2 derived from the CAR endodomain (and potentially signal 3 in fourth generation constructs), they would still be expected to require an inflamed TME with high levels of chemoattractant chemokines necessary for efficient trafficking and tumor entry. While the exogenous delivery of pro-inflammatory DC-derived cytokines (such as IL-12) can to some extent restore or enhance an adaptive immune response, OVs, which can commonly induce the local release of type I IFNs and may also be armed with matching chemokines, may prove to be the optimal partners for combination immunotherapy utilizing CAR T-cells.

The central role of the IFN-inducible C-X-C motif chemokine ligands CXCL-9 and CXCL-10 in mediating effective anti-tumor immunity is further evidenced by their repeated detection in "inflamed" immunophenotypic signatures seen using tumor transcriptome analysis, which are often prognostic and in the case of anti-PD-1/PD-L1 ICB predictive of long-term response.[232,233] This has been confirmed in a number of studies involving patients with a variety of solid tumors including high grade serous ovarian cancer, breast and colorectal cancer. In a preclinical study involving combination therapy with an HSV strain OV and OVA-specific T-cells derived from OT-I TCR transgenic mice, ACT-mediated control of PDAC tumors in immunocompetent and incompetent mice was improved with OV infection and was associated with a chemokine profile in the TME enriched for CXCL-9 and CXCL-10.[63] This profile was associated with increased T-cell migration and persistence within the TME, conferring a significantly enhanced therapeutic response. Similarly, OVs have been engineered to express chemokines designed to augment the trafficking of CTLs or other immune cells (such as DCs) into the TME. For example, utilizing a syngeneic murine mesothelioma model, direct intratumoral delivery of a double-deleted vaccinia OV strain (vvDD) armed with CXCL-11 demonstrated an ability to enhance the infiltration of tumor-specific $CD8^+$ T-cells and was associated with both a reduction in the expression of a number of immunosuppressive molecules including COX2, TGFβ and CCL-22 (a Treg chemoattractant) and a concomitant increase in local mediators of cytotoxicity such as granzyme B.[234] Moreover, the approach was also able to foster systemic immunity, evidenced by an increase in the numbers of tumor-specific IFNγ-producing $CD8^+$ effector T-cells present in splenic and lymphoid tissue, and an overall survival benefit in treated animals. A second research group have examined the potential synergy of combining CXCL-11 with ACT using mesothelin-redirected CAR T-cells in both syngeneic and human mesothelioma xenograft mouse model.[235] Here, CXCL-11 was delivered to subcutaneous tumors either via local injection of a CXCL-11-armed vaccinia virus or via expression in adoptively transferred T-cells transduced with a bicistronic lentiviral transgene cassette encoding both CXCL-11 and a second generation 4-1BB containing anti-mesothelin CAR (SS1BBz). While the use of either strategy could increase CXCL-11 protein levels in tumor, only vaccinia-delivered CXCL-11 could enhance the anti-tumor efficacy of subsequent ACT utilizing mesothelin-redirected CAR T-cells or endogenous T-cells following vaccination with an adenoviral E7 cancer vaccine (Ad.E7).

Similarly, a vvDD vaccinia OV strain armed with CCL-5 (otherwise known as regulated on activation, normal T cell expressed and secreted (RANTES)) was found to augment adaptive immunity compared to an unarmed control in a syngeneic murine colorectal cancer model.[236] Here, treatment with vvCCL-5 could arrest Treg infiltration, while facilitating a skewed Th2-predominant immune infiltrate. The addition of the chemokine was also associated with an increase in viral persistence in vivo. Responses could be improved further by combining vvCCL-5 with a DC1 vaccination schedule. This was able to elicit Th1 skewing with high surface expression of CCR5 on activated $CD8^+$ T-cells within the TME. Utilizing a conditionally replicating adenovirus selective for TERT positive tumor cells and armed with both CCL-20, a chemoattractant molecule for iDCs, and the co-stimulatory ligand CD40L, which can foster the phenotypic maturation ofDCs and induce the expression ofTh1-supportive cytokines as well as death receptor ligands, Liu et al. were able to show enhanced CTL-mediated tumor cytotoxicity in vitro.[237] In vivo efficacy remains to be evaluated but the combined approach would seem to promising. The cytokine CCL-19, through its interaction with CCR7 on mature DCs, early/central memory T-cells and cytokine induced killer (CIK) cells, is able to elicit coordinated lymph node homing. A similar vvDD strain armed with CCL-19 demonstrated an ability to enhance DC and T-cell infiltration in syngeneic mouse tumor models.[238] CCL-19 carriage conferred enhanced safety with more rapid clearance of armed virus from normal tissues. Use of this OV may prove particularly efficacious with ACT utilizing early/central memory T-cells that highly express CCR7 or with engineered/flow-sorted CAR T-cells.

The delivery of certain chemokines to the TME is to a certain extent tempered by controversy—while the IFNγ-induced CXC chemokines have been associated with improved prognosis in some studies (e.g., CXCL-9 and 10 in advanced high grade serous ovarian cancer[239] and CXCL-9 in CRC[240]), other studies have implied a poor prognosis (e.g., CXCL-10 and its receptor CXCR3 in stage II CRC[241] and serum CXCL-9 in nasopharyngeal carcinoma[242]). Indeed, in certain circumstances tumor-associated CXCL-11 may induce infiltration by TAM and is associated with poor prognosis in colorectal cancer.[243] Fractalkine (otherwise known as CX3CL-1), has also been associated with an increased propensity to invade and metastasize in several solid tumors, including lung, ovarian and prostate cancer. The cognate receptor CX3CR1 has been shown to be a negative prognostic marker in ovarian cancer, where overexpression inversely correlates with survival in post-menopausal patients with advanced disease.[244]

In certain contexts, different chemokine receptor isoforms can exhibit functional differences, potentially explaining some of these disparities. For example, there are at least three known isoforms of the chemokine receptor CXCR3—CXCR3A, CXCR3B and CXCR3alt, which may be expressed to varying degrees by tumors and immune cells.[245] On effector T-cells, CXCR3A is the main isoform that elicits chemotaxis; while tumor cell CXCR3A is known to be pathogenic and may drive invasion and metastasis following engagement with CXCL-9 and activation of downstream PI3K and MAPK signaling pathways. Expression of CXCR3B, on the other hand, has been shown to be anti-angiogenic with engagement by CXCL-9 having a negative impact on endothelial cell proliferation and tumor angiogenesis. CXCL3-alt, generated via alternative splicing of mRNA, is the least well characterized isoform and is unable to signal in response to either CXCL-9 or CXCL-10. This isoform may therefore have a negative regulatory role by acting as a sink for excess chemokine. Due to the complexity of these pleiotropic effects on both tumor cells and immune cells, the enforced expression of certain chemokines could be expected to have unexpected deleterious consequences. Some examples of potential pitfalls with armed OVs are suggested from preclinical studies highlighting the context-dependent role of certain chemokines in different tumor models. Neuroendocrine differentiation in a murine model of colorectal cancer has been shown to drive the recruitment of immunosuppressive TAMs into the TME due to an upregulation in local CXCL-10 and CXCL-11, leading to immune evasion and poorer outcomes.[243] Likewise, CXCL-9 has been implicated in the infiltration by Tregs of certain tumors and may be able to induce $CXCR3^+$ B-cells to polarize immunosuppressive M2b TAMs.[246] In this context, OVs have also been designed to counter the pro-tumorigenic role of certain chemokine/chemokine receptors interactions involved in promoting tumor cell extravasation and metastasis. For example, a vaccinia strain engineered to disrupt the CXCL-12/CXCR4 axis using a peptide antagonist of CXCR4 fused to an IgG-derived Fc fragment has been evaluated in immune competent mice with orthotopically delivered mammary tumors.[247] Prior evidence has implicated this axis in tumor immune escape due to a direct effect on tumor cell survival, invasion and stemness as well an indirect effect elicited through the recruitment of a myeloid-rich "distal stroma" and angiogenesis.[248] In this model the delivery of this armed virus led to the destruction of tumor-associated vasculature and was able to both mediate primary tumor regression and limit the formation of spontaneous metastases.

5.4 OV expression of bispecific T-cell engagers

The ability of OV infection to resculpt the TME and engender enhanced adaptive immunity has been extended by combining them with agents designed to engage effector T-cells via the TCR complex. Such agents are typically bispecific adaptor proteins whereby one arm is able to engage with a cell surface TAA and the other with CD3ε.[249] One prototypic bispecific T-cell engager (BiTE) with dual targeting of CD19 and CD3, namely Blinatumomab, has entered routine clinical practice for the treatment of Philadelphia chromosome-negative relapsed or refractory B-cell precursor ALL.[250]

A number of groups have explored potential synergies between OVs and BiTEs by either delivering them in combination or arming the OV itself with a BiTE that can be delivered directly to the TME due to the OV's tumor tropism and specificity. An example of the latter has been demonstrated by Wang et al., who have developed a novel T-cell engager armed vaccinia virus (TEA-VV) that is able to redirect T-cells to target the TAA EphA2.[251] Viral infection was thus able to elicit T-cell mediated bystander killing of non-infected tumor cells. Furthermore, in combination with a HER2-redirected CAR this virus could enhance the killing of dual $HER2^{+}$/$EphA2^{+}$ A549 cells in vitro, suggesting that such an approach may be particularly beneficial in addressing tumoral heterogeneity and preventing CAR-mediated antigen escape. Indeed, with the ability of the BiTE to redirect endogenous effector T-cells and the capacity of both the OV and CAR to induce an inflammatory TME, this combination is likely be particularly beneficial for cold tumors. The potential for dual delivery as a combination therapy may be further enhanced by the finding that TEA-VV could itself replicate in HER2-redirected CAR T-cells, suggesting that the latter may provide a means of delivering the former while shielding it from host elimination.

Separately, this approach may be applied to the tumor stroma, a potent physical and immunosuppressive barrier to T-cell-mediated anti-tumoral responses. Here, a TEA-VV encoding a secretory bi-specific T-cell engager consisting of two single-chain variable fragments specific for murine CD3 and fibroblast activation protein (FAP) was evaluated in vitro and in an immunocompetent B16 melanoma mouse model.[252] mFAP-TEA-VV was able to elicit bystander killing of non-infected FAP^+ stromal cells in the presence of murine T-cells in co-cultures and in vivo mFAP-TEA-VV infection led to higher viral titers compared to wild-type VV control and was associated to a lower metastatic tumor burden.

Similar strategies have been applied to adenoviral OVs. For example, the oncolytic adenovirus ICOVIR-15K has been engineered to express an EGFR-targeting BiTE (each arm derived from Cetuximab and Blinatumomab, respectively) under the control of the viral late promoter. Viral replication itself is dependent upon dysregulation of the retinoblastoma pathway. Incorporation of the BiTE led to robust T-cell activation, expansion, and bystander cell-mediated cytotoxicity in co-cultures.[253] Enhanced in vivo efficacy was seen in human NSCLC and colorectal cancer cell line derived xenografts (namely A549 and HCT116, respectively) compared to parental control virus and could be augmented further if used in combination with ACT using PBMCs or purified T-cells.

This same EGFR BiTE-armed adenovirus ICOVIR15K-cBiTE (or OAd-BiTE) has also been utilized in combination with CAR T-cell therapy with the aim of enhancing CAR-mediated cytotoxicity at low antigen site density. In this case, a second generation ICOS-costimulated anti-FRα CAR was investigated in combination with or without OAd-BiTE.[254] Alone, delivery of CAR T-cells to NSG mice bearing SKOV tumors was able to elicit transient disease control but ultimately tumor escape in the majority of treated animals. Post-mortem analysis of tumors revealed evidence of heterogeneous FRα expression but homogenous EGFR expression. Combination therapy, however, in NSG mice with subcutaneous HCT116 tumors was associated with a significant reduction in tumor volume and improved survival when compared to either agent utilized on its own. Further experiments utilizing tumor cell lines with low or very low FRα expression were able to show significant expansion of both FR-directed CAR T-cells and indeed irrelevant CAR T-cells in combination with OAd-BiTE.

A similar approach has been utilized to re-engineer the oncolytic group B adenovirus enadenotucirev by arming it with a BiTE cross-linking EpCAM expressed on tumor cells with CD3 on T-cells.[255] Transcription of the BiTE was placed under the control of the virus major late promoter, ensuring that BiTE expression was localized to tumor cells permissive to viral replication. Utilizing primary pleural effusions and peritoneal malignant ascites, the approach was found to enhance enadenotucirev-mediated cytotoxicity via both $CD4^+$ and $CD8^+$ T-cells.

An alternative strategy may be to utilize a bispecific engager with specificity for an OV-associated antigen. An early example of this approach was described by Haas et al. Here, a recombinant bispecific single-chain antibody was designed to dually target NDV-expressing hemagglutinin-neuraminidase (HN) and either CD3 or CD28.[256] Effector T-cells were thus redirected to NDV-infected tumor cells, leading to enhanced polyclonal T-cell-mediated cytotoxicity at nanomolar concentrations in vitro. Furthermore, sustained efficacy was seen if purified T-cells were co-incubated with this agent for 3 days and then serially transferred to new NDV-infected tumor cell monolayers.

5.5 CAR carriage of OV

The systemic delivery of OV may be compromised by a number of factors including poor tumor targeting and infectivity, sequestration at non-tumor sites and elimination by innate and adaptive immune responses. Toxicity, for example, mediated by cytokine release following systemic administration of adenoviral vectors may also be a concern.[257] The repeated administration of certain OV strains is liable to induce antibody and complement-mediated neutralization, particularly in individuals previously vaccinated against the vector (e.g., vaccinia). For these reasons and with the aim of leveraging the dual capabilities of OV and ACT at recontextualizing the TME and inducing epitope spreading, a number of groups have investigated the strategy of infecting tumor-targeting T-cells with OVs and administering these infected T-cells systemically as a combined product.[77,258,259] While such an approach would be expected to shield the OVs from immune recognition, the combined product remains subject to the inherent barriers that compromise the entry into the TME of both OVs and T-cells. Alternative strategies for optimizing the systemic delivery of OVs include administering cytotoxic agents such as cyclophosphamide or specifically ablating B-cells with rituximab. Complement has been inhibited in preclinical models with compstatin[69] or sublethal doses of cobra venom.[260] Particular OV strains may have intrinsic resistance to antibody or complement-mediated clearance such as extracellular enveloped virus (EEV)-high vaccinia strains.[261] Modification to the viral coat, e.g., with multi-layer ionic polymers may also render viral particles more resistant to immune evasion.[262]

Attempts have been made to improve ACT and indeed CAR T-cell therapy utilizing vaccination-based strategies to potentially enhance in vivo expansion and anti-tumor efficacy. In particular, several groups have attempted to utilize virus-specific (e.g., CMV[263] or EBV[183]) T-cells for CAR transduction targeting a number of solid tumors and hematological malignancies. Here, in an allogeneic setting the selection of clonal anti-viral T-cells may reduce GvHD and, more generally in the case of previously infected patients, the interaction of CAR T-cells with professional DCs presenting viral epitopes may aid their in vivo expansion and persistence. Likewise, attempts have been made to develop oncolytic virus-specific CAR T-cells for ACT. JX-594 (pexastimogene devacirepvec) is a TK-deleted Copenhagen-strain vaccinia virus dependent upon aberrant EGFR-Ras pathway activation and (like T-VEC) armed with recombinant GM-CSF, that has been investigated primarily in patients with advanced HCC. By analyzing PBMCs obtained from patients receiving intratumoral JX-594 and healthy donors previously vaccinated against vaccinia, Ngo et al. were able to identify VV-specific T-cells which could then be transduced with a HER2-directed CAR.[184] Anti-HER2 VV-specific CAR T-cells were able to target both VV peptide pulsed activated T cells and HER2-expressing tumor cell lines. Conceptually, dual treatment with JX-594 and ACT using HER2-redirected VV-specific T-cells may be anticipated to synergistically improve anti-tumoral efficacy. However, the viability of this approach is tempered by data suggesting that dual activation of the TCR and CAR on the same cell may in fact be detrimental. Both in a murine allogeneic hematopoietic stem cell transplantation model utilizing second generation CD28-costimulated CD19-redirected CAR[264] and in an immunocompetent syngeneic mouse model of $CD19^+$ B-cell ALL,[265] engagement of the endogenous TCR with its target antigen (OVA in the case of the former and the male minor histocompatibility antigen (HY) in the case of the latter) was found to impair CAR functionality and induce anergy and exhaustion. In the latter, this was particularly evident in the case of $CD8^+$ rather than $CD4^+$ CAR T-cells, which underwent AICD and exhibited poor persistence in vivo.

5.6 Chemokine matching

Attempts have been made to enhance CAR T-cell migration and entry into solid tumors by enforcing the expression of chemokine receptors specific to chemokines secreted by tumor cells and/or associated cells within the TME.

In the case of malignant pleural mesothelioma, mesothelin-targeting CAR T-cells have been re-engineered to co-express CCR2b, the cognate chemokine receptor ligand being CCL-2, which was found to be consistently overexpressed in the conditioned media from a panel of 11 human MPM-derived cell lines. Such an approach was found to enhance the ability of mesoCARs to traffic into MPM xenografts and aid tumor eradication.[266] Outside of MPM, such an approach may be expected to be synergistic with dual infection of a tumor-tropic OV armed with recombinant human CCL-2.

Separately in non-CAR models of ACT, this approach has been investigated over many years. For example, Kershaw et al. identified elevated levels of Growth-Regulated Oncogene-a (Gro-a or CXCL1) in a number of tumor cell lines as well as in the FNAs of melanoma samples from patients.[173] Activated T-cells themselves were found not to secrete CXCL1. The transduction of activated T-cells with a retroviral vector encoding the CXCL1 receptor CXCR2 facilitated the chemotaxis of transduced T-cells in response to tumor-derived CXCL1 and also interestingly induced T-cell secretion of IFNγ. Similarly, Di Stasi et al. have shown that co-expressing CCR4 with an anti-CD30 CAR on activated T-cells can elicit enhanced control of human Hodgkin lymphoma xenografts.[172] In this context CARs are able to circumvent the immunosuppressive role of the CCR4 ligands thymus and activation-regulated chemokine/CC chemokine ligand 17 (TARC/CCL-17) and macrophage-derived chemokine (MDC/CCL-22) that are both produced by Reed Sternberg cells and are able to mediate the local recruitment of $CCR4^+$ Th2 and Tregs.

In a preclinical model of hematopoietic stem cell transplant preconditioning, Arai et al. have shown that coexpression of CXCR4 with an anti-c-kit CAR led to enhanced bone marrow trafficking of CAR T-cells, improved depletion of resident c-kit^+ BM cells, enhanced engraftment of donor BM and improved chimerism.[174] Outside of this context, however, engagement of tumor or stroma-derived CXCL-12 with endogenous CXCR4 on T-cells has been shown to mediate the exclusion of effector T-cells from a number of solid tumors including PDAC.[267] By pharmacologically inhibiting CXCR4 with either AMD3100 (plerixafor, an agent utilized for the mobilization of BM-resident hematopoietic stem cells prior to transplantation) in combination with anti-PD-L1 in a PDAC model[268] or the l-RNA-aptamer NOX-A12 (olaptesed pegol) in combination with anti-PD-1 in a syngeneic colorectal cancer model,[269] led to increased intra-tumoral T-cell infiltration and enhanced disease control. Similarly, use of BQ-788, a specific endothelin B receptor (ETBR) inhibitor peptide is able to restore aberrant adhesion of T-cells to endothelium, which may occur in the context of impaired extravasation through tumor associated vasculature.[270] Here, T-cell adhesion appears to be compromised by the ability of ETBR to induce nitric oxide-mediated endothelial ICAM-1 downregulation and relocalization.

Finally, T-cells have been armed with the fractalkine (CX3CL1) receptor CX3CR1, thus improving their ability to home to fractalkine-rich tumor xenografts.[271] Caveats remain, however, when using such an approach due to the overexpression of many chemokine family members by non-tumoral tissues potentially leading to aberrant homing and toxic off-target effects. For example, fractalkine protein is very widely expressed at a moderate level in many organs including the lung, brain and GI tract and at least at an mRNA level at a very high level in neural cells present in the cerebral cortex.[272]

The ex vivo activation and expansion of T-cells (including CAR T-cells), whether it be with anti-CD3/CD28 dynabeads or with phytohemagglutinin (PHA) is associated with a significant upregulation of cell surface CXCR3, the main receptor for the IFNγ-dependent chemokine ligands CXCL-9, CXCL-10 and CXCL-11. This process mirrors the endogenous upregulation of CXCR3 in T-cells following their priming by DCs in the lymph node. $CXCR3^+$ effector T-cells are subsequently drawn into locally inflamed IFNγ-rich tissue, where they exert their cytotoxic function and further expand. CAR T-cell therapy may therefore be expected to contribute to a CXCR3-driven amplification loop, whereby their migration into the TME and activation following target engagement will further elevate local IFNγ levels, inducing additional secretion of CXCL-9/10, the further recruitment of endogenous T-cells (and indeed other CAR T-cells) and ultimately a highly inflamed TME susceptible to epitope spreading and bystander cytotoxicity. Due to the IFNγ-inducible upregulation of immune checkpoint ligands such as PD-L1, this positive feedback loop would be expected to benefit further from combination therapy with PD-1/PD-L1 axis blockade. In addition the potential benefits of further elevating CXCR3 levels on CAR T-cells has been demonstrated by Newick et al. Here, mesothelin-redirected CAR T-cells were further engineered to express a "regulatory subunit I anchoring disruptor" (RIAD) that interferes with the ability of ezrin and PKA to colocalize, thus blocking the ability of adenosine or PGE2 to negatively regulate TCR activation.[273] Enhanced in vivo efficacy, however, was found to be dependent upon enhanced CXCR3 expression, highlighting the importance of this molecule in driving adoptively transferred T-cells into the TME.

5.7 Delivery of pro-T-cell cytokines and chemokines in combination

It would be anticipated that the provision of pro-inflammatory cytokine(s) in combination with T-cell attractant chemokine(s) by an armed recombinant OV may be particularly favorable for effective CAR T-cell therapy targeting solid tumors. An example of this strategy combining an adenoviral OV (Ad5Δ24) armed with CCL-5 and IL-15 with third generation GD2-redirected CAR T-cells (incorporating tandem CD28 and OX40 endodomains) has been explored targeting neuroblastoma in a xenogenic model using immune-incompetent NSG mice.[274] Here, the intra-tumoral delivery of both CCL-5 and IL-15 was found to attract and promote the survival of CAR T-cells in the TME, while viral infection itself was associated with an increase in caspase-dependent cell death by tumor cells. Such an approach may also be expected to reduce potential toxicity mediated by the systemic delivery of pro-inflammatory cytokines such as IL-15. Naturally, the full synergy of such a strategy can only be partially evaluated using immune-incompetent animal models. Understanding the ability of both OVs and CAR T-cells to modulate the TME (particularly with regard to immunosuppressive players such as Tregs, MDSCs, TAMs and CAFs) will require the use of either murine CAR-cells or ideally the development of an optimized humanized mouse model.[275]

5.8 Delivery of immune co-stimulatory molecules

Various OVs with large genomes (such as vaccinia or adenovirus) have been engineered to express co-stimulatory ligands, thereby rendering infected tumor cells capable of providing activating signals to effector immune cells within the TME. The majority of such OVs have expressed TNF receptor ligand superfamily members such as OX40L,[276] 4-1BBL[277] and GITRL,[278] all of which are known to provide co-stimulatory signals to effector T-cells, with modulating effects on other immune cells depending upon cognate receptor expression. CD40L (either in membrane-bound or soluble form[279,280]) has also been utilized in various OVs and may elicit pleiotropic anti-tumoral effects due to the broad expression of CD40 by various immune cells, including DCs, B-cells and $CD4^+$ helper T-cells. Indeed, CD40/CD40L-mediated interactions between $CD4^+$ and $CD8^+$ T-cells have been shown to be supportive of an effector memory phenotype[281,282] and the activation of this pathway may prove to be particularly attractive in the context of CAR T-cell therapy. An oncolytic adenovirus engineered with trimeric membrane-bound CD40L (TMZ-CD40L) was evaluated in vitro and in vivo and found to induce the activation of myeloid and endothelial cells.[280] In a human PDAC xenograft nude beige mouse model, TMZ-CD40L infection was associated with phenotypic skewing of the macrophage compartment in tumor from M2 to M1. And when utilizing murine CD40L in immune competent pmel-1 TCR transgenic mice harboring gp100$^+$ B16F10 melanoma tumors, viral infection was associated with enhanced expansion and migration into the TME of adoptively transferred T-cells. Such an approach may prove to be particularly attractive when targeting tumors exhibiting a highly immunosuppressive M2-skewed microenvironment such as PDAC or GBM. Indeed recently released early phase clinical trial data has suggested promise in PDAC using a similar approach by combining cytotoxic chemotherapy with an agonistic monoclonal antibody targeting CD40 (APX005M).[283] Similarly, CAR T-cells have been engineered to constitutively express CD40L on their surface.[120] These modified CAR T-cells are able to enhance DC maturation and local secretion of Th1 pro-inflammatory cytokines including IL-12 (Fig. 1).

Similarly, OVs have been engineered with CD28 ligand family members including B7-1 (CD80), which when expressed by mature DCs is able to facilitate antigen cross-presentation to CD28-expressing effector T-cells. For example, a modified vaccinia Ankara (MVA) strain expressing B7-1 alone or in combination with two cell adhesion molecules, namely ICAM-1 (CD54) and LFA-1 (CD58)—the triplet termed TRICOM or TRIad of CO-stimulatory Molecules, is able to elicit potent cytotoxic T-cell activation and has been utilized in various vaccination strategies with specific TAAs (such as PSA,[284] CEA and/or MUC-1[285]) also engineered into the viral backbone. While early phase clinical trials have been promising in a number of tumor models with excellent tolerability, a recently reported phase III trial (PROSPECT) evaluating PSA-TRICOM (rilimogene galvacirepvec or PROSTVAC) in patients with metastatic castration-resistant prostate cancer did not demonstrate a survival benefit.[286]

5.9 Delivery of molecules targeting immune checkpoints

In a manner akin to endogenous T-cell exhaustion, it is becoming clear that ACT, whether comprising CAR T-cell, engineered TCR or TIL therapy, can be highly susceptible to the immunosuppressive effects exerted by inhibitory immune checkpoint molecules in the TME. A large number of immune checkpoints have been elucidated, many with pleiotropic, highly

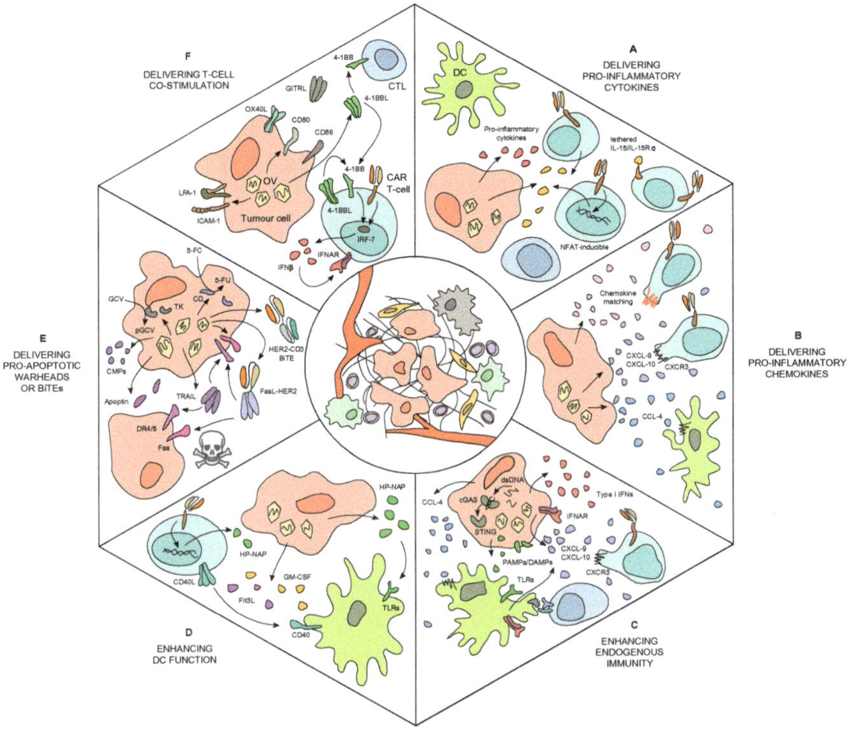

Fig. 1 Oncolytic virus and CAR T-cell-mediated immune stimulation. Oncolytic viruses and CAR T-cells may act in concert to promote an inflamed TME, enhancing and broadening host adaptive anti-tumoral immunity. Furthermore, the rational combination of next generation OVs and CARs may prove particularly efficacious due to their ability to enhance and support each other's function in a synergistic fashion. For example: (A) both OVs and CARs may be engineered to secrete pro-inflammatory cytokines, constitutively or in a controlled manner; (B) OVs may be designed to deliver chemokines capable of attracting CAR T-cells or endogenous immune cells into the TME, whereas the enforced expression of chemokine receptors on CAR T-cells may aid tumor homing; (C) OV-mediated release of PAMPs and DAMPs, coupled with the triggering of the cGAS-STING cytosolic DNA sensing pathway, can foster enhanced innate and adaptive immune responses ultimately leading to epitope spreading and a profound bystander effect; similarly, CAR T-cell activation and the killing of tumor cells may result in local inflammation, the release of type I and II interferons and a resculpting of the TME; (D) OVs and CARs may be further engineered to directly engage with professional antigen presenting DCs, for example by secreting cytokines (such as GM-CSF or Flt3), TLR-triggering agents such as HP-NAP or by expressing co-stimulatory ligands such as CD40L; (E) armed OVs have been engineered to deliver pro-apoptotic "warheads" such as death receptor ligands or fusion proteins, apoptin or CMPs, or alternatively bispecific T-cell engagers, which couple an antigen targeting moiety with a CD3-binding domain; and (F) both may provide effector T-cell co-stimulation—for instance, CAR T-cells armed with cell surface 4-1BBL may elicit 4-1BB-mediated co-stimulation in *trans* while a variety of OVs have been designed to express co-stimulatory ligands such as OX40L, 4-1BBL, GITRL, CD80 or CD86 as well as cell adhesion molecules such as LFA-1 or ICAM-1.

dynamic, or context-dependent effects; others remain poorly characterized with an incomplete understanding of their binders or indeed their role in effector T-cell modulation. The two main target pathways that have featured in the clinic comprise the interactions between CTLA-4 and B7-1/B7-2 and between PD-1 and PD-L1/PD-L2. Additional immune checkpoints that have recently been targeted within clinical trials include LAG-3, TIM-3, B7-H3, VISTA, CEACAM1, CD200 and TIGIT.[287,288] Other potential targets remain the subject of pre-clinical investigation and include HVEM/BTLA/CD160, KLRG-1 and LAIR-1.[289]

A number of studies have demonstrated that CAR and TCR-engineered T-cells commonly overexpress PD-1 following their adoptive transfer and are thus prone to PD-L1/PD-L2-mediated exhaustion and hypofunction in the TME.[290,291] Utilizing a mesothelioma xenograft model in NSG mice, mesothelin-redirected CAR TIL hypofunction was found to be associated with the upregulation of a number of inhibitory enzymes including diacylglycerol kinase (DGK) and SHP-1 and correlated with enhanced surface expression of a number of inhibitory immune checkpoint molecules including PD-1, LAG-3, TIM3 and 2B4.[292] Similarly, adoptively transferred T-cells may also be liable to AICD via interactions between death receptors and their ligands, which may often be overexpressed in the TME. Furthermore, CAR T-cells may exhibit an intrinsic propensity to undergo AICD.[293] In a study utilizing HER2 and CEA-redirected first and second generation CARs, all constructs studied were found liable to progressively upregulate Fas, FasL, DR5 and TRAIL, leading to programmed cell death independent of antigen-mediated TCR or CAR activation. Additional preclinical studies utilizing TCR-clonal (anti-HY or EBV-specific) CAR T-cells as well as clinical trials evaluating allogeneic CD19 CARs have, on the other hand, suggested that CAR T-cells may be particularly prone to TCR-mediated anergy and AICD.[264,265]

A number of clinical trials have been set-up to examine the potential of combining CAR T-cell therapy with immune checkpoint inhibitors (such as monoclonal antibodies targeting PD-1, PD-L1 or CTLA-4).[294] However, it is conceivable that combination therapy utilizing CAR T-cells with OVs engineered to express immune checkpoint blocking agents (ICBs) may prove more fruitful due to the enhanced, pleiotropic innate and adaptive immunomodulatory effects provided by the latter and the ability of OVs to potentially prevent the emergence of acquired resistance to ICBs by broadening the neoantigen targeting repertoire of endogenous effector T-cells. Furthermore, the tumor tropism of OVs would be expected to ensure that ICBs are delivered in a more targeted and potentially less toxic fashion. As such a number of ICB-armed OVs have undergone preclinical evaluation. These include an adenoviral OV engineered to express a full length IgG2 anti-hCTLA4 mAb[295] and a Western Reserve (WR) vaccinia strain armed with different ICBs against murine PD-1.[296] The latter was targeted utilizing either a full-length hamster IgG mAb or with Fab or scFv versions. The viruses were evaluated in immune competent murine syngeneic melanoma and fibrosarcoma models. Following direct intra-tumoral delivery, these ICB-armed OVs were able to induce a very robust infiltrate of both $CD8^+$ and $CD4^+$ lymphocytes with a reduction in the proportion of Tregs. OVs armed with whole mAb or scFv were able to confer a significant survival advantage in treated mice compared to unarmed OV controls. In a similar fashion, one may also conceive of OVs able to induce the local secretion of ICBs targeting other immune checkpoints or potentially downregulate tumor cell surface inhibitory ligands such as PD-L1, B7-H3 or CD47 (a "don't eat me" signal designed to inhibit phagocytosis by binding SIRPa on macrophages) by utilizing a number of strategies such inhibitory short hairpin RNAs (shRNAs), ribozyme switches, intrabodies (intracellular antibodies) with or without ER-retention motifs.

Combination immunotherapeutic strategies can have unexpected outcomes due to the complex interplay between key players in the system. Using immune competent syngeneic murine models of colorectal cancer and renal cell carcinoma, Rojas et al. have evaluated various combination approaches using various vaccinia OV strains with an anti-CTLA-4 blocking antibody or Treg-depleting anti-CD25 mAb.[297] Whereas initiation of anti-CTLA-4 therapy 4 days after viral delivery led to robust efficacy, simultaneous delivery of both agents revealed no efficacy due to rapid viral clearance mediated by virus-specific CTLs. The role of Treg-mediated immunosuppression was found to be uncertain in these models given that combination therapy with a CD25-targeting antibody failed to confer any therapeutic advantage to vaccinia virotherapy. However, modification of the vaccinia strain by deleting a secreted type I IFN binding protein (encoded by the B18R gene) leading to an "immune activated" phenotype was able to increase the potency of the combination strategy significantly.

Taking these strategies a step further, the use of an OV armed with an anti-PD-L1 minibody (comprising an anti-PD-L1 scFv fused to the hinge, CH2 and CH3 domains of human IgG1, driven off a CMV promoter) has recently been evaluated in combination with a HER2-redirected second generation CD28 endodomain-containing CAR in a human prostate cancer xenograft model using NSG mice.[298] Here, combination therapy utilizing the anti-PD-L1 minibody-armed OV outperformed CAR T-cell monotherapy, combined treatment with an unarmed control adenoviral vector and the latter in conjunction with an infused mAb targeting PD-L1, highlighting the benefits of producing the minibody directly within the TME. This approach has been further investigated in a head and neck squamous cell carcinoma xenograft model, whereby an intra-tumorally delivered adenoviral OV vector armed with both the aforementioned anti-PD-L1 minibody and a pro-inflammatory cytokine (heterodimeric IL-12p70) was combined with systemically-delivered HER2-redirected CAR T-cell therapy.[299] Combined treatment with both modalities was able to prolong survival in NSG mice beyond 100 days versus approximately 25 days when either approach was utilized alone as a monotherapy. Orthotopic models incorporating locoregional nodal metastases could also be targeted using this approach. However, distant disease could only be eradicated by CAR T-cell therapy in the presence of the dually-armed OV. Due to the locoregional delivery of the OV-armed transgenes, this approach is also likely to have benefits in terms of safety and indeed serum concentration of IL-12 was found to be 20–100 times lower compared to tumor levels.

A similar approach has been undertaken with regard to CAR T-cell therapy, which is also highly amenable to bespoke engineering and customization. As such, CAR T-cells have been designed to secrete antibodies targeting PD-L1[300] or scFvs targeting PD-1.[126] However, in the absence of a tumor-homing or regulatory mechanism the constitutive expression of secreted ICBs by CARs may prove liable to induce systemic toxicity in a similar fashion to the mAbs currently available in the clinic. Potentially more nuanced approaches include: utilizing CRISPR/Cas9 to knock out inhibitory receptors expressed by CAR T-cells (such as PD-1[301] or Fas[302]); co-expressing cell surface dominant negative receptors such as dominant-negative PD-1,[123] TGFβR[124] or Fas[122]; or by using "switch receptors" capable of converting negative signals into positive outputs, e.g., PD-1/CD28,[164] PD-1/4-1BB, CD200R/CD28[303] or Fas/4-1BB.[304] Alternatively, downstream inhibitory circuits within CAR T-cells can themselves be rewired. For example, the efficacy of murine mesothelin-redirected CAR T-cells have been shown to be greatly enhanced in transgenic mice deficient in both isoforms of DGK (dgkα/dgkζ double knock out), an inhibitory enzyme highly expressed by effector T-cells and able to limit downstream Ras/ERK activation by depleting the second messenger diacylglycerol (DAG).[305] Similarly, EGFRvIII-redirected CAR T-cells with DGK specifically knocked out using CRISPR/Cas9 have demonstrated enhanced tumor infiltration and CAR T-cell effector function in a GBM xenograft murine model.[306]

5.10 Delivery of molecules targeting immunosuppressive cells

Should CAR T-cells succeed in trafficking into solid tumors, commonly they continue to encounter significant obstacles in the form of various highly immunosuppressive immune cells, capable of stifling CTL activation and in situ expansion. Many solid tumors, to varying degrees, harbor an abundance of Tregs, TAMS, MDSCs, CAFs and dysfunctional endothelial cells.[307] Many OVs have now been designed to target these immunosuppressive cell populations in various ways and would be anticipated to have synergistic effects when combined with ACT, including CAR T-cell therapy.

Regulatory T-cells and TGFβ, a cytokine supportive of the maintenance and proliferation of the former while simultaneously suppressing the activation, maturation and differentiation of both CTLs and DCs are important mediators of immunosuppression in the TME. Tregs themselves are potent secretors of TGFβ (as well as other immunosuppressive cytokines such as IL-10), inducing autocrine and paracrine activation in a feedback loop manner.[308] Despite numerous attempts, no clinical candidate mAbs or small molecule inhibitors targeting the TGFβ pathway have yet demonstrated sufficient safety and efficacy in randomized phase III clinical trials.[309] However, a novel TGFβ trap/anti-PD-1 fusion protein (bintrafusp alfa or M7824) has recently emerged in the clinic, exhibiting promising overall response rates and evidence of a good therapeutic index in several phase I/II trials.[310] Currently, this agent is being compared head to head in a large randomized phase II trial with single agent pembrolizumab in patients with treatment-naïve PD-L1 high NSCLC (NCT03631706) and is also being advanced in patients with other solid tumors, particularly cervical SCC. An adenoviral vector has been used to target the TGFβ pathway by being armed with decorin, a small leucine-rich cellular or pericellular matrix proteoglycan family member that is able to block TGFβ-mediated gene expression.[311] This OV, also armed with the pro-inflammatory cytokine IL-12, was able to limit the infiltration by Tregs of mammary tumors and draining lymph nodes, while enhancing the entry and expansion of $CD8^+$ effector T-cells. Similarly, an adenoviral OV engineered to express both decorin and a soluble Wnt decoy peptide (sLRP6E1E2), chemically conjugated to a pegylated neurotensin-based targeting moiety, was able to demonstrate improved anti-tumor efficacy in an orthotopic PDAC nude mouse model and was facilitated by both enhanced ECM degradation and the depletion of a number of Wnt signaling-related factors (including Wnt, β-catenin, and/or vimentin) in the TME.[50] Furthermore, by targeting the OV to neurotensin receptor (NTR) positive tumor cells, the OV was able to demonstrate improved efficacy and tumor tropism.

Tumor-associated TGFβ is also implicated in the epigenetic transition of normal fibroblasts into so-called cancer associated fibroblasts (CAFs), which can not only modulate tumor-associated stroma, tumor stiffness and osmotic pressure within the extracellular matrix but can, via paracrine TGFβ secretion, induce local tumor cells to undergo epithelial-mesenchymal transition (EMT),[312] further mediating immune evasion via numerous mechanisms (including downregulation of MHC class I and/or death receptors, upregulated PD-L1 expression, increased heterogeneity or intrinsic resistance to CTL-mediated cytotoxicity).[313] CAFs, themselves, may be prone to infection by OVs, thus acting as potential sinks that can protect associated tumor cells. However, their inadvertent infection may also lead to enhanced secretion of fibroblast growth factor 2 (FGF2), a cytokine that in turn may render tumor cells more susceptible to OV infection.[314] Likewise, a Maraba MG1 OV armed with human FGF2 demonstrated enhanced tumor tropism and oncolytic potential.

Glucocorticoid-induced TNFR-related protein (GITR) is a costimulatory receptor and member of the tumor necrosis factor receptor superfamily. In effector T-cells it is rapidly upregulated from a low basal state following NFκB-mediated activation. Subsequent binding of GITR ligand (GITRL) is then able to foster proliferation, maturation, pro-inflammatory cytokine release[315] and resistance to Treg-mediated suppression.[316] Tregs themselves, on the other hand, exhibit high level constitutive expression of GITR[317] but are liable to be depleted or rendered dysfunctional in the presence of agonist antibodies. The use of a murine anti-GITR agonist antibody in several mouse tumor models has been shown to impair the accumulation of intra-tumoral Tregs, while enhancing $CD8^+$ effector T-cell function and anti-tumor immune responses.[318] Similarly, a humanized antibody (MK-4166) has been shown to inhibit the generation of Tregs in mixed lymphocyte reactions and downregulate Foxp3 expression in ex vivo tumor culture TIL populations.[319] This agent is currently undergoing evaluation in early phase clinical trials (NCT02132754 and NCT03707457). Similarly, an adenoviral OV has been armed with either membrane-bound or soluble GITRL and delivered by direct intra-tumoral injection into syngeneic murine melanomas, leading to a significantly enhanced local infiltrate of both $CD4^+$ and $CD8^+$ T-cells.[320] Tumor growth retardation was maximized in the group receiving the OV armed with soluble GITRL.

Attempts have also been made to deplete Tregs utilizing various pre-conditioning agents prior to OV infection. Similar lymphodepleting conditioning regimens have been utilized in ACT and have proved particularly efficacious in enhancing CAR T-cell expansion in patients with hematological malignancies.[146] In one example, the systemic administration of a VSV OV strain in a syngeneic murine melanoma model was preceded with a non-toxic low dose of IL-2, resulting in a mild NK-dependent vascular leak-type syndrome followed by enhanced Treg depletion.[321] IL-2 conditioning led to enhanced extravasation of VSV from the circulation, improved tumor localization, tumor-specific viral replication and ultimately animal survival.

In addition, Treg inhibition and potentially depletion has been highlighted in many murine models of anti-CTLA-4 antibody therapy.[322] Tregs constitutively express CTLA-4, which is able to mediate target cell immune suppression and tolerance by sequestering B7 ligands on professional DCs and activated effector T-cells. Ipilimumab is a monoclonal antibody targeting CTLA-4 that is currently approved for the management of patients with malignant melanoma either alone or in combination with the anti-PD-1 mAb nivolumab.[323] The combination has also been recently approved for patients with intermediate or poor risk treatment-naïve advanced RCC[324] as well as in previously treated patients with MSI-H-/dMMR metastatic colorectal cancer.[325] The agent, however, is associated with significant risks of systemic autoimmune toxicity, leading to a number of second-generation agents designed to improve its safety profile. These include a nonfucosylated version with increased ADCC activity and a probody format designed to be selectively cleaved and rendered functional in the TME.[326] Alternatively, tumor tropic OVs may be able to deliver anti-CTLA-4 agents preferentially to the TME. In this way, use of an oncolytic adenovirus (Ad5/3-Δ24aCTLA4) engineered to express a full-length human IgG2 subtype anti-CTLA-4 mAb has demonstrated a more favorable tumor:plasma concentration ratio, suggestive that this approach may raise the therapeutic index of this class of agents.[295]

Other potential approaches to target Treg-mediated immune suppression with recombinant OVs may involve arming the latter with agents that block Treg-attractive chemokine ligands (such as CCL-1, CCL-17 or CCL-22)[327] or their cognate receptors (such as CCR4 or CCR8); agents that block or reverse the effects of Treg-supportive cytokines in addition to TGFβ such as IL-10 or IL-35[328]; and agents that disrupt other important Treg pathways such as the neuropilin 1/semaphorin axis.[329]

Myeloid cells, including prototypic MDSCs, TAMs and neutrophils may also prove highly immunosuppressive and can help engender a metastatic niche favorable to immune evasion.[330] Such tumors, which commonly exhibit a "mesenchymal" gene expression pattern (e.g., consensus molecular subtype 4 (CMS4) in colorectal cancer[331]), are typically resistant to standard forms of immunotherapy utilizing ICBs and are also proving particularly difficult to treat with ACT. While various agents capable of targeting MDSCs are well known in the clinic, including bisphosphonates, various cytotoxic chemotherapeutic agents (such as Gemcitabine or 5-FU), c-kit and/or VEGFR-targeting TKIs such as sunitinib, retinoic acid receptor ligands such as ATRA, non-specific disruptors of iNOS and ARG1 (including sildenafil), vitamin D3 and COX2 inhibitors such as celecoxib, more selective targeted agents, such as small molecule inhibitors of arginase, CSF-1 or PFKFB3 inhibitors remain experimental.[332] Here, OVs may prove to be particularly efficacious if able to pre-empt early MDSC-mediated clearance. One such OV, a vaccinia strain engineered to express the PGE2-inactivating enzyme 15-prostaglandin dehydrogenase (HPGD) has been shown to induce a reduction in local CXCL-12 levels, leading to a reduction in tumor infiltrating granulocyte-like MDSCs.[66] However, the importance of tailoring OV therapy to the particular context is demonstrated by the failure of a similar vaccinia strain armed with mGM-CSF to elicit effective anti-tumor adaptive immunity in similar tumor models exhibiting high basal infiltration by G-MDSCs.

In a similar vein, treatment of neuroblastoma xenograft-bearing immune compromised mice with third generation anti-GD2 CAR-transduced T-cells (incorporating tandem CD28, OX40 and CD3ζ activating endodomains) could be enhanced when combined with all-trans retinoic acid (ATRA), currently used in the clinic in patients with APML. Here, ATRA was able to deplete monocytic MDSCs and limit G-MDSC-mediated immunosuppression, presumably by impairing the proliferation of bone-marrow derived $CD34^+$ progenitor cells and impacting upon MDSC myelopoiesis.[333] Similarly, ATRA has been incorporated as an adjuvant therapy with an adenoviral OV armed with a TK suicide switch.[334] One can envisage further approaches focused upon delivering ATRA to the TME, potentially utilizing enzymes involved in ATRA biosynthesis, such as RDH10, RALDH2 and CRABP2.[335]

5.11 Delivery of molecules targeting immunosuppressive metabolic pathways within the tumor microenvironment

OV replication and, to an even greater extent CAR T-cell function, may be also be compromised due to metabolic reprogramming in the TME. Indeed, many studies have shown that endogenous T-cell activation and differentiation may be rendered dysfunctional by so called "metabolic checkpoints,"[336] which include but are not limited to: tumor hypoxia, low pH, low glucose, oxidative stress and tryptophan depletion (coupled with high kynurenine levels) due to overexpression of IFNγ-induced indoleamine 2,3-dioxygenase 1 (IDO1).[337] Other interrelated factors that might be expected to induce T-cell (and indeed CAR T-cell) hypofunction include a preponderance of immunosuppressive soluble small molecules including inhibitory prostaglandins (e.g., PGE2) or adenosine, electrolyte imbalances (especially local hyperkalemia caused by the release of intracellular potassium by dying tumor cells), or the depletion of other key amino acids such as arginine or cysteine. The infection of tumor cells by OVs designed to reverse metabolic dysfunction in the TME may therefore prove synergistic with CAR T-cell therapy.

The COX2/PGE2 pathway is one such metabolic checkpoint that has been targeted by researchers working with engineered OVs. PGE2 has been demonstrated to exert a broad range of pleiotropic effects on immune cells in the TME and has been implicated as a mediator of intrinsic or acquired resistance to ICBs.[338] More specifically, it can suppress the effector functions of macrophages and neutrophils, negatively regulating Th1 effector cells while supporting Th2, Th17, and Treg responses.[339] Furthermore, via altering patterns of chemokine expression, PGE2 can negatively impact the trafficking of pro-inflammatory cells, while promoting the local accumulation of MDSCs and Tregs. A Western Reserve vaccinia OV armed with HPGD (WR.TK-HPGD$^+$), an enzyme capable of degrading PGE2, has been shown enhance local T-cell activation and limit premature MDSC-induced immune shutdown—a commonly encountered limiting factor of OV-mediated anti-tumor efficacy.[66] With potential relevance for clinical practice, WR.TK-HPGD$^+$ was able to sensitize tumors typically resistant to ICB in an immune competent murine syngeneic kidney cancer model. Similarly, an oncolytic strain of HSV-1 armed with HPGD was able to reverse PGE2-mediated immunosuppression and significantly inhibit the development and accumulation of metastatic breast cancer in an immune competent 4T1/BALB/c mouse model.[340]

Another pathway that has been subject to considerable interest in the field of cancer immunotherapy is the L-tryptophan to *N*-formyl-L-kynurenine catabolic pathway, the first step involving the related enzymes indoleamine 2,3-dioxygenase 1/2 (IDO1/2) and the closely related tryptophan 2,3-dioxygenase (TDO). The dual depletion of tryptophan and upregulation of kynurenine within the TME is known to confer a number of immunosuppressive effects, including: enhanced Treg function[341]; reduced effector T-cell activation[342]; impaired TCR-mediated signaling[343]; and poor CTL expansion.[344] In a human GBM model, tumor cells infected with an oncolytic neurotropic HSV strain (KOS), with or without deletion of ICP0—a virulence gene able to counteract IFNγ-mediated viral clearance—were found unable to upregulate IDO expression in response to local IFNγ.[345] While IDO/TDO inhibition utilizing OVs may have potential for combination therapy with ACT, the field has suffered a significant knock back with the late phase clinical trial failure of epacadostat, an orally available inhibitor of IDO1, in combination with the anti-PD-1 agent pembrolizumab in patients with unresectable or metastatic melanoma.[346] This trial, ECHO-301, demonstrated almost entirely identical PFS and OS rates in two groups of over 350 treatment-naïve or post-BRAF inhibitor patients receiving either first line pembrolizumab alone (standard of care) or pembrolizumab in combination with epacadostat. Several other inhibitors of IDO or, moreover TDO as well as novel tryptophan-mimetics continue to undergo clinical evaluation.[347]

Adenosine represents a further soluble factor capable of inducing multiple immunosuppressive effects on a variety of immune cells within the TME. Here, adenosine is predominantly synthesized from the catabolism of extracellular ATP, which is first converted to ADP and then AMP by CD39 and then to adenosine by CD73, a second ectonucleotidases that is highly expressed by many solid tumors including malignant melanoma, TNBC, PDAC and HNSCC.[348] Furthermore, CD73 expression on tumor cells is often correlated with disease stage and may have prognostic relevance.[349] Adenosine is able to both enhance Treg function while negatively regulating the effector function of CTLs. Furthermore, it can block the activation of M1 macrophages and can promote skewing of the TAM compartment toward an M2 phenotype with an associated upregulation of anti-inflammatory cytokines.[350] While small molecule inhibitors of CD39, CD73 and the target receptor A2AR are currently undergoing pre-clinical and clinical evaluation (most commonly with anti-PD-1/PD-L1 axis ICBs),[351] no reports have yet been published describing OVs capable of circumventing this immunosuppressive pathway. However, such an approach is liable to be efficacious for CAR T-cell therapy, as evidenced by the increased efficacy seen in CAR T-cells following the knock-down of A2AR using either a small inhibitory hairpin RNA (shRNA) or with pharmacological inhibition.[352] These CARs were able to demonstrate enhanced CAR-mediated activation and cytokine production, with improvements in efficacy that were particularly pronounced when combined with an anti-PD-1 ICB. Similarly, albeit using NK-92 cells transduced with a piggyBac transposon system encoding an NKG2D-based CAR incorporating DAP10 and CD3ζ activating domains, the use of a monoclonal blocking antibody targeting CD73 has been shown to significantly enhance CAR-mediated anti-tumor efficacy in xenographic NSCLC (A549) and GBM mouse models.[353]

Disturbances in extracellular electrolyte levels within tumors have also been highlighted as potential mediators of T-cell hypofunction, particularly in the context of TIL-based ACT. Specifically, increased levels of potassium can impair T-cell metabolism and induce autophagy, leading in turn to the depletion of nucleocytosolic acetyl-coenzyme A (AcCoA) and methionine intermediates, impaired epigenetic reprogramming and ultimately the maintenance of a stemness-like state.[354] In this population, effector function could be restored in T-cells transduced to constitutively express acyl-CoA synthetase short-chain family member Acss1, an enzyme responsible for mitochondrial and cytoplasmic AcCoA metabolism. Furthermore, these T-cells exhibited many of the hallmarks of high extracellular potassium-conditioned T-cells with significantly improved persistence and enhanced effector function in an established syngeneic B16 melanoma model. In a separate study, T-cells transduced to overexpress the potassium channel Kv1.3 were able to augment potassium efflux, leading to improvements in effector function in vitro and in vivo and ultimately improving the survival of melanoma-bearing mice in an immune competent TCR-transgenic Pmel-1 model.[355]

Finally, physical changes within the TME can have profound metabolic effects on infiltrating immune cells, often to their detriment. These changes may also impact upon OV infection of tumors. For example, tumor associated hypoxia, particularly in the centrally necrotic cores of many tumors, can reduce the replicative capacity of OVs, particularly group C adenoviruses, by preventing the optimal translation of viral proteins.[356] Despite this, several OVs are capable of functioning and even thriving in the midst of a hypoxic TME. For example, NDV strains commonly exhibit enhanced oncolytic potential in hypoxic conditions and can counteract survival signals mediated by hypoxia-induced proteins such as HIF-1α.[357] Accordingly, NDVs may prove to be particularly efficacious partners with CARs when attempting to target hypoxic tumors. Low pH can impair the regeneration of T-cell NAD^+, which is important in driving the mitochondrial TCA cycle forward to generate ATP.[337] As mentioned earlier, an adenovirus coated with a pH-sensitive and bio-reducible polymer (PPCBA) demonstrated improved infective capacity in hypoxic tumors, while also benefiting from the ability to be internalized via micropinocytosis, rather than relying upon coxsackie and adenovirus receptor (CAR) expression, which may be reduced or absent in many solid tumors.[49]

5.12 Delivery of pro-apoptotic or cytotoxic molecules

While our understanding of OV-induced tumor regression has shifted to focus on the ability of these viruses to enhance the cross-presentation of tumor associated antigens and improve adaptive T-cell-mediated immune responses, oncolysis in itself remains a powerful mechanism by which tumors may be destroyed following viral infection. There is thus a complex interaction in an immune competent host between the factors that engender OV clearance, tumor infection, oncolysis and anti-tumor immune modulation. Several strategies may be adopted with the aim of skewing this dynamic interplay toward a more favorable outcome. These include: optimizing OV treatment schedules; applying techniques designed to mitigate OV clearance by the host immune system; attempting to exert regulatory control on OV infection and oncolysis; or engineering OVs to incorporate additional "suicide" transgenes designed to enhance oncolysis and/or bystander cytotoxicity.

Given that CAR T-cells exert their anti-tumor effects by redirecting TCR-mediated cytotoxicity in a targeted fashion, there is a clear rationale for combination therapy with tumor-tropic OVs that exhibit enhanced oncolytic potential. A large number of OVs have been engineered to express pro-apoptotic molecules capable of inducing bystander cytotoxicity of non-infected tumor cells. Example payloads include recombinant TRAIL[358] or synthetic death receptor ligand fusion molecules (such as FasL-aHER2, combining a secreted form of a HER2 single chain antibody linked to a self-multimerizing Fas ligand extracellular domain[359]); pro-cytotoxic cytokines such as type I IFNs or TNFα[360]; tumor-specific pro-apoptotic viral molecules such as the chicken anemia virus protein apoptin[67] or the adenovirus death protein (ADP), which can enhance oncolysis when overexpressed in the adenoviral E3 locus[40]; cationic antimicrobial peptides (CMPs) with lipid disrupting and tumoricidal effects—examples include lactoferricin (a CMP derived from cow's milk following acid-pepsin hydrolysis of lactoferrin)[361] or lactaptin (a pro-apoptotic proteolytic fragment of kappa-casein derived from human breast milk)[362]; or enzymes designed to render tumor cells more susceptible to cytotoxic chemotherapy, such as bacterial cytosine deaminase (CD) which is able to convert the pro-drug 5-fluorocytosine into the cytotoxic molecule 5-fluorouracil (5-FU).[363] In a similar vein, enhanced oncolysis may be subject to spatial control by using host tissue-enriched or tissue-specific promoters that facilitate the preferential expression of suicide genes within tumors. For example, by engineering an adenoviral OV to express the HSV-1 TK gene under the control of the osteocalcin promoter, osteosarcoma cells may be rendered more susceptible to ganciclovir-induced DNA synthesis arrest and apoptotic cell death.[364] Additional combination strategies exist with other pro-apoptotic compounds such as second mitochondria-derived activator of caspases (Smac) mimetics (SMCs). By antagonizing the effects of inhibitor of apoptosis proteins (IAPs), these agents are able to enhance the sensitivity of tumor cells to pro-apoptotic signaling induced by inflammatory cytokines and have demonstrated an ability to induce significant bystander cytotoxicity when used in combination with oncolytic virotherapy.[365] One would envisage that such an approach could be enhanced further by utilizing OVs armed with type I interferons, TNFα, TRAIL or other death receptor ligands. Interestingly, genetically engineered mesenchymal stromal/stem cells (MSCs) (rather than T-cells) have been engineered to express both a truncated CAR targeting GD2 (incorporating an anti-GD2 scFv tethered to a hinge and TMD derived of CD8α) and the pro-apoptotic death receptor ligand TRAIL to successfully target $GD2^+$ human GBM cell lines.[366]

5.13 Delivery of molecules capable of structurally altering the tumor microenvironment

Many solid tumors feature a dense fibrotic stroma composed of extracellular matrix proteoglycans (including elastins, fibronectins and collagens) interspersed with CAFs and immunosuppressive myeloid cells (such as TAMs and MDSCs).[367] Indeed, in certain diseases, such as PDAC, this dense stroma may predominate and surround small islets of tumor cells, which are to an extent shielded from immune surveillance. Stroma within the TME can thus act as a formidable barrier to both OV infection and CAR

T-cell entry. Several agents with the capacity to modulate and potentially break down tumor-associated stroma have already entered the clinic. For example, a pegylated form of recombinant human hyaluronidase (PEGPH20) delivered in intravenous form has been shown capable of structurally altering the desmoplastic TME present in patients with PDAC.[368] In the HALO-202 phase II clinical trial a total of 279 patients with untreated metastatic PDAC were randomly assigned to standard of care doublet cytotoxic chemotherapy (Gemcitabine and nab-Paclitaxel) with or without PEGPH20. Administration of the latter was associated with a PFS benefit in the entire patient population, but was most striking in patients with hyaluronan-high tumors (9.2 versus 5.2 months, HR 0.51) (Fig. 2).[369]

In the context of OV infection, the use of a bacterial collagenase capable of degrading collagen in the TME has been shown to enhance the spread and anti-tumor efficacy of an oncolytic HSV. Likewise, the re-engineering of an oncolytic adenovirus to express human sperm hyaluronidase (PH20) under the control of the adenoviral major late promoter (MLP) led to a degradation of TME-associated hyaluronan, improved OV dissemination and enhanced therapeutic efficacy in a human melanoma xenograft model.[370] This approach was further extended by incorporating PH20 into a novel oncolytic adenovirus designed to selectively kill pRb-pathway defective tumor cells (ICOVIR15). This virus (ICOVIR17) showed significantly improved anti-tumor efficacy delivered either intra-tumorally or systemically in mice bearing melanoma or PDAC xenografts. Similarly, an oncolytic adenovirus has been designed to express matrix metalloproteinase 8 (MMP-8), a member of the zinc- and calcium-dependent endopeptidases capable of degrading tumor-related ECM.[371] Armed with MMP-8 this virus demonstrated improved spread in human PDAC and NSCLC xenograft models. In addition, infection of PDAC tumors in nude mice with a recombinant vaccinia strain (GLV-1h255) armed with MMP-9, a related zinc-metalloproteinase, was associated with a reduction in type IV collagen within the TME, enhanced viral tumor load and accelerated disease regression.[372]

While freshly isolated T-cells have been shown to express heparanase, an enzyme able to degrade heparan sulfate proteoglycans present in tumor ECM, long-term ex vivo expanded (LTE) T-cells have been shown to downregulate heparanase mRNA with a concomitant inability to degrade ECM in a Matrigel-based cell invasion assay.[125] This finding has implications for the successful delivery of CAR T-cells targeting solid tumors. The same group were able to restore the ECM-degrading capacity in GD2-redirected LTE CAR T-cells by transducing the latter with a retroviral cassette

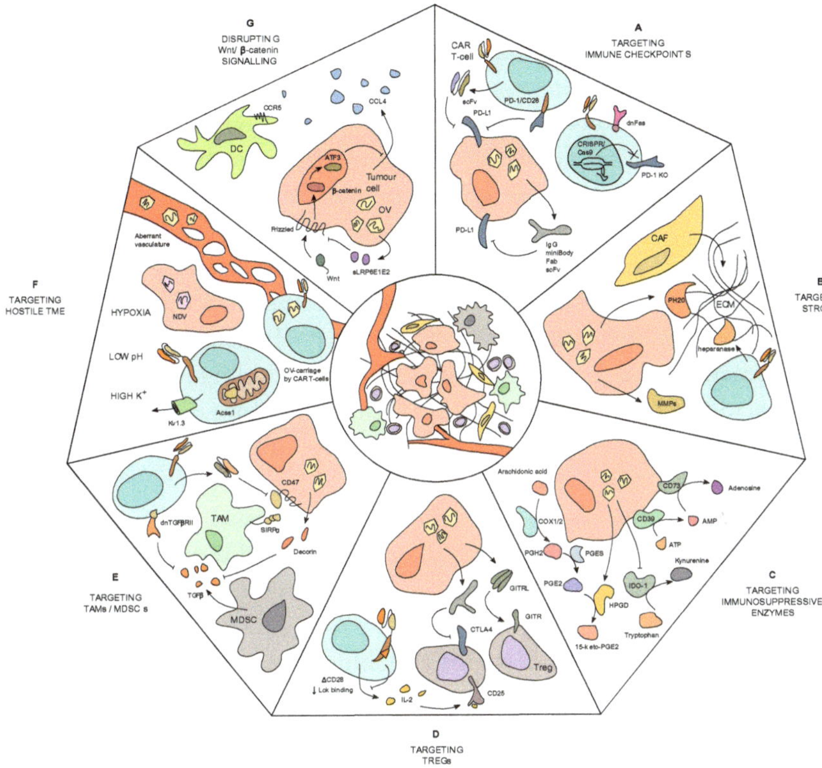

Fig. 2 Oncolytic virus and CAR T-cell-mediated reversal of immunosuppression. Many tumors have an immunosuppressive microenvironment that is not conducive to the emergence of an effective immune response. Indeed, numerous physical, metabolic and immunological barriers will need to be circumvented to optimize both OV and CAR T-cell efficacy in solid tumors. The synergistic combination of engineered OVs with CAR T-cells may counter many of these obstacles. For example: (A) tumor cells may be shielded from attack by the local upregulation of immune checkpoint molecules such as PD-L1 or CTLA-4; both OVs and CAR T-cells may be engineered to secrete antibodies or antibody fragments capable of inhibiting these pathways; (B) tumor-associated stroma may trap or exclude CAR T-cell entry, but can potentially be targeted by the co-delivery of OVs engineered to secrete ECM-degrading enzymes such as PH20 or MMPs; alternatively the CAR T-cells themselves may be engineered to secrete heparanase; (C) several immunosuppressive enzymatic pathways that can contribute to poor CAR T-cell function in the TME may be disrupted utilizing OVs with specific innate or engineered functionality—for example, OVs have been designed to degrade PGE2 utilizing the enzyme HPGD, while certain HSV strains exhibit a natural propensity to downregulate IDO-1; (D) Tregs are known to mediate immunosuppression in many tumor models and may be targeted by the delivery of depleting antibodies or by modulating the release of IL-2 following CAR engagement; (E) M2-skewed TAMs and MDSCs may also exert a formidable immunosuppressive effect in many tumors; this, in part, may be mediated by their secretion of cytokines such as IL-10 and TGFβ—the latter may be neutralized by

(Continued)

incorporating human heparanase. These CAR T-cells demonstrated increased GD2-specific cytotoxicity when neuroblastoma cells were incorporated into the Matrigel-based invasion assay and conferred a survival advantage in mice bearing human neuroblastoma xenografts. However, such CAR T-cells still face a considerable challenge trafficking through dense stroma to reach tumor cells and the disruption of tumor-related ECM using an armed OV in advance of ACT may in fact be preferential. A minor caveat may also relate to the fact that many collagenases (particularly members of the MMP family) have also been shown to facilitate cancer progression, invasion and metastasis[373] with MMPs-2 and 9 emerging as negative biomarkers in numerous solid tumors including breast, NSCLC, PDAC, bladder and CRC.[374] Caution should therefore be exercised when deploying OVs engineered to express these proteins in future clinical trials with or without ACT.

In addition to the use of ECM-degrading enzymes, certain OVs have been shown to harbor an intrinsic ability to modulate the tumor microvasculature and potentially enhance vascular permeability. These attributes may prove useful for combination therapy with ACT, including CAR T-cells. For instance, several vaccinia strains are able to exert an anti-tumor effect following systemic delivery by demonstrating a propensity to infect endothelial cells and modulate cancer-associated endothelium.[57] Several other OVs including HSV and VSV are also tropic to tumor-associated endothelial cells.[375] Separately, OVs may also be armed with anti-angiogenic molecules such as endothelial disrupting molecules (see below), inhibitors of vascular endothelial growth factors (VEGFs) or their receptors (VEGFRs) (e.g., ONYX-015).[376] Certain cytokines are known to exhibit anti-angiogenic properties, including IL-12,[377] IL-18,[378] IL-24[379] or IL-27.[380] These may be incorporated into OVs to exert an impact on cancer-associated

Fig. 2—cont'd co-expressing a dominant negative TGFβ receptor or by co-delivery of an OV incorporating a TGFβ pathway antagonist. TAMs may be enlisted in the effort to eradicate tumors by CAR T-cells armed with inhibitors of the SIRPα-CD47 axis; (F) tumor-associated aberrant microvasculature may be targeted by using certain vaccinia OV strains while CAR T-cells may themselves be utilized to shield OVs from being cleared by anti-viral antibodies or complement; in addition, certain OV species may be particularly resistant to hypoxia (such as NDV), while CAR T-cells may be engineered to exhibit resistance to immunosuppressive electrolyte shifts; (G) Wnt/β-catenin signaling has been shown to negatively regulate innate and adaptive immune responses in several tumor models and may be targeted by OVs armed with Wnt pathway inhibitors or with pro-inflammatory chemokines such as CCL-4 or CXCL-9/10.

microvasculature. Contrarily, pro-angiogenic cytokines such as VEGF or IL-8 may be blocked by engineering OVs with small interfering ribonucleic acids (siRNAs).[381,382] In addition, OVs may incorporate anti-angiogenic chemokines such as platelet factor 4,[383] or induce or encode angiostatic chemokines including members of the C-X-C motif family such as CXCL-10.[384] Alternatively, endogenous angiogenesis inhibitors such as angiostatin,[385] endostatin,[386] vasculostatin[387] or Kringle 5 of human plasminogen,[388] or novel agents such as GLAF-1 (a single chain antibody targeting murine and human VEGF)[389] may be incorporated into OV vectors. Furthermore, such OVs are likely to exhibit pleiotropic effects in the TME that, when combined with ACT in the form of CAR T-cell therapy may outperform direct combination with VEGF-pathway targeting agents already present in the clinic such as the anti-VEGF-A mAb bevacizumab or the anti-VEGFR2 mAb, ramucirumab.

6. Conclusions

The great promise heralded by the rational design of CARs 30 years ago[105] has finally been realized in the clinic for many patients with advanced hematological malignancy. However, no transformational results have yet been seen using CARs designed to target solid tumors. A great number of physical, metabolic and immunosuppressive obstacles will need to be overcome before CAR T-cells can become a standard of care for these patients. The immune checkpoint blockade era has had a dramatic impact on our understanding of cancer immune evasion and has changed the way that we treat many cancers. Combination strategies, in concert with the development of innovative next generation CAR constructs, are likely to provide solutions to many of the problems that we continue to encounter in the clinic. After a period of relative obscurity, oncolytic virotherapy has re-emerged as a potent form of immunotherapy and one that has considerable synergistic potential with both ICB and ACT, including CAR T-cell therapy. The ability to further engineer both OVs and CAR T-cells with customizable transgenes further extends the scope of their combined use and presages a time when a practically limitless number of patient-specific bespoke cellular and viral products (potentially composed of modular component parts) are delivered routinely to patients with cancer.

Acknowledgments

Not applicable.

Ethical approval and consent to participate

Not applicable.

Consent for publication

Not applicable.

Availability of supporting data

Not applicable.

Competing interests

J.M. is chief scientific officer of Leucid Bio, a spinout company focussed on CAR T-cell and gamma delta T-cell immunotherapies for malignant disease. A.A. does not have any conflicts of interest to declare.

Funding

Research in JM's laboratory is supported by the Medical Research Council, Cancer Research UK, Bayer HealthCare, Bloodwise, British Lung Foundation, Breast Cancer Now, the King's Health Partners Research and Development Fund, British Council Newton Fund Institutional Links Award, the Experimental Cancer Medicine Centre at King's College London, the Cancer Research UK Centre at King's Health Partners and by the National Institute for Health Research (NIHR) Biomedical Research Centre based at Guy's and St Thomas' NHS Foundation Trust and King's College London. The views expressed are those of the authors and not necessarily those of the NHS, the NIHR or the Department of Health.

Authors' contributions

Both authors wrote the manuscript. Both authors read and approved the final manuscript.

References

1. Khalil DN, et al. The future of cancer treatment: immunomodulation, CARs and combination immunotherapy. *Nat Rev Clin Oncol.* 2016;13:273.
2. Ribas A. Releasing the brakes on cancer immunotherapy. *N Engl J Med.* 2015;373(16): 1490–1492.
3. Hargadon KM, Johnson CE, Williams CJ. Immune checkpoint blockade therapy for cancer: an overview of FDA-approved immune checkpoint inhibitors. *Int Immunopharmacol.* 2018;62:29–39.
4. Kaiser J, Couzin-Frankel J. Cancer immunotherapy sweeps Nobel for medicine. *Science.* 2018;362(6410):13.
5. Andtbacka RH, et al. Talimogene laherparepvec improves durable response rate in patients with advanced melanoma. *J Clin Oncol.* 2015;33(25):2780–2788.
6. Maude SL, et al. Tisagenlecleucel in children and young adults with B-cell lymphoblastic leukemia. *N Engl J Med.* 2018;378(5):439–448.
7. Boyiadzis MM, et al. Chimeric antigen receptor (CAR) T therapies for the treatment of hematologic malignancies: clinical perspective and significance. *J Immunother Cancer.* 2018;6(1):137.

8. Bu X, Mahoney KM, Freeman GJ. Learning from PD-1 resistance: new combination strategies. *Trends Mol Med.* 2016;22(6):448–451.
9. Scarfo I, Maus MV. Current approaches to increase CAR T cell potency in solid tumors: targeting the tumor microenvironment. *J Immunother Cancer.* 2017;5:28.
10. Mullen JT, Tanabe KK. Viral oncolysis. *Oncologist.* 2002;7(2):106–119.
11. Hamid O, et al. Oncolytic immunotherapy: unlocking the potential of viruses to help target cancer. *Cancer Immunol Immunother.* 2017;66(10):1249–1264.
12. Endo Y, et al. Virus-mediated oncolysis induces danger signal and stimulates cytotoxic T-lymphocyte activity via proteasome activator upregulation. *Oncogene.* 2008;27(17):2375–2381.
13. Achard C, et al. Induction of immunogenic tumor cell death by attenuated oncolytic measles virus. *J Clin Cell Immunol.* 2015;6(1):291.
14. Kaufman HL, Kohlhapp FJ, Zloza A. Oncolytic viruses: a new class of immunotherapy drugs. *Nat Rev Drug Discov.* 2015;14(9):642–662.
15. Gesundheit B, Rosenzweig JP. Editorial: oncolytic viruses—genetically engineering the future of cancer therapy. *Front Oncol.* 2017;7:271.
16. Eissa IR, et al. The current status and future prospects of oncolytic viruses in clinical trials against melanoma, glioma, pancreatic, and breast cancers. *Cancers (Basel).* 2018;10(10): pii: E356.
17. Ajina A, Maher J. Prospects for combined use of oncolytic viruses and CAR T-cells. *J Immunother Cancer.* 2017;5(1):90.
18. Ries S, Korn WM. ONYX-015: mechanisms of action and clinical potential of a replication-selective adenovirus. *Br J Cancer.* 2002;86(1):5–11.
19. Bennett JJ, et al. Comparison of safety, delivery, and efficacy of two oncolytic herpes viruses (G207 and NV1020) for peritoneal cancer. *Cancer Gene Ther.* 2002;9(11): 935–945.
20. Cripe TP, et al. Phase 1 study of intratumoral Pexa-Vec (JX-594), an oncolytic and immunotherapeutic vaccinia virus, in pediatric cancer patients. *Mol Ther.* 2015;23(3):602–608.
21. Chan WM, Rahman MM, McFadden G. Oncolytic myxoma virus: the path to clinic. *Vaccine.* 2013;31(39):4252–4258.
22. Pol J, et al. Trial watch—oncolytic viruses and cancer therapy. *Oncoimmunology.* 2016;5(2):e1117740.
23. Geletneky K, et al. Oncolytic H-1 parvovirus shows safety and signs of immunogenic activity in a first phase I/IIa glioblastoma trial. *Mol Ther.* 2017;25(12):2620–2634.
24. Annels NE, et al. Oncolytic immunotherapy for bladder cancer using coxsackie A21 virus. *Mol Ther Oncolytics.* 2018;9:1–12.
25. Holl EK, et al. Recombinant oncolytic poliovirus, PVSRIPO, has potent cytotoxic and innate inflammatory effects, mediating therapy in human breast and prostate cancer xenograft models. *Oncotarget.* 2016;7(48):79828–79841.
26. Burke MJ. Oncolytic seneca valley virus: past perspectives and future directions. *Oncolytic Virother.* 2016;5:81–89.
27. Mahalingam D, et al. A phase II study of REOLYSIN((R)) (pelareorep) in combination with carboplatin and paclitaxel for patients with advanced malignant melanoma. *Cancer Chemother Pharmacol.* 2017;79(4):697–703.
28. Lundstrom K. Oncolytic alphaviruses in cancer immunotherapy. *Vaccine.* 2017;5(2):9.
29. Aref S, Bailey K, Fielding A. Measles to the rescue: a review of oncolytic measles virus. *Viruses.* 2016;8(10):294.
30. Zamarin D, Palese P. Oncolytic Newcastle disease virus for cancer therapy: old challenges and new directions. *Future Microbiol.* 2012;7(3):347–367.
31. Bishnoi S, et al. Oncotargeting by vesicular stomatitis virus (VSV): advances in cancer therapy. *Viruses.* 2018;10(2): pii: E90.

32. Pol JG, et al. Development and applications of oncolytic Maraba virus vaccines. *Oncolytic Virother.* 2018;7:117–128.
33. Wei D, et al. Fighting cancer with viruses: oncolytic virus therapy in China. *Hum Gene Ther.* 2017;29(2):151–159.
34. Kohlhapp FJ, Kaufman HL. Molecular pathways: mechanism of action for talimogene laherparepvec, a new oncolytic virus immunotherapy. *Clin Cancer Res.* 2016;22(5): 1048–1054.
35. Garber K. China approves World's first oncolytic virus therapy for cancer treatment. *J Natl Cancer Inst.* 2006;98(5):298–300.
36. Chalikonda S, Bartlett DL. Vaccinia and pox-virus. In: Hunt KK, Vorburger SA, Swisher SG, eds. *Gene Therapy for Cancer.* Totowa, NJ: Humana Press; 2007:73–85. ISBN: 978-1-59745-222-9.
37. Jhawar SR, et al. Oncolytic viruses—natural and genetically engineered cancer immunotherapies. *Front Oncol.* 2017;7(202).
38. Maroun J, et al. Designing and building oncolytic viruses. *Future Virol.* 2017;12(4): 193–213.
39. McCart JA, et al. Systemic cancer therapy with a tumor-selective vaccinia virus mutant lacking thymidine kinase and vaccinia growth factor genes. *Cancer Res.* 2001;61(24):8751–8757.
40. Doronin K, et al. Tumor-specific, replication-competent adenovirus vectors overexpressing the adenovirus death protein. *J Virol.* 2000;74(13):6147–6155.
41. Richardson A, et al. THER-30. Combination Toca 511 & 5-FC significantly extends survival in a murine orthotopic model of intracerebellar medulloblastoma. *Neuro Oncol.* 2019;21(Suppl 2):ii120.
42. Giunco S, et al. Telomere and telomerase in carcinogenesis: their role as prognostic biomarkers. *Curr Pathobiol Rep.* 2015;3(4):315–328.
43. Kawashima T, et al. Telomerase-specific replication-selective virotherapy for human cancer. *Clin Cancer Res.* 2004;10(1):285–292.
44. Machitani M, et al. Enhanced oncolytic activities of the telomerase-specific replication-competent adenovirus expressing short-hairpin RNA against dicer. *Mol Cancer Ther.* 2017;16(1):251–259.
45. Bradley S, et al. Applications of coxsackievirus A21 in oncology. *Oncolytic Virother.* 2014;3:47–55.
46. Ruiz AJ, Russell SJ. MicroRNAs and oncolytic viruses. *Curr Opin Virol.* 2015;13:40–48.
47. Kelly E, Hadac E, Russell SJ. 52. Targeting tissue tropism of coxsackievirus A21, an oncolytic enterovirus. *Mol Ther.* 2007;15:S22.
48. Kuhn I, et al. OvAd1, a novel, potent, and selective chimeric oncolytic virus developed for ovarian cancer by 3D-directed evolution. *Mol Ther Oncolytics.* 2017;4:55–66.
49. Moon CY, et al. Dual tumor targeting with pH-sensitive and bioreducible polymer-complexed oncolytic adenovirus. *Biomaterials.* 2015;41:53–68.
50. Na Y, et al. Potent antitumor effect of neurotensin receptor-targeted oncolytic adenovirus co-expressing decorin and Wnt antagonist in an orthotopic pancreatic tumor model. *J Control Release.* 2015;220(Pt B):766–782.
51. Pikor LA, Bell JC, Diallo J-S. Oncolytic viruses: exploiting cancer's deal with the devil. *Trends Cancer.* 2015;1(4):266–277.
52. Mansour M, Palese P, Zamarin D. Oncolytic specificity of Newcastle disease virus is mediated by selectivity for apoptosis-resistant cells. *J Virol.* 2011;85(12):6015–6023.
53. Bischoff JR, et al. An adenovirus mutant that replicates selectively in p53-deficient human tumor cells. *Science.* 1996;274(5286):373–376.
54. Farassati F, et al. Ras signaling influences permissiveness of malignant peripheral nerve sheath tumor cells to oncolytic herpes. *Am J Pathol.* 2008;173(6):1861–1872.

55. Coffey MC, et al. Reovirus therapy of tumors with activated Ras pathway. *Science.* 1998;282(5392):1332–1334.
56. Parato KA, et al. The oncolytic poxvirus JX-594 selectively replicates in and destroys cancer cells driven by genetic pathways commonly activated in cancers. *Mol Ther.* 2012;20(4):749–758.
57. Breitbach CJ, et al. Oncolytic vaccinia virus disrupts tumor-associated vasculature in humans. *Cancer Res.* 2013;73(4):1265–1275.
58. Zhang KX, et al. Down-regulation of type I interferon receptor sensitizes bladder cancer cells to vesicular stomatitis virus-induced cell death. *Int J Cancer.* 2010;127(4):830–838.
59. Huang PY, Guo JH, Hwang LH. Oncolytic Sindbis virus targets tumors defective in the interferon response and induces significant bystander antitumor immunity in vivo. *Mol Ther.* 2012;20(2):298–305.
60. Matveeva OV, Chumakov PM. Defects in interferon pathways as potential biomarkers of sensitivity to oncolytic viruses. *Rev Med Virol.* 2018;28(6):e2008.
61. Ahmed M, et al. Ability of the matrix protein of vesicular stomatitis virus to suppress beta interferon gene expression is genetically correlated with the inhibition of host RNA and protein synthesis. *J Virol.* 2003;77(8):4646–4657.
62. Su C, Zhan G, Zheng C. Evasion of host antiviral innate immunity by HSV-1, an update. *Virol J.* 2016;13:38.
63. Fu X, et al. An HSV-2 based oncolytic virus can function as an attractant to guide migration of adoptively transferred T cells to tumor sites. *Oncotarget.* 2015;6(2):902–914.
64. Todo T, et al. Oncolytic herpes simplex virus vector with enhanced MHC class I presentation and tumor cell killing. *Proc Natl Acad Sci USA.* 2001;98(11):6396–6401.
65. Jason John L, et al. Correlation of WNT/β-catenin pathway activation with immune exclusion across most human cancers. In: *ASCO Annual Meeting*; 2016.
66. Hou W, et al. Oncolytic virus-mediated targeting of PGE2 in the tumor alters the immune status and sensitizes established and resistant tumors to immunotherapy. *Cancer Cell.* 2016;30(1):108–119.
67. Kochneva G, et al. Apoptin enhances the oncolytic properties of vaccinia virus and modifies mechanisms of tumor regression. *Oncotarget.* 2014;5(22):11269–11282.
68. Ferguson MS, Lemoine NR, Wang Y. Systemic delivery of oncolytic viruses: hopes and hurdles. *Adv Virol.* 2012;2012:805629.
69. Evgin L, et al. Complement inhibition prevents oncolytic vaccinia virus neutralization in immune humans and cynomolgus macaques. *Mol Ther.* 2015;23(6):1066–1076.
70. Miller AC, Russell SJ. Heterogeneous delivery is a barrier to the translational advancement of oncolytic virotherapy for treating solid tumors. *Virus Adapt Treat.* 2014;6(1):11–31.
71. Conry RM, et al. Talimogene laherparepvec: first in class oncolytic virotherapy. *Hum Vaccin Immunother.* 2018;14(4):839–846.
72. Rudin CM, et al. LBA40 Phase Ib KEYNOTE-200: a study of an intravenously delivered oncolytic virus, coxsackievirus A21 in combination with pembrolizumab in advanced NSCLC and bladder cancer patients. *Ann Oncol.* 2018;29(Suppl 8): viii732.
73. Mell LK, et al. Phase I trial of intravenous oncolytic vaccinia virus (GL-ONC1) with cisplatin and radiotherapy in patients with locoregionally advanced head and neck carcinoma. *Clin Cancer Res.* 2017;23(19):5696–5702.
74. Wong HH, Lemoine NR, Wang Y. Oncolytic viruses for cancer therapy: overcoming the obstacles. *Viruses.* 2010;2(1):78–106.
75. Peng KW, et al. Using clinically approved cyclophosphamide regimens to control the humoral immune response to oncolytic viruses. *Gene Ther.* 2013;20(3):255–261.

76. Yotnda P, et al. Targeted delivery of adenoviral vectors by cytotoxic T cells. *Blood.* 2004;104(8):2272–2280.
77. VanSeggelen H, et al. Chimeric antigen receptor-engineered T cells as oncolytic virus carriers. *Mol Ther Oncolytics.* 2015;2:15014.
78. Marelli G, et al. Oncolytic viral therapy and the immune system: a double-edged sword against cancer. *Front Immunol.* 2018;9:866.
79. Guo ZS, et al. Oncolytic immunotherapy: conceptual evolution, current strategies, and future perspectives. *Front Immunol.* 2017;8:555.
80. Robilotti EV, et al. Viral oncolytic immunotherapy in the war on cancer: infection control considerations. *Infect Control Hosp Epidemiol.* 2019;40(3):350–354.
81. June CH, Sadelain M. Chimeric antigen receptor therapy. *N Engl J Med.* 2018;379(1):64–73.
82. Sadelain M, Brentjens R, Riviere I. The basic principles of chimeric antigen receptor design. *Cancer Discov.* 2013;3(4):388–398.
83. Ormhøj M, et al. CARs in the lead against multiple myeloma. *Curr Hematol Malig Rep.* 2017;12(2):119–125.
84. Ghobadi A. Chimeric antigen receptor T cell therapy for non-Hodgkin lymphoma. *Curr Res Transl Med.* 2018;66(2):43–49.
85. Ramos CA, et al. CD30-chimeric antigen receptor (CAR) T cells for therapy of hodgkin lymphoma (HL). *Biol Blood Marrow Transplant.* 2019;25(3):S63.
86. Hofmann S, et al. Chimeric antigen receptor (CAR) T cell therapy in acute myeloid leukemia (AML). *J Clin Med.* 2019;8(2): pii: E200.
87. Scarfò I, Frigault MJ, Maus MV. CAR-based approaches to cutaneous T-cell lymphoma. *Front Oncol.* 2019;9:259.
88. Knochelmann HM, et al. CAR T cells in solid tumors: blueprints for building effective therapies. *Front Immunol.* 2018;9:1740.
89. Schmidts A, Maus MV. Making CAR T cells a solid option for solid tumors. *Front Immunol.* 2018;9:2593.
90. Wang LC, et al. Targeting fibroblast activation protein in tumor stroma with chimeric antigen receptor T cells can inhibit tumor growth and augment host immunity without severe toxicity. *Cancer Immunol Res.* 2014;2(2):154–166.
91. Rosenberg SA, Restifo NP. Adoptive cell transfer as personalized immunotherapy for human cancer. *Science.* 2015;348(6230):62–68.
92. McGranahan N, et al. Allele-specific HLA loss and immune escape in lung cancer evolution. *Cell.* 2017;171(6):1259–1271.e11.
93. Restifo NP, et al. Loss of functional beta 2-microglobulin in metastatic melanomas from five patients receiving immunotherapy. *J Natl Cancer Inst.* 1996;88(2):100–108.
94. Leone P, et al. MHC class I antigen processing and presenting machinery: organization, function, and defects in tumor cells. *J Natl Cancer Inst.* 2013;105(16):1172–1187.
95. Sharma P, et al. Primary, adaptive, and acquired resistance to cancer immunotherapy. *Cell.* 2017;168(4):707–723.
96. Mansoor W, et al. Engineering T cells for cancer therapy. *Br J Cancer.* 2005;93(10):1085–1091.
97. Rossig C, et al. Carbohydrate targets for CAR T cells in solid childhood cancers. *Front Oncol.* 2018;8:513.
98. Liu H, et al. Targeting alpha-fetoprotein (AFP)-MHC complex with CAR T-cell therapy for liver cancer. *Clin Cancer Res.* 2017;23(2):478–488.
99. Bausch-Fluck D, et al. A mass spectrometric-derived cell surface protein atlas. *PLoS One.* 2015;10(3):e0121314.
100. Wu CY, et al. Remote control of therapeutic T cells through a small molecule-gated chimeric receptor. *Science.* 2015;350(6258):aab4077.

101. Roybal KT, et al. Engineering T cells with customized therapeutic response programs using synthetic notch receptors. *Cell*. 2016;167(2):419–432.e16.
102. Cho JH, Collins JJ, Wong WW. Universal chimeric antigen receptors for multiplexed and logical control of T cell responses. *Cell*. 2018;173(6):1426–1438.e11.
103. Davies DM, Maher J. Gated chimeric antigen receptor T-cells: the next logical step in reducing toxicity? *Transl Cancer Res*. 2016;5(1):S61–S65.
104. Gross G, et al. Generation of effector T cells expressing chimeric T cell receptor with antibody type-specificity. *Transplant Proc*. 1989;21(1 Pt 1):127–130.
105. Gross G, Waks T, Eshhar Z. Expression of immunoglobulin-T-cell receptor chimeric molecules as functional receptors with antibody-type specificity. *Proc Natl Acad Sci USA*. 1989;86(24):10024–10028.
106. Karlsson H, et al. Evaluation of intracellular signaling downstream chimeric antigen receptors. *PLoS One*. 2015;10(12):e0144787.
107. Salter AI, et al. Phosphoproteomic analysis of chimeric antigen receptor signaling reveals kinetic and quantitative differences that affect cell function. *Sci Signal*. 2018;11:544.
108. Park JH, Brentjens RJ. Are all chimeric antigen receptors created equal? *J Clin Oncol*. 2015;33(6):651–653.
109. van der Stegen SJC, Hamieh M, Sadelain M. The pharmacology of second-generation chimeric antigen receptors. *Nat Rev Drug Discov*. 2015;14:499.
110. Abate-Daga D, Davila ML. CAR models: next-generation CAR modifications for enhanced T-cell function. *Mol Ther Oncolytics*. 2016;3:16014.
111. Ramello MC, et al. An immunoproteomic approach to characterize the CAR interactome and signalosome. *Sci Signal*. 2019;12(568):eaap9777.
112. Kagoya Y, et al. A novel chimeric antigen receptor containing a JAK-STAT signaling domain mediates superior antitumor effects. *Nat Med*. 2018;24(3):352–359.
113. Kloss CC, et al. Combinatorial antigen recognition with balanced signaling promotes selective tumor eradication by engineered T cells. *Nat Biotechnol*. 2013;31(1):71–75.
114. Feucht J, et al. Calibration of CAR activation potential directs alternative T cell fates and therapeutic potency. *Nat Med*. 2019;25(1):82–88.
115. Avanzi MP, et al. Engineered tumor-targeted T cells mediate enhanced anti-tumor efficacy both directly and THROUGH activation of the endogenous immune system. *Cell Rep*. 2018;23(7):2130–2141.
116. Hurton LV, et al. Tethered IL-15 augments antitumor activity and promotes a stem-cell memory subset in tumor-specific T cells. *Proc Natl Acad Sci USA*. 2016;113(48): E7788–E7797.
117. Craddock JA, et al. Enhanced tumor trafficking of GD2 chimeric antigen receptor T cells by expression of the chemokine receptor CCR2b. *J Immunother*. 2010;33(8):780–788.
118. Muller N, et al. Engineering NK cells modified with an EGFRvIII-specific chimeric antigen receptor to overexpress CXCR4 improves immunotherapy of CXCL12/SDF-1alpha-secreting glioblastoma. *J Immunother*. 2015;38(5):197–210.
119. Zhao Z, et al. Structural design of engineered costimulation determines tumor rejection kinetics and persistence of CAR T cells. *Cancer Cell*. 2015;28(4):415–428.
120. Curran KJ, et al. Enhancing antitumor efficacy of chimeric antigen receptor T cells through constitutive CD40L expression. *Mol Ther*. 2015;23(4):769–778.
121. Rataj F, et al. PD1-CD28 fusion protein enables CD4+ T cell help for adoptive T cell therapy in models of pancreatic cancer and non-hodgkin lymphoma. *Front Immunol*. 2018;9:1955.
122. Yamamoto TN, et al. T cells genetically engineered to overcome death signaling enhance adoptive cancer immunotherapy. *J Clin Invest*. 2019;129(4):1551–1565.

123. Chen N, et al. CAR T-cell intrinsic PD-1 checkpoint blockade: a two-in-one approach for solid tumor immunotherapy. *Oncoimmunology*. 2016;6(2):e1273302.
124. Kloss CC, et al. Dominant-negative TGF-beta receptor enhances PSMA-targeted human CAR T cell proliferation and augments prostate cancer eradication. *Mol Ther* 2018;26(7):1855–1866.
125. Caruana I, et al. Heparanase promotes tumor infiltration and antitumor activity of CAR-redirected T lymphocytes. *Nat Med*. 2015;21(5):524–529.
126. Rafiq S, et al. Targeted delivery of a PD-1-blocking scFv by CAR-T cells enhances anti-tumor efficacy in vivo. *Nat Biotechnol*. 2018;36(9):847–856.
127. Pegram HJ, Purdon TJ, Brentjens RJ. 770. Blocking CD47 improves CAR T cell therapy. *Mol Ther*. 2014;22:S297.
128. Choi BD, Yu X, Castano AP, et al. BiTE-armored CARs overcome antigen escape in EGFRvIII-targeted therapy for glioblastoma. In: *American Association for Cancer Research (AACR) Annual Meeting 2019*; 2019. Atlanta, GA, USA.
129. Condomines M, et al. 655. Constitutive 4-1BBL expression dramatically increases proliferation, survival and in vivo persistence of CD19-targeted T cells, promoting complete tumor elimination. *Mol Ther*. 2010;18:S255.
130. Eyquem J, et al. Targeting a CAR to the TRAC locus with CRISPR/Cas9 enhances tumour rejection. *Nature*. 2017;543(7643):113–117.
131. MacLeod DT, et al. Integration of a CD19 CAR into the TCR alpha chain locus streamlines production of allogeneic gene-edited CAR T cells. *Mol Ther*. 2017;25(4):949–961.
132. Bachmann M. The UniCAR system: a modular CAR T cell approach to improve the safety of CAR T cells. *Immunol Lett*. 2019;211:13–22.
133. Mata M, et al. Inducible activation of MyD88 and CD40 in CAR T cells results in controllable and potent antitumor activity in preclinical solid tumor models. *Cancer Discov*. 2017;7(11):1306–1319.
134. Patel S, et al. Beyond CAR T cells: other cell-based immunotherapeutic strategies against cancer. *Front Oncol*. 2019;9:196.
135. FDA News Service. *FDA Approval Brings First Gene Therapy to the United States*. 2017, https://www.fda.gov/news-events/press-announcements/fda-approval-brings-first-gene-therapy-united-states.
136. FDA News Service. *FDA Approves Tisagenlecleucel for Adults With Relapsed or Refractory Large B-Cell Lymphoma*. 2018, https://www.fda.gov/drugs/resources-information-approved-drugs/fda-approves-tisagenlecleucel-adults-relapsed-or-refractory-large-b-cell-lymphoma.
137. FDA News Service. *FDA Approves CAR-T Cell Therapy to Treat Adults With Certain Types of Large B-Cell Lymphoma*. 2017, https://www.fda.gov/news-events/press-announcements/fda-approves-car-t-cell-therapy-treat-adults-certain-types-large-b-cell-lymphoma.
138. Neelapu SS, et al. Axicabtagene ciloleucel CAR T-cell therapy in refractory large B-cell lymphoma. *N Engl J Med*. 2017;377(26):2531–2544.
139. Lu P, et al. Which is better in CD19 CAR-T treatment of r/r B-ALL, CD28 or 4-1BB? A parallel trial under the same manufacturing process. *J Clin Oncol*. 2018;36(Suppl 15):3041.
140. Doan A, Pulsipher MA. Hypogammaglobulinemia due to CAR T-cell therapy. *Pediatr Blood Cancer*. 2018;65(4):e26914.
141. Lamers CH, et al. Treatment of metastatic renal cell carcinoma with CAIX CAR-engineered T cells: clinical evaluation and management of on-target toxicity. *Mol ther*. 2013;21(4):904–912.
142. Irving M, et al. Engineering chimeric antigen receptor T-cells for racing in solid tumors: don't forget the fuel. *Front Immunol*. 2017;8:267.

143. Adusumilli PS, et al. Regional delivery of mesothelin-targeted CAR T cell therapy generates potent and long-lasting CD4-dependent tumor immunity. *Sci Transl Med.* 2014;6(261):261ra151.
144. O'Rourke DM, et al. A single dose of peripherally infused EGFRvIII-directed CAR T cells mediates antigen loss and induces adaptive resistance in patients with recurrent glioblastoma. *Sci Transl Med.* 2017;9:399.
145. Hombach AA, Abken H. Of chimeric antigen receptors and antibodies: OX40 and 41BB costimulation sharpen up T cell-based immunotherapy of cancer. *Immunotherapy.* 2013;5(7):677–681.
146. Shalabi H, et al. Intensification of lymphodepletion optimizes CAR re-treatment efficacy. *Blood.* 2017;130(Suppl 1):3889.
147. Poh A. JCAR015 in ALL: a root-cause investigation. *Cancer Discov.* 2018;8(1):4–5.
148. Dholaria BR, Bachmeier CA, Locke F. Mechanisms and management of chimeric antigen receptor T-cell therapy-related toxicities. *BioDrugs.* 2019;33(1):45–60.
149. Cummins KD, Gill S. Anti-CD123 chimeric antigen receptor T-cells (CART): an evolving treatment strategy for hematological malignancies, and a potential ace-in-the-hole against antigen-negative relapse. *Leuk Lymphoma.* 2018;59(7):1539–1553.
150. Ruella M, et al. Induction of resistance to chimeric antigen receptor T cell therapy by transduction of a single leukemic B cell. *Nat Med.* 2018;24(10):1499–1503.
151. Bedoya F, Frigault MJ, Maus MV. The flipside of the power of engineered T cells: observed and potential toxicities of genetically modified T cells as therapy. *Mol Ther.* 2017;25(2):314–320.
152. Morgan RA, et al. Case report of a serious adverse event following the administration of T cells transduced with a chimeric antigen receptor recognizing ERBB2. *Mol Ther.* 2010;18(4):843–851.
153. Gargett T, Brown MP. The inducible caspase-9 suicide gene system as a "safety switch" to limit on-target, off-tumor toxicities of chimeric antigen receptor T cells. *Front Pharmacol.* 2014;5:235.
154. Oliveira G, et al. Use of TK-cells in haploidentical hematopoietic stem cell transplantation. *Curr Opin Hematol.* 2012;19(6):427–433.
155. Sakemura R, et al. A Tet-On inducible system for controlling CD19-chimeric antigen receptor expression upon drug administration. *Cancer Immunol Res.* 2016;4(8): 658–668.
156. Wang X, et al. A transgene-encoded cell surface polypeptide for selection, in vivo tracking, and ablation of engineered cells. *Blood.* 2011;118(5):1255–1263.
157. Philip B, et al. A highly compact epitope-based marker/suicide gene for easier and safer T-cell therapy. *Blood.* 2014;124(8):1277–1287.
158. Ma JSY, et al. Versatile strategy for controlling the specificity and activity of engineered T cells. *Proc Natl Acad Sci U S A.* 2016;113(4):E450–E458.
159. Raj D, et al. Switchable CAR-T cells mediate remission in metastatic pancreatic ductal adenocarcinoma. *Gut.* 2019;68(6):1052–1064.
160. DeRenzo C, Gottschalk S. Genetic modification strategies to enhance CAR T cell persistence for patients with solid tumors. *Front Immunol.* 2019;10:218.
161. Jennings RN, Grayson JM, Barton ES. Type I interferon signaling enhances $CD8^+$ T cell effector function and differentiation during murine gammaherpesvirus 68 infection. *J Virol.* 2014;88(24):14040–14049.
162. Curtsinger JM, et al. Type I IFNs provide a third signal to CD8 T cells to stimulate clonal expansion and differentiation. *J Immunol.* 2005;174(8):4465–4469.
163. Garg AD, et al. Molecular and translational classifications of DAMPs in Immunogenic cell death. *Front Immunol.* 2015;6:588.
164. Kobold S, et al. Impact of a new fusion receptor on PD-1-mediated immunosuppression in adoptive T cell therapy. *J Natl Cancer Inst.* 2015;107(8):djv146.

165. Walker R, et al. Enhancing synergy of CAR T cell therapy and oncolytic virus therapy for pancreatic cancer. *bioRxiv*. 2016;055988.
166. Inoue H, Tani K. Multimodal immunogenic cancer cell death as a consequence of anticancer cytotoxic treatments. *Cell Death Differ*. 2014;21(1):39–49.
167. Corrales L, et al. Innate immune signaling and regulation in cancer immunotherapy. *Cell Res*. 2017;27(1):96–108.
168. Zaretsky JM, et al. Mutations associated with acquired resistance to PD-1 blockade in melanoma. *N Engl J Med*. 2016;375(9):819–829.
169. Parker BS, Rautela J, Hertzog PJ. Antitumour actions of interferons: implications for cancer therapy. *Nat Rev Cancer*. 2016;16(3):131–144.
170. Dumitriu IE, et al. Release of high mobility group box 1 by dendritic cells controls T cell activation via the receptor for advanced glycation end products. *J Immunol*. 2005;174(12):7506–7515.
171. Moser B, et al. Receptor for advanced glycation end products expression on T cells contributes to antigen-specific cellular expansion in vivo. *J Immunol*. 2007; 179(12):8051–8058.
172. Di Stasi A, et al. T lymphocytes coexpressing CCR4 and a chimeric antigen receptor targeting CD30 have improved homing and antitumor activity in a Hodgkin tumor model. *Blood*. 2009;113(25):6392–6402.
173. Kershaw MH, et al. Redirecting migration of T cells to chemokine secreted from tumors by genetic modification with CXCR2. *Hum Gene Ther*. 2002;13(16): 1971–1980.
174. Arai Y, et al. *Enhanced Expression of CXCR4 Facilitates C-Kit-Targeted CAR-T Cell Trafficking to Bone Marrow and Enables Donor Stem Cell Engraftment*. Biology of Blood and Marrow Transplantation. 2018;24:S311.
175. Thorne SH. Adding STING to the tale of oncolytic virotherapy. *Trends Cancer*. 2016;2(2):67–68.
176. Xia T, et al. Deregulation of STING signaling in colorectal carcinoma constrains DNA damage responses and correlates with tumorigenesis. *Cell Rep*. 2016;14(2):282–297.
177. Xia T, Konno H, Barber GN. Recurrent loss of STING signaling in melanoma correlates with susceptibility to viral oncolysis. *Cancer Res*. 2016;76(22):6747–6759.
178. Harrington KJ, et al. Preliminary results of the first-in-human (FIH) study of MK-1454, an agonist of stimulator of interferon genes (STING), as monotherapy or in combination with pembrolizumab (Pembro) in patients with advanced solid tumors or lymphomas. *Ann Oncol*. 2018;29(Suppl 8):viii712.
179. Haile ST, Conner J, Mackall C. Abstract 2304: attenuated oncolytic virus HSV1716 increases in vivo expansion of GD2- targeting CAR T cells in murine solid tumor models. *Cancer Res*. 2016;76(Suppl 14):2304.
180. Chen X, et al. A combinational therapy of EGFR-CAR NK cells and oncolytic herpes simplex virus 1 for breast cancer brain metastases. *Oncotarget*. 2016;7(19):27764–27777.
181. Tahtinen S, et al. Adenovirus improves the efficacy of adoptive T-cell therapy by recruiting immune cells to and promoting their activity at the tumor. *Cancer Immunol Res*. 2015;3(8):915–925.
182. Ilett E, et al. Prime-boost using separate oncolytic viruses in combination with checkpoint blockade improves anti-tumour therapy. *Gene Ther*. 2017;24(1):21–30.
183. Pule MA, et al. Virus-specific T cells engineered to coexpress tumor-specific receptors: persistence and antitumor activity in individuals with neuroblastoma. *Nat Med*. 2008;14(11):1264–1270.
184. Ngo MC, et al. Combining oncolytic vaccinia virotherapy with adoptive T cell therapy. *Blood*. 2015;118(21):4042.
185. Rehman H, et al. Into the clinic: talimogene laherparepvec (T-VEC), a first-in-class intratumoral oncolytic viral therapy. *J Immunother Cancer*. 2016;4:53.

186. Iankov ID, Haralambieva IH, Galanis E. Immunogenicity of attenuated measles virus engineered to express Helicobacter pylori neutrophil-activating protein. *Vaccine.* 2011;29(8):1710–1720.
187. Ramachandran M, et al. Vector-encoded Helicobacter pylori neutrophil-activating protein promotes maturation of dendritic cells with Th1 polarization and improved migration. *J Immunol.* 2014;193(5):2287–2296.
188. Jin C, et al. *CD19 CAR T-Cells With Induced Secretion of Helicobacter pylori Neutrophil-Activating Protein (HP-NAP) Yields Improved Anti-Tumor Activity and Reduced Immunosuppression.* 2016. Retrieved from http://urn.kb.se/resolve?urn=urn:nbn:se:uu:diva-300198.
189. Ramachandran M. Cancer immunotherapy: evolving oncolytic viruses and CAR T-cells. In: *Digital Comprehensive Summaries of Uppsala Dissertations from the Faculty of Medicine.* Uppsala: Acta Universitatis Upsaliensis; 2016:77. PhD thesis.
190. Berraondo P, et al. Cytokines in clinical cancer immunotherapy. *Br J Cancer.* 2019; 120(1):6–15.
191. Rose S. Bempegaldesleukin stimulates immune response. *Cancer Discov.* 2019; 9(6):OF1.
192. Klein C, et al. Cergutuzumab amunaleukin (CEA-IL2v), a CEA-targeted IL-2 variant-based immunocytokine for combination cancer immunotherapy: overcoming limitations of aldesleukin and conventional IL-2-based immunocytokines. *Oncoimmunology.* 2017; 6(3):e1277306.
193. Miller JS, et al. A first-in-human phase I study of subcutaneous outpatient recombinant human IL15 (rhIL15) in adults with advanced solid tumors. *Clin Cancer Res.* 2018;24(7):1525–1535.
194. Wrangle JM, et al. ALT-803, an IL-15 superagonist, in combination with nivolumab in patients with metastatic non-small cell lung cancer: a non-randomised, open-label, phase 1b trial. *Lancet Oncol.* 2018;19(5):694–704.
195. Naing A, et al. PEGylated IL-10 (Pegilodecakin) induces systemic immune activation, CD8(+) T cell invigoration and polyclonal T cell expansion in cancer patients. *Cancer Cell.* 2018;34(5):775–791.e3.
196. Schmidt H, et al. Safety and clinical effect of subcutaneous human interleukin-21 in patients with metastatic melanoma or renal cell carcinoma: a phase I trial. *Clin Cancer Res.* 2010;16(21):5312–5319.
197. Algazi AP, et al. Immune monitoring outcomes of patients with stage III/IV melanoma treated with a combination of pembrolizumab and intratumoral plasmid interleukin 12 (pIL-12). *J Clin Oncol.* 2017;35(Suppl 7S). abstract 78.
198. Yeku OO, et al. Armored CAR T cells enhance antitumor efficacy and overcome the tumor microenvironment. *Sci Rep.* 2017;7(1):10541.
199. Durham NM, et al. Oncolytic VSV primes differential responses to immuno-oncology therapy. *Mol Ther.* 2017;25(8):1917–1932.
200. Havunen R, et al. Abscopal effect in non-injected tumors achieved with cytokine-armed oncolytic adenovirus. *Mol Ther Oncolytics.* 2018;11:109–121.
201. Cervera-Carrascon V, et al. TNFa and IL-2 armed adenoviruses enable complete responses by anti-PD-1 checkpoint blockade. *OncoImmunology.* 2018;7(5):e1412902.
202. Post DE, et al. Targeted cancer gene therapy using a hypoxia inducible factor dependent oncolytic adenovirus armed with interleukin-4. *Cancer Res.* 2007;67(14): 6872–6881.
203. Leoni V, et al. A fully-virulent retargeted oncolytic HSV armed with IL-12 elicits local immunity and vaccine therapy towards distant tumors. *PLoS Pathog.* 2018;14(8): e1007209.
204. Kowalsky SJ, et al. Superagonist IL-15-armed oncolytic virus elicits potent antitumor immunity and therapy that are enhanced with PD-1 blockade. *Mol Ther.* 2018;26(10):2476–2486.

205. Yang C, et al. Oncolytic adenovirus expressing interleukin-18 improves antitumor activity of dacarbazine for malignant melanoma. *Drug Des Devel Ther.* 2016;10: 3755–3761.
206. Wang X, Lin Y. Tumor necrosis factor and cancer, buddies or foes? *Acta Pharmacol Sin.* 2008;29(11):1275–1288.
207. Meisen WH, Kaur B. How can we trick the immune system into overcoming the detrimental effects of oncolytic viral therapy to treat glioblastoma? *Expert Rev Neurother.* 2013;13(4):341–343.
208. Choi IK, et al. Oncolytic adenovirus co-expressing IL-12 and IL-18 improves tumor-specific immunity via differentiation of T cells expressing IL-12Rbeta2 or IL-18Ralpha. *Gene Ther.* 2011;18(9):898–909.
209. Bai FL, et al. Genetically engineered Newcastle disease virus expressing interleukin-2 and TNF-related apoptosis-inducing ligand for cancer therapy. *Cancer Biol Ther.* 2014;15(9):1226–1238.
210. Patel MR, et al. Vesicular stomatitis virus expressing interferon-beta is oncolytic and promotes antitumor immune responses in a syngeneic murine model of non-small cell lung cancer. *Oncotarget.* 2015;6(32):33165–33177.
211. Sturm JB, et al. Functional hyper-IL-6 from vaccinia virus-colonized tumors triggers platelet formation and helps to alleviate toxicity of mitomycin C enhanced virus therapy. *J Transl Med.* 2012;10:9.
212. Chard LS, Lemoine NR, Wang Y. New role of interleukin-10 in enhancing the antitumor efficacy of oncolytic vaccinia virus for treatment of pancreatic cancer. *Oncoimmunology.* 2015;4(9):e1038689.
213. Barnard Z, et al. Expression of FMS-like tyrosine kinase 3 ligand by oncolytic herpes simplex virus type I prolongs survival in mice bearing established syngeneic intracranial malignant glioma. *Neurosurgery.* 2012;71(3):741–748.
214. Lasek W, Zagożdżon R, Jakobisiak M. Interleukin 12: still a promising candidate for tumor immunotherapy? *Cancer Immunol Immunother.* 2014;63(5):419–435.
215. Zhang L, et al. Tumor-infiltrating lymphocytes genetically engineered with an inducible gene encoding interleukin-12 for the immunotherapy of metastatic melanoma. *Clin Cancer Res.* 2015;21(10):2278–2288.
216. Pegram HJ, et al. Tumor-targeted T cells modified to secrete IL-12 eradicate systemic tumors without need for prior conditioning. *Blood.* 2012;119(18):4133–4141.
217. Markley JC, Sadelain M. IL-7 and IL-21 are superior to IL-2 and IL-15 in promoting human T cell-mediated rejection of systemic lymphoma in immunodeficient mice. *Blood.* 2010;115(17):3508–3519.
218. Watanabe K, et al. Pancreatic cancer therapy with combined mesothelin-redirected chimeric antigen receptor T cells and cytokine-armed oncolytic adenoviruses. *JCI Insight.* 2018;3:7.
219. Afferni C, et al. The pleiotropic immunomodulatory functions of IL-33 and its implications in tumor immunity. *Front Immunol.* 2018;9:2601.
220. Chen Y, Lu B. Guided delivery of the "alarming" cytokine IL-33 to tumor by chimeric antigen receptor T cells. *J Immunol.* 2017;198(Suppl 1):204.23.
221. Mardiana S. Novel strategies to enhance the therapeutic efficacy of chimeric antigen receptor T cell therapy against solid cancer. In: *Centre for Cancer Research.* University of Melbourne; 2018. PhD thesis.
222. Weist MR, et al. PET of adoptively transferred chimeric antigen receptor T cells with (89)Zr-oxine. *J Nucl Med.* 2018;59(10):1531–1537.
223. Slaney CY, Kershaw MH, Darcy PK. Trafficking of T cells into tumors. *Cancer Res.* 2014;74(24):7168–7174.
224. Sridhar P, Petrocca F. Regional delivery of chimeric antigen receptor (CAR) T-cells for cancer therapy. *Cancers (Basel).* 2017;9:7.

225. van Schalkwyk MC, et al. Design of a phase I clinical trial to evaluate intratumoral delivery of ErbB-targeted chimeric antigen receptor T-cells in locally advanced or recurrent head and neck cancer. *Hum Gene Ther Clin Dev.* 2013;24(3):134–142.
226. Adusumilli, P.S., et al., CT036—A Phase I Clinical Trial of Malignant Pleural Disease Treated With Regionally Delivered Autologous Mesothelin-Targeted CAR T Cells: Safety and Efficacy in AACR Annual Meeting 2019. 2019: Atlanta, GA, USA.
227. Lanitis E, et al. Mechanisms regulating T-cell infiltration and activity in solid tumors. *Ann Oncol.* 2017;28(Suppl 12):xii18–xii32.
228. Roberts EW, et al. Critical role for CD103(+)/CD141(+) dendritic cells bearing CCR7 for tumor antigen trafficking and priming of T cell immunity in Melanoma. *Cancer Cell.* 2016;30(2):324–336.
229. Spranger S, et al. Tumor-residing Batf3 dendritic cells are required for effector T cell trafficking and adoptive T cell therapy. *Cancer Cell.* 2017;31(5):711–723.e4.
230. Spranger S, Zha Y, Gajewski T. Effect of melanoma intrinsic β-catenin signaling on immune exclusion and resistance to immunotherapies. *J Clin Oncol.* 2015;33(Suppl 15):9014.
231. Yao C, et al. Cross-dressing and antigen transfer after DC targeting induce robust T cell response. *J Immunol.* 2017;198(Suppl 1):209.2.
232. Ayers M, et al. IFN-gamma-related mRNA profile predicts clinical response to PD-1 blockade. *J Clin Invest.* 2017;127(8):2930–2940.
233. Roelands J, et al. Immunogenomic classification of colorectal cancer and therapeutic implications. *Int J Mol Sci.* 2017;18(10):2229.
234. Liu Z, et al. CXCL11-armed oncolytic poxvirus elicits potent antitumor immunity and shows enhanced therapeutic efficacy. *Oncoimmunology.* 2016;5(3):e1091554.
235. Moon EK, et al. Intra-tumoral delivery of CXCL11 via a vaccinia virus, but not by modified T cells, enhances the efficacy of adoptive T cell therapy and vaccines. *OncoImmunology.* 2018;7(3):e1395997.
236. Li J, et al. Chemokine expression from oncolytic vaccinia virus enhances vaccine therapies of cancer. *Mol Ther.* 2011;19(4):650–657.
237. Liu GY, et al. Enhanced growth suppression of TERT-positive tumor cells by oncolytic adenovirus armed with CCL20 and CD40L. *Int Immunopharmacol.* 2015;28(1): 487–493.
238. Li J, et al. Expression of CCL19 from oncolytic vaccinia enhances immunotherapeutic potential while maintaining oncolytic activity. *Neoplasia.* 2012;14(12):1115–1121.
239. Bronger H, et al. CXCL9 and CXCL10 predict survival and are regulated by cyclooxygenase inhibition in advanced serous ovarian cancer. *Br J Cancer.* 2016;115(5):553–563.
240. Wu Z, et al. The chemokine CXCL9 expression is associated with better prognosis for colorectal carcinoma patients. *Biomed Pharmacother.* 2016;78:8–13.
241. Bai M, Chen X, Ba YI. CXCL10/CXCR3 overexpression as a biomarker of poor prognosis in patients with stage II colorectal cancer. *Mol Clin Oncol.* 2016;4(1):23–30.
242. Hsin LJ, et al. Serum CXCL9 levels are associated with tumor progression and treatment outcome in patients with nasopharyngeal carcinoma. *PLoS One.* 2013;8(11):e80052.
243. Zeng YJ, et al. Neuroendocrine-like cells-derived CXCL10 and CXCL11 induce the infiltration of tumor-associated macrophage leading to the poor prognosis of colorectal cancer. *Oncotarget.* 2016;7(19):27394–27407.
244. Gurler Main H, et al. Emergent role of the fractalkine axis in dissemination of peritoneal metastasis from epithelial ovarian carcinoma. *Oncogene.* 2017;36(21):3025–3036.
245. Korniejewska A, et al. Expression and agonist responsiveness of CXCR3 variants in human T lymphocytes. *Immunology.* 2011;132(4):503–515.
246. Viola A, et al. The pros and cons of chemokines in tumor immunology. *Trends Immunol.* 2012;33(10):496–504.

247. Komorowski MP, et al. Reprogramming antitumor immunity against chemoresistant ovarian cancer by a CXCR4 antagonist-armed viral oncotherapy. *Mol Ther Oncolytics*. 2016;3:16034.
248. Meng W, Xue S, Chen Y. The role of CXCL12 in tumor microenvironment. *Gene*. 2018;641:105–110.
249. Suryadevara CM, et al. Are BiTEs the "missing link" in cancer therapy? *Oncoimmunology*. 2015;4(6):e1008339.
250. Kantarjian H, et al. Blinatumomab versus chemotherapy for advanced acute lymphoblastic leukemia. *N Engl J Med*. 2017;376(9):836–847.
251. Wang X, Gottschalk S, Song X-T. Synergistic antitumor effects of chimeric antigen receptor-modified T cells and oncolytic virotherapy. *Blood*. 2014;124(21):5808.
252. Yu F, Hong B, Song XT. A T-cell engager-armed oncolytic vaccinia virus to target the tumor stroma. *Cancer Transl Med*. 2017;3:122–132.
253. Fajardo CA, et al. Oncolytic adenoviral delivery of an EGFR-targeting T-cell engager improves antitumor efficacy. *Cancer Res*. 2017;77(8):2052–2063.
254. Wing A, et al. Improving CART-cell therapy of solid tumors with oncolytic virus-driven production of a bispecific T-cell engager. *Cancer Immunol Res*. 2018;6(5): 605–616.
255. Freedman JD, et al. Oncolytic adenovirus expressing bispecific antibody targets T-cell cytotoxicity in cancer biopsies. *EMBO Mol Med*. 2017;9(8):1067–1087.
256. Haas C, et al. T-cell triggering by CD3- and CD28-binding molecules linked to a human virus-modified tumor cell vaccine. *Vaccine*. 2005;23(19):2439–2453.
257. Zhang Y, et al. Acute cytokine response to systemic adenoviral vectors in mice is mediated by dendritic cells and macrophages. *Mol Ther*. 2001;3(5 Pt 1):697–707.
258. Cole C, et al. Tumor-targeted, systemic delivery of therapeutic viral vectors using hitchhiking on antigen-specific T cells. *Nat Med*. 2005;11(10):1073–1081.
259. Willmon C, et al. Cell carriers for oncolytic viruses: fed Ex for cancer therapy. *Mol Ther*. 2009;17(10):1667–1676.
260. Magge D, et al. Inhibitors of C5 complement enhance vaccinia virus oncolysis. *Cancer Gene Ther*. 2013;20(6):342–350.
261. Kirn DH, et al. Enhancing poxvirus oncolytic effects through increased spread and immune evasion. *Cancer Res*. 2008;68(7):2071–2075.
262. Nosaki K, et al. A novel, polymer-coated oncolytic measles virus overcomes immune suppression and induces robust antitumor activity. *Mol Ther Oncolytics*. 2016;3:16022.
263. Ahmed NM, et al. Autologous HER2 CMV bispecific CAR T cells for progressive glioblastoma: results from a phase I clinical trial. *J Clin Oncol*. 2015;33(Suppl 15):3008.
264. Ghosh A, et al. Donor CD19 CAR T cells exert potent graft-versus-lymphoma activity with diminished graft-versus-host activity. *Nat Med*. 2017;23(2):242–249.
265. Yang Y, et al. TCR engagement negatively affects CD8 but not CD4 CAR T cell expansion and leukemic clearance. *Sci Transl Med*. 2017;9:417.
266. Moon EK, et al. Expression of a functional CCR2 receptor enhances tumor localization and tumor eradication by retargeted human T cells expressing a mesothelin-specific chimeric antibody receptor. *Clin Cancer Res*. 2011;17(14):4719–4730.
267. Joyce JA, Fearon DT. T cell exclusion, immune privilege, and the tumor microenvironment. *Science*. 2015;348(6230):74–80.
268. Feig C, et al. Targeting CXCL12 from FAP-expressing carcinoma-associated fibroblasts synergizes with anti-PD-L1 immunotherapy in pancreatic cancer. *Proc Natl Acad Sci USA*. 2013;110(50):20212–20217.
269. Zboralski D, et al. Increasing tumor-infiltrating T cells through inhibition of CXCL12 with NOX-A12 synergizes with PD-1 blockade. *Cancer Immunol Res*. 2017;5(11):950–956.
270. Buckanovich RJ, et al. Endothelin B receptor mediates the endothelial barrier to T cell homing to tumors and disables immune therapy. *Nat Med*. 2008;14(1):28–36.

271. Siddiqui I, et al. Enhanced recruitment of genetically modified CX3CR1-positive human T cells into fractalkine/CX3CL1 expressing tumors: importance of the chemokine gradient. *J Immunother Cancer*. 2016;4:21.
272. Hulshof S, et al. CX3CL1 and CX3CR1 expression in human brain tissue: non-inflammatory control versus multiple sclerosis. *J Neuropathol Exp Neurol*. 2003;62(9): 899–907.
273. Newick K, et al. Augmentation of CAR T-cell trafficking and antitumor efficacy by blocking protein kinase A localization. *Cancer Immunol Res*. 2016;4(6):541–551.
274. Nishio N, Dotti G. Oncolytic virus expressing RANTES and IL-15 enhances function of CAR-modified T cells in solid tumors. *Oncoimmunology*. 2015;4(2):e988098.
275. Lan P, et al. Reconstitution of a functional human immune system in immunodeficient mice through combined human fetal thymus/liver and CD34+ cell transplantation. *Blood*. 2006;108(2):487–492.
276. Jiang H, et al. Oncolytic adenovirus and tumor-targeting immune modulatory therapy improve autologous cancer vaccination. *Cancer Res*. 2017;77(14):3894–3907.
277. Kim HS, et al. Host lymphodepletion enhances the therapeutic activity of an oncolytic vaccinia virus expressing 4-1BB ligand. *Cancer Res*. 2009;69(21):8516–8525.
278. Rivera-Molina YA, et al. Abstract 4565: forced expression of GITRL in cancer cells enhances adenovirus-mediated in situ vaccination. *Cancer Res*. 2017;77(Suppl 13):4565.
279. Wise M, et al. 643. Various forms of CD40L encoded as an immune plasmid adjuvant generate unique anti-cancer DNA vaccine induced responses. *Mol Ther*. 2016;24:S255.
280. Eriksson E, et al. Activation of myeloid and endothelial cells by CD40L gene therapy supports T-cell expansion and migration into the tumor microenvironment. *Gene Ther*. 2017;24(2):92–103.
281. Xydia M, et al. CD40L co-stimulation from CD8+ to CD4+ effector memory T cells supports CD4+ expansion. *Immunol Cell Biol*. 2011;89(6):670–680.
282. Shugart JA, et al. A self-help program for memory CD8+ T cells: positive feedback via CD40-CD40L signaling as a critical determinant of secondary expansion. *PLoS One*. 2013;8(5):e64878.
283. Mark H. O'Hara, Eileen M. O'Reilly, Mick Rosemarie, Gauri Varadhachary, Zev A. Wainberg, Andrew Ko, George A. Fisher, Jr., Osama Rahma, Jaclyn P. Lyman, Christopher R. Cabanski, Erica L. Carpenter, Travis Hollmann, Pier Federico Gherardini, Lacey Kitch, Cheryl Selinsky, Theresa LaVallee, Ovid C. Trifan, Ute Dugan, Vanessa M. Hubbard-Lucey, Robert H. Vonderheide. CT004—A Phase Ib Study of CD40 Agonistic Monoclonal Antibody APX005M Together With Gemcitabine (Gem) and Nab-Paclitaxel (NP) With or Without Nivolumab (Nivo) in Untreated Metastatic Ductal Pancreatic Adenocarcinoma (PDAC) Patients in AACR Annual Meeting 2019. 2019. Conference abstract, AACR, Atlanta, USA.
284. Gulley JL, et al. Immune impact induced by PROSTVAC (PSA-TRICOM), a therapeutic vaccine for prostate cancer. *Cancer Immunol Res*. 2014;2(2):133–141.
285. Gulley JL, et al. Pilot study of vaccination with recombinant CEA-MUC-1-TRICOM poxviral-based vaccines in patients with metastatic carcinoma. *Clin Cancer Res*. 2008;14(10):3060–3069.
286. Bavarian Nordic, Bavarian Nordic Announces Acceptance of Abstracts and Oral Presentation at the American Society of Clinical Oncology (ASCO) Annual Meeting. 2018: Copenhagen, Denmark.
287. De Sousa Linhares A, et al. Not all immune checkpoints are created equal. *Front Immunol*. 2018;9:1909.
288. Salama AKS, Moschos SJ. Next steps in immuno-oncology: enhancing antitumor effects through appropriate patient selection and rationally designed combination strategies. *Ann Oncol*. 2016;28(1):57–74.
289. Śledzińska A, et al. Negative immune checkpoints on T lymphocytes and their relevance to cancer immunotherapy. *Mol Oncol*. 2015;9(10):1936–1965.

290. Abate-Daga D, et al. Expression profiling of TCR-engineered T cells demonstrates overexpression of multiple inhibitory receptors in persisting lymphocytes. *Blood.* 2013;122(8):1399–1410.
291. Kalos M, et al. T cells with chimeric antigen receptors have potent antitumor effects and can establish memory in patients with advanced leukemia. *Sci Transl Med.* 2011;3(95): 95ra73.
292. Moon EK, et al. Multifactorial T-cell hypofunction that is reversible can limit the efficacy of chimeric antigen receptor-transduced human T cells in solid tumors. *Clin Cancer Res.* 2014;20(16):4262–4273.
293. Tschumi BO, et al. CART cells are prone to Fas- and DR5-mediated cell death. *J Immunother Cancer.* 2018;6(1):71.
294. Yoon DH, et al. Incorporation of immune checkpoint blockade into chimeric antigen receptor T cells (CAR-Ts): combination or built-in CAR-T. *Int J Mol Sci.* 2018;19(2):340.
295. Dias JD, et al. Targeted cancer immunotherapy with oncolytic adenovirus coding for a fully human monoclonal antibody specific for CTLA-4. *Gene Ther.* 2012;19(10):988–998.
296. Kleinpeter P, et al. Vectorization in an oncolytic vaccinia virus of an antibody, a Fab and a scFv against programmed cell death -1 (PD-1) allows their intratumoral delivery and an improved tumor-growth inhibition. *Oncoimmunology.* 2016;5(10): e1220467.
297. Rojas JJ, et al. Defining effective combinations of immune checkpoint blockade and oncolytic virotherapy. *Clin Cancer Res.* 2015;21(24):5543–5551.
298. Tanoue K, et al. Armed oncolytic adenovirus-expressing PD-L1 mini-body enhances antitumor effects of chimeric antigen receptor T cells in solid tumors. *Cancer Res.* 2017;77(8):2040–2051.
299. Rosewell Shaw A, et al. Adenovirotherapy delivering cytokine and checkpoint inhibitor augments CAR T cells against metastatic head and neck cancer. *Mol Ther.* 2017;25(11):2440–2451.
300. Suarez ER, et al. Chimeric antigen receptor T cells secreting anti-PD-L1 antibodies more effectively regress renal cell carcinoma in a humanized mouse model. *Oncotarget.* 2016;7(23):34341–34355.
301. Welstead G, et al. 572. Successful generation of CAR + PD-1-primary T cells using Cas9-mediated genome editing. *Mol Ther.* 2016;24:S229.
302. Ren J, et al. A versatile system for rapid multiplex genome-edited CAR T cell generation. *Oncotarget.* 2017;8(10):17002–17011.
303. Oda SK, et al. A CD200R-CD28 fusion protein appropriates an inhibitory signal to enhance T-cell function and therapy of murine leukemia. *Blood.* 2017;130(22): 2410–2419.
304. Oda SK, et al. Cheating death: a Fas-41BB immunomodulatory fusion protein obviates a death signal to enhance T cell function and adoptive therapy targeting leukemia and solid tumors. *J Immunol.* 2018;200(Suppl 1):179.11.
305. Riese MJ, et al. Enhanced effector responses in activated CD8 + T cells deficient in diacylglycerol kinases. *Cancer Res.* 2013;73(12):3566–3577.
306. Jung IY, et al. CRISPR/Cas9-mediated knockout of DGK improves antitumor activities of human T cells. *Cancer Res.* 2018;78(16):4692–4703.
307. Pitt JM, et al. Targeting the tumor microenvironment: removing obstruction to anticancer immune responses and immunotherapy. *Ann Oncol.* 2016;27(8):1482–1492.
308. Oh SA, Li MO. TGF-β: guardian of T cell function. *J Immunol.* 2013;191(8): 3973–3979.
309. Haque S, Morris JC. Transforming growth factor-β: a therapeutic target for cancer. *Hum Vaccin Immunother.* 2017;13(8):1741–1750.

310. Strauss J, et al. Phase I trial of M7824 (MSB0011359C), a bifunctional fusion protein targeting PD-L1 and TGFbeta, in advanced solid tumors. *Clin Cancer Res.* 2018;24(6):1287–1295.
311. Oh E, et al. Oncolytic adenovirus coexpressing interleukin-12 and decorin overcomes Treg-mediated immunosuppression inducing potent antitumor effects in a weakly immunogenic tumor model. *Oncotarget.* 2017;8(3):4730–4746.
312. Yu Y, et al. Cancer-associated fibroblasts induce epithelial-mesenchymal transition of breast cancer cells through paracrine TGF-beta signalling. *Br J Cancer.* 2014;110(3): 724–732.
313. Terry S, et al. New insights into the role of EMT in tumor immune escape. *Mol Oncol.* 2017;11(7):824–846.
314. Ilkow CS, et al. Reciprocal cellular cross-talk within the tumor microenvironment promotes oncolytic virus activity. *Nat Med.* 2015;21(5):530–536.
315. Ronchetti S, et al. Glucocorticoid-induced TNFR-related protein lowers the threshold of CD28 costimulation in CD8+ T cells. *J Immunol.* 2007;179(9):5916–5926.
316. Mitsui J, et al. Two distinct mechanisms of augmented antitumor activity by modulation of immunostimulatory/inhibitory signals. *Clin Cancer Res.* 2010;16(10):2781–2791.
317. Knee DA, Hewes B, Brogdon JL. Rationale for anti-GITR cancer immunotherapy. *Eur J Cancer.* 2016;67:1–10.
318. Cohen AD, et al. Agonist anti-GITR monoclonal antibody induces melanoma tumor immunity in mice by altering regulatory T cell stability and intra-tumor accumulation. *PLoS One.* 2010;5(5):e10436.
319. Sukumar S, et al. Characterization of MK-4166, a clinical agonistic antibody that targets human GITR and inhibits the generation and suppressive effects of T regulatory cells. *Cancer Res.* 2017;77(16):4378–4388.
320. Calmels B, et al. Bypassing tumor-associated immune suppression with recombinant adenovirus constructs expressing membrane bound or secreted GITR-L. *Cancer Gene Ther.* 2005;12(2):198–205.
321. Kottke T, et al. Treg depletion-enhanced IL-2 treatment facilitates therapy of established tumors using systemically delivered oncolytic virus. *Mol Ther.* 2008;16(7): 1217–1226.
322. Simpson TR, et al. Fc-dependent depletion of tumor-infiltrating regulatory T cells co-defines the efficacy of anti-CTLA-4 therapy against melanoma. *J Exp Med.* 2013;210(9):1695–1710.
323. Hodi FS, et al. Nivolumab plus ipilimumab or nivolumab alone versus ipilimumab alone in advanced melanoma (CheckMate 067): 4-year outcomes of a multicentre, randomised, phase 3 trial. *Lancet Oncol.* 2018;19(11):1480–1492.
324. Motzer RJ, et al. Nivolumab plus Ipilimumab versus sunitinib in advanced renal-cell carcinoma. *N Engl J Med.* 2018;378(14):1277–1290.
325. Overman MJ, et al. Durable clinical benefit with nivolumab plus ipilimumab in DNA mismatch repair-deficient/microsatellite instability-high metastatic colorectal cancer. *J Clin Oncol.* 2018;36(8):773–779.
326. Korman AJ, et al. Abstract SY09-01: next-generation anti-CTLA-4 antibodies. *Cancer Res.* 2017;77(Suppl 13):SY09-01.
327. Mailloux AW, Young MRI. Regulatory T-cell trafficking: from thymic development to tumor-induced immune suppression. *Crit Rev Immunol.* 2010;30(5):435–447.
328. Turnis ME, et al. Interleukin-35 limits anti-tumor immunity. *Immunity.* 2016;44(2):316–329.
329. Delgoffe GM, et al. Stability and function of regulatory T cells is maintained by a neuropilin-1-semaphorin-4a axis. *Nature.* 2013;501(7466):252–256.
330. Fleming V, et al. Targeting myeloid-derived suppressor cells to bypass tumor-induced immunosuppression. *Front Immunol.* 2018;9:398.

331. Isella C, et al. Selective analysis of cancer-cell intrinsic transcriptional traits defines novel clinically relevant subtypes of colorectal cancer. *Nat Commun.* 2017;8:15107.
332. Chesney JA, Mitchell RA, Yaddanapudi K. Myeloid-derived suppressor cells—a new therapeutic target to overcome resistance to cancer immunotherapy. *J Leukoc Biol.* 2017;102(3):727–740.
333. Long AH, et al. Reduction of MDSCs with all-trans retinoic acid improves CAR therapy efficacy for sarcomas. *Cancer Immunol Res.* 2016;4(10):869–880.
334. Chen W, et al. ATRA enhances bystander effect of suicide gene therapy in the treatment of prostate cancer. *Urol Oncol.* 2008;26(4):397–405.
335. Gyöngyösi A, et al. RDH10, RALDH2, and CRABP2 are required components of PPARγ-directed ATRA synthesis and signaling in human dendritic cells. *J Lipid Res.* 2013;54(9):2458–2474.
336. Shevchenko I, Bazhin AV. Metabolic checkpoints: novel avenues for immunotherapy of cancer. *Front Immunol.* 2018;9:1816.
337. Scharping NE, Delgoffe GM. Tumor microenvironment metabolism: a new checkpoint for anti-tumor immunity. *Vaccine.* 2016;4(4):46.
338. Zelenay S, et al. Cyclooxygenase-dependent tumor growth through evasion of immunity. *Cell.* 2015;162(6):1257–1270.
339. Kalinski P. Regulation of immune responses by prostaglandin E2. *J Immunol.* 2012;188(1):21–28.
340. Walker JD, Sehgal I, Kousoulas KG. Oncolytic herpes simplex virus 1 encoding 15-prostaglandin dehydrogenase mitigates immune suppression and reduces ectopic primary and metastatic breast cancer in mice. *J Virol.* 2011;85(14):7363–7371.
341. Curti A, et al. Modulation of tryptophan catabolism by human leukemic cells results in the conversion of CD25- into CD25+ T regulatory cells. *Blood.* 2007;109(7):2871–2877.
342. Munn DH, et al. GCN2 kinase in T cells mediates proliferative arrest and anergy induction in response to indoleamine 2,3-dioxygenase. *Immunity.* 2005;22(5):633–642.
343. Fallarino F, et al. The combined effects of tryptophan starvation and tryptophan catabolites down-regulate T cell receptor zeta-chain and induce a regulatory phenotype in naive T cells. *J Immunol.* 2006;176(11):6752–6761.
344. Prendergast GC, et al. Discovery of IDO1 inhibitors: from bench to bedside. *Cancer Res.* 2017;77(24):6795–6811.
345. Reinhart B, et al. Inhibition of indoleamine-2,3-dioxygenase (IDO) in glioblastoma cells by oncolytic herpes simplex virus. *Adv Virol.* 2012;2012:815465.
346. Long GV, et al. Epacadostat (E) plus pembrolizumab (P) versus pembrolizumab alone in patients (pts) with unresectable or metastatic melanoma: results of the phase 3 ECHO-301/KEYNOTE-252 study. *J Clin Oncol.* 2018;36(Suppl 15):108.
347. Labadie BW, Bao R, Luke JJ. Reimagining IDO pathway inhibition in cancer immunotherapy via downstream focus on the tryptophan–kynurenine–aryl hydrocarbon axis. *Clin Cancer Res.* 2019;25(5):1462–1471.
348. Allard B, et al. The ectonucleotidases CD39 and CD73: novel checkpoint inhibitor targets. *Immunol Rev.* 2017;276(1):121–144.
349. Monteiro I, et al. CD73 expression and clinical significance in human metastatic melanoma. *Oncotarget.* 2018;9(42):26659–26669.
350. Morello S, et al. Myeloid cells in the tumor microenvironment: role of adenosine. *Oncoimmunology.* 2015;5(3):e1108515.
351. Schindler U, et al. Abstract 2640: small-molecule inhibitors of CD73, CD39 and $A_{2a}R$: three anti-cancer targets in the ATP/adenosine signaling pathway. *Cancer Res.* 2017;-77(Suppl 13):2640.
352. Beavis PA, et al. Targeting the adenosine 2A receptor enhances chimeric antigen receptor T cell efficacy. *J Clin Invest.* 2017;127(3):929–941.

353. Wang J, et al. Purinergic targeting enhances immunotherapy of CD73+ solid tumors with piggyBac-engineered chimeric antigen receptor natural killer cells. *J Immunother Cancer*. 2018;6(1):136.
354. Vodnala SK, et al. T cell stemness and dysfunction in tumors are triggered by a common mechanism. *Science*. 2019;363(6434):eaau0135.
355. Eil R, et al. Ionic immune suppression within the tumour microenvironment limits T cell effector function. *Nature*. 2016;537:539.
356. Wojton J, Kaur B. Impact of tumor microenvironment on oncolytic viral therapy. *Cytokine Growth Factor Rev*. 2010;21(2-3):127–134.
357. Abd-Aziz N, Stanbridge EJ, Shafee N. Newcastle disease virus degrades HIF-1alpha through proteasomal pathways independent of VHL and p53. *J Gen Virol*. 2016;97(12):3174–3182.
358. Zhou W, et al. Telomerase-specific oncolytic adenovirus expressing TRAIL suppresses peritoneal dissemination of gastric cancer. *Gene Ther*. 2017;24(4):199–207.
359. Loya SM, Zhang X. Enhancing the bystander killing effect of an oncolytic HSV by arming it with a secretable apoptosis activator. *Gene Ther*. 2015;22(3):237–246.
360. Hirvinen M, et al. Immunological effects of a tumor necrosis factor alpha-armed oncolytic adenovirus. *Hum Gene Ther*. 2015;26(3):134–144.
361. Mader JS, et al. Bovine lactoferricin selectively induces apoptosis in human leukemia and carcinoma cell lines. *Mol Cancer Ther*. 2005;4(4):612–624.
362. Kochneva G, et al. Engineering of double recombinant vaccinia virus with enhanced oncolytic potential for solid tumor virotherapy. *Oncotarget*. 2016;7(45):74171–74188.
363. Yamada S, et al. Oncolytic herpes simplex virus expressing yeast cytosine deaminase: relationship between viral replication, transgene expression, prodrug bioactivation. *Cancer Gene Ther*. 2012;19(3):160–170.
364. Kubo H, et al. Phase I dose escalation clinical trial of adenovirus vector carrying osteocalcin promoter-driven herpes simplex virus thymidine kinase in localized and metastatic hormone-refractory prostate cancer. *Hum Gene Ther*. 2003;14(3):227–241.
365. Dobson CC, et al. Oncolytic virus synergizes with Smac mimetic compounds to induce rhabdomyosarcoma cell death in a syngeneic murine model. *Oncotarget*. 2017;8(2): 3495–3508.
366. Golinelli G, et al. Targeting GD2-positive glioblastoma by chimeric antigen receptor empowered mesenchymal progenitors. *Cancer Gene Ther*. 2018;1–13.
367. Lu P, Weaver VM, Werb Z. The extracellular matrix: a dynamic niche in cancer progression. *J Cell Biol*. 2012;196(4):395–406.
368. Infante JR, et al. Phase 1 trials of PEGylated recombinant human hyaluronidase PH20 in patients with advanced solid tumours. *Br J Cancer*. 2018;118(2):153–161.
369. Hingorani SR, et al. HALO 202: randomized phase II study of PEGPH20 plus Nab-Paclitaxel/gemcitabine versus Nab-Paclitaxel/gemcitabine in patients with untreated, metastatic pancreatic ductal adenocarcinoma. *J Clin Oncol*. 2018;36(4):359–366.
370. Guedan S, et al. Hyaluronidase expression by an oncolytic adenovirus enhances its intratumoral spread and suppresses tumor growth. *Mol Ther*. 2010;18(7):1275–1283.
371. Cheng J, et al. Human matrix metalloproteinase-8 gene delivery increases the oncolytic activity of a replicating adenovirus. *Mol Ther*. 2007;15(11):1982–1990.
372. Schäfer S, et al. Vaccinia virus-mediated intra-tumoral expression of matrix metalloproteinase 9 enhances oncolysis of PC-3 xenograft tumors. *BMC Cancer*. 2012;12:366.
373. Gialeli C, Theocharis AD, Karamanos NK. Roles of matrix metalloproteinases in cancer progression and their pharmacological targeting. *FEBS J*. 2011;278(1):16–27.
374. Roy R, Yang J, Moses MA. Matrix metalloproteinases as novel biomarkers and potential therapeutic targets in human cancer. *J Clin Oncol*. 2009;27(31):5287–5297.

375. Toro Bejarano M, Merchan JR. Targeting tumor vasculature through oncolytic virotherapy: recent advances. *Oncolytic Virother.* 2015;4:169–181.
376. Tysome JR, Lemoine NR, Wang Y. Update on oncolytic viral therapy—targeting angiogenesis. *Onco Targets Ther.* 2013;6:1031–1040.
377. Strasly M, et al. IL-12 Inhibition of endothelial cell functions and angiogenesis depends on lymphocyte-endothelial cell cross-talk. *J Immunol.* 2001;166(6):3890–3899.
378. Cao R, et al. Interleukin-18 acts as an angiogenesis and tumor suppressor. *FASEB J.* 1999;13(15):2195–2202.
379. Persaud L, et al. Mechanism of action and applications of interleukin 24 in immunotherapy. *Int J Mol Sci.* 2016;17(6):869.
380. Shimizu M, et al. Antiangiogenic and antitumor activities of IL-27. *J Immunol.* 2006;176(12):7317–7324.
381. Yoo JY, et al. VEGF-specific short hairpin RNA-expressing oncolytic adenovirus elicits potent inhibition of angiogenesis and tumor growth. *Mol Ther.* 2007;15(2): 295–302.
382. Yoo JY, et al. Short hairpin RNA-expressing oncolytic adenovirus-mediated inhibition of IL-8: effects on antiangiogenesis and tumor growth inhibition. *Gene Ther* 2008;15(9):635–651.
383. Liu T-C, et al. Oncolytic HSV armed with platelet factor 4, an antiangiogenic agent, shows enhanced efficacy. *Mol Ther.* 2006;14(6):789–797.
384. Carew JS, et al. Oncolytic reovirus inhibits angiogenesis through induction of CXCL10/IP-10 and abrogation of HIF activity in soft tissue sarcomas. *Oncotarget.* 2017;8(49):86769–86783.
385. Zhang W, et al. Combination of oncolytic herpes simplex viruses armed with angiostatin and IL-12 enhances antitumor efficacy in human glioblastoma models. *Neoplasia.* 2013;15(6):591–599.
386. Zhu G, et al. Glioma stem cells targeted by oncolytic virus carrying endostatin-angiostatin fusion gene and the expression of its exogenous gene in vitro. *Brain Res.* 2011;1390:59–69.
387. Tomita Y, et al. Oncolytic herpes virus armed with vasculostatin in combination with bevacizumab abrogate glioma invasion via the CCN1 and AKT signaling pathways. *Mol Cancer Ther.* 2019;.
388. Bui Nguyen TM, et al. Adeno-associated virus-mediated delivery of kringle 5 of human plasminogen inhibits orthotopic growth of ovarian cancer. *Gene Ther.* 2010;17(5):606–615.
389. Frentzen A, et al. Anti-VEGF single-chain antibody GLAF-1 encoded by oncolytic vaccinia virus significantly enhances antitumor therapy. *Proc Natl Acad Sci USA.* 2009;106(31):12915–12920.

CHAPTER EIGHT

Application of CAR T cells for the treatment of solid tumors

Jonathan F. Khan[a,b,†], **Abdul Salam Khan**[c,†], **Renier J. Brentjens**[a,b,c,*]
[a]Department of Pharmacology, Weill Cornell Graduate School of Medical Sciences, Weill Cornell Medicine, New York, NY, United States
[b]Molecular Pharmacology Program, Sloan Kettering Institute, Memorial Sloan Kettering Cancer Center, New York, NY, United States
[c]Department of Medicine, Memorial Sloan Kettering Cancer Center, New York, NY, United States
*Corresponding author: e-mail address: brentjer@mskcc.org

Contents

1. Introduction 294
2. CAR T cell therapy for solid tumors 297
   2.1 Glioblastoma 297
   2.2 Neuroblastoma 299
   2.3 Mesothelioma 300
   2.4 Ovarian cancer 301
3. Overcoming challenges to CAR T cell therapy of solid tumors 302
   3.1 Antigen heterogeneity 302
   3.2 Suboptimal CAR T cell trafficking 305
   3.3 Immunosuppressive tumor microenvironment 309
   3.4 CAR T cell persistence 315
   3.5 Site-specific CAR integration into the T cell genome 316
4. Concluding remarks 317
References 318

Abstract

CAR T cell therapy of cancers promises to revolutionize oncology by harnessing the powers of synthetic biology and immunotherapy in a single agent. CARs are synthetic receptors composed of an extracellular antigen binding domain and one or more intracellular signaling domains which act in concert to activate the T cell upon antigen recognition. CARs targeting B cell associated CD19 demonstrated robust in vivo cytolytic activity, expansion, and persistence upon antigen exposure paving the way for clinical application of this technology and ultimately FDA approval for pediatric and young adult acute lymphoblastic leukemia as well as patients with relapsed or refractory diffuse large B cell lymphoma. However, these successes have not yet been replicated in the

† These authors contributed equally.

Progress in Molecular Biology and Translational Science, Volume 164
ISSN 1877-1173
https://doi.org/10.1016/bs.pmbts.2019.07.004

arena of solid tumors. Unlike hematologic malignancies, solid tumors present numerous challenges in the form of an immunosuppressive tumor microenvironment. In this chapter, we will highlight clinical application of CAR T cells in solid tumors, discuss hurdles that have impeded CAR T cell function in these malignancies, and propose methods to overcome these limitations.

1. Introduction

T cell activation is predicated upon T cell receptor (TCR) mediated recognition of a cognate peptide presented by the major histocompatibility complex (MHC) of a target antigen presenting cell (APC). This event, termed "signal 1," in the presence of a costimulatory "signal 2" provided by CD28 (or other molecules), initiates an efficient T cell response, which drives the T cell to kill the APC, proliferate, and produce cytokines enhancing proliferation and immune response[1–3] (Fig. 1A). Antitumor efficacy of T cells has been observed in the context of leukemia patients receiving bone marrow transplants; however this "graft-versus-leukemia" effect is not targeted against a specific antigen. Furthermore, for the minority of tumor antigens that are presented in the form of a peptide-MHC complex, very few have been matched with a corresponding TCR thereby limiting the therapeutic efficacy of the approach.

One T cell-based therapy that promises to overcome these and other challenges is chimeric antigen receptor (CAR) T cell therapy. CAR T cell therapy of cancers promises to revolutionize oncology by harnessing the powers of synthetic biology and immunotherapy in a single agent. CARs are synthetic receptors composed of an extracellular antigen binding domain and one or more intracellular signaling domains which act in concert to activate the T cell upon antigen recognition.[10] The antigen binding domain is most commonly presented as an antibody-derived single chain fragment variable (scFv), but may also take the form of another antigen binding polypeptide, such as a mutein,[11] thus providing a single treatment strategy for all cancers expressing a shared antigen. Additionally, recognition of antigens in an MHC independent fashion facilitates targeting tumor-associated antigens (TAAs) present in the form of extracellular proteins, complex carbohydrates, or lipids, thereby expanding the repertoire of antigens targetable by a T cell (Fig. 1A).

Initial efforts toward applying this technology to cancer utilized CAR constructs composed of a scFv and a CD3ζ intracellular domain,[12] termed

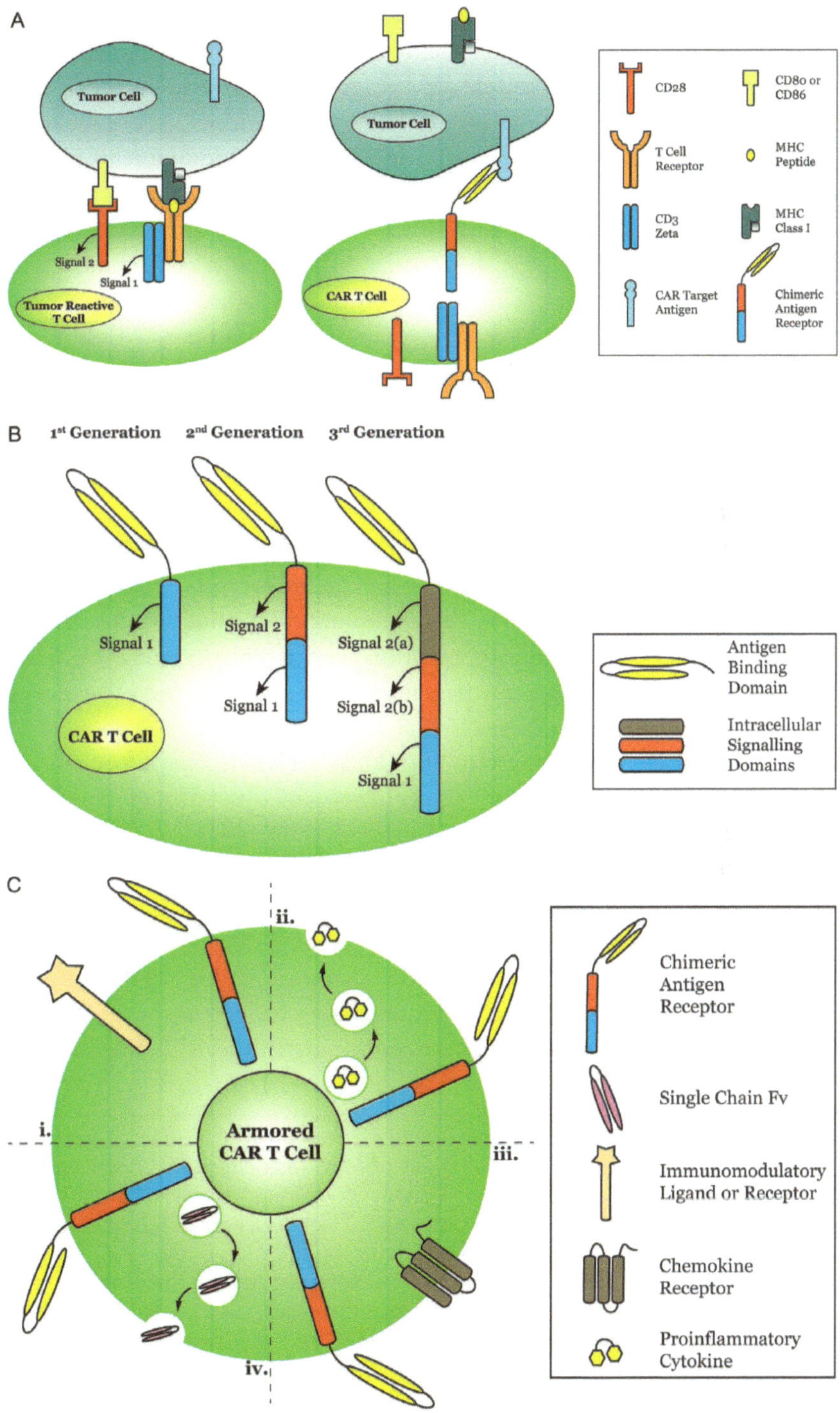

Fig. 1 See legend on next page.

"first-generation" CARs (Fig. 1B). T cells transduced with first-generation constructs demonstrated antigen specific cytolytic activity but were susceptible to activation-induced cell death[13,14] (ACID). Subsequent work demonstrated that ACID in the context of first-generation CAR T cells was due to a lack of costimulation,[15] and led to the development of second-generation CAR constructs wherein a costimulatory signal is delivered *in cis* by addition most commonly of either CD28[16] or 4-1BB[17] intracellular domains within the CAR construct (Fig. 1B). Second-generation CARs targeting B cell associated CD19 demonstrated robust *in vivo* cytolytic activity, expansion, and persistence upon antigen exposure,[16] thus paving the way for clinical application of this technology[18–20] and ultimately FDA approval for pediatric and young adult acute lymphoblastic leukemia[19] (ALL) as well as patients with relapsed or refractory diffuse large B cell lymphoma[20] (DLBCL). Third-generation CARs combine multiple costimulatory domains within the construct (Fig. 1B) but have demonstrated inconsistent preclinical antitumor capacities as compared to second-generation CAR T cells, demonstrating greater efficacy in some models[21–23] while equal efficacy in others.[24] Clinical evaluation is underway to compare antitumor efficacy of second-generation versus third-generation CAR T cells (NCT01853631). Fourth-generation CARs, termed "armored" CARs (Fig. 1C), aim to bolster CAR T cell function through recruitment of the endogenous immune system via additional modification of the CAR T cells with immunomodulatory molecules.[5–7,9,25,26]

To date, the clinical response of patients treated with CD19 targeted T cells has led to the approval of two separate CD19 targeted CAR T cell products by the FDA, and the data from BCMA-targeted CAR T cells is encouraging and may soon lead to FDA approval as well.[27] However, these successes have not been replicated with solid tumors. Unlike hematologic malignancies, solid tumors present numerous challenges in the form of an immunosuppressive tumor microenvironment (TME). The TME is

Fig. 1 (A) T cells recognize antigens presented in the context of MHC while CAR T cells are able to recognize antigens independent of MHC restriction. (B) CARs have been engineered with different intracellular signaling domains to confer differing effector functions. (C) Armored CARs aim to bolster CAR T cell function through recruitment of the endogenous immune system via additional modification of the CAR T cells with immunomodulatory molecules such as (i) a coexpressed immunomodulatory ligand such as CD40L[4]; (ii) secreted proinflammatory cytokines such as Il-12[5,6] or IL-18[7]; (iii) coexpression of chemokine receptors such as CCR2[8]; and (iv) a secreted immunomodulatory scFv.[9]

characterized by poor vascularization, hypoxia, immunosuppressive cell types such as tumor-associated macrophages (TAMs), tumor-associated neutrophils (TANs), myeloid derived suppressor cells (MDSCs), regulatory T cells (Tregs), soluble factors such as transforming growth factor β (TGFβ), adenosine,[28] interleukin 10 (IL-10), and the upregulation of immunoinhibitory checkpoint ligands such as programmed death ligand 1[29] (PD-L1) and programmed death ligand 2 (PD-L2).[30,31] In this chapter, we will highlight clinical application of CAR T cells in solid tumors, discuss hurdles that have impeded CAR T cell function in these malignancies, and propose methods to overcome these limitations.

2. CAR T cell therapy for solid tumors

While CAR T cell therapy of hematologic malignancies has been proven efficacious, there is little meaningful clinical data on CAR T cell therapy in solid tumors. Most promising data are still in preclinical animal models and await translation to the clinical setting. We discuss below clinical trials of CAR T cells for selected solid tumors that have provided insight into specific challenges that must be overcome if such therapies are to be effective.

2.1 Glioblastoma

2.1.1 Clinical efforts

Glioblastoma multiforme (GBM) is the most prevalent form of primary brain tumor, with a dismal prognosis. Presenting a 95% 5-year mortality rate, there exists an urgent need for novel therapies as recent trials have all failed to improve overall survival.[32] CAR T cell therapy of GBM for patients that have demonstrated no or limited response to prior therapy has been investigated primarily by targeting interleukin-13 receptor alpha 2 (IL-13rα2) or epidermal growth factor receptor variant three (EGFRvIII). To date, there have been five clinical trials, each demonstrating a different facet of the GBM TME. IL-13rα2 is a potential target for CAR T cell therapy of GBM as it is overexpressed on more than 50% of GBM and has limited expression on healthy brain tissue.[33] This antigen is expressed by both stem-like and established tumor cells,[34] thus targeting this antigen may prevent tumor relapse in addition to addressing overt disease. A trial targeting this antigen utilized a "zetakine" composed of an IL-13 mutein (E13Y) and a CD3ζ domain.[35] Three patients were treated with escalating doses of CAR

T cells over 11, 12, and 12 + 5 infusions after excision of the tumor, and were monitored for toxicity and relapse. One patient experienced neurologic toxicity that was hypothesized to be due to the level of antigen present on the tumor, while the other two patients experienced mild or no toxicity. All patients relapsed, two of which had disease at sites distinct from the original mass, while the third relapsed with an antigen-low tumor proximal to the original mass. These results demonstrated that IL-13rα2 targeted CAR T cells may confer a prolonged antitumor response but also stimulate antigen escape. A second trial was conducted by the same group utilizing an improved IL-13Rα2 targeting zetakine engineered to include a 4-1BB costimulatory domain. T cells transduced to express the construct were administered to a single patient presenting multifocal disease over 16 infusions intracranially or intraventricularly. Clinical efficacy was observed as a complete regression of all eight tumors, and this effect was durable for 7.5 months before disease relapse at sites nonadjacent to the original tumors.[36] This trial demonstrated that multifocal GBM can be therapeutically targeted by CAR T cells and that administration of the cells to the CSF may be more effective than to the site of a resected tumor. However, the significance of these findings is limited by the study's sample size of a single patient.

EGFRvIII results from the deletion of exons two through seven and the introduction of a new glycine residue between exons one and eight of the EGFR gene,[37] which creates a cancer specific neoantigen. This improvement on safety as compared to IL-13rα2 comes at the cost of prevalence as this antigen is only present on approximately 30% of gliomas.[37,38] The third completed trial of CAR T cells for GBM utilized a 4-1BBζ construct targeting EGFRvIII. Ten patients each received a single intravenous administration of CAR modified T cells expressing this construct, in the absence of lymphodepleting chemotherapy, and an extended effect was seen in one patient whose disease remained stable until resection after 3 months. Following resection, this patient has not experienced disease progression as of their latest clinical follow-up, more than 18 months after CAR T cell therapy. Together these observations indicate that the infused CAR T cells may have demonstrated control of the tumor, as an inclusion criterion for enrollment onto the trial was progressive disease.[38] Importantly, this study demonstrated that CAR T cells can cross the blood-brain-barrier and exert a therapeutic response. Additionally, induction of adaptive resistance in the form of upregulation of immune-inhibitory checkpoints in response to CAR T cell therapy was observed.

2.1.2 Challenges

Immunohistochemistry (IHC) analysis of tumor tissue before and after CAR T cell therapy in patients from the trial targeting EGFRvIII revealed important information about the GBM TME. Most striking was the relatively rapid upregulation of multiple immune-inhibitory checkpoint pathways in response to CAR T cell therapy, which may have contributed to an eventual abrogation of efficacy. Specifically, upregulation of indoleamine-pyrrole 2,3-dioxygenase (IDO1), tryptophan 2,3-dioxygenase (TDO), and PD-L1 were observed on the tumor by IHC, in addition to increased staining for IL-10, TGFβ, and FoxP3+ Tregs.[38] Additionally, the first trial targeting IL-13rα2 and the trial targeting EGFRvIII both demonstrated some form of antigen loss,[35,38] though in the latter case this effect may be the consequence of EGFRvIII's inherent biology, as this antigen is not stably expressed.[39,40] As GBM patients almost always relapse, infused CAR T cells must persist for an extended duration in order to provide continued anti-tumor immunity. None of the trials discussed in this section demonstrated long-term persistence of the infused CAR T cells, which is a concern for the long-term efficacy of this therapeutic modality in GBM.

2.2 Neuroblastoma

2.2.1 Clinical overview

Neuroblastoma is a pediatric malignancy of neuroendocrine origin.[41] To date, two clinical trials utilizing CAR T cells targeting either ganglioside G2[42,43] (GD2) or L1 cell adhesion molecule[44] (L1CAM) have been completed. The trial targeting GD2[42,43] utilized a first-generation construct transduced into activated T cells (ATC) or into Epstein-Barr virus (EBV) specific T cells. Nineteen patients received a single intravenous dose of cells composed equally of ATC CAR T cells and EBV CAR T cells, after which they were evaluated for therapeutic efficacy and CAR T cell persistence. Of the 19 patients treated, 11 presented with active disease at the time of infusion and 4 of these 11 patients demonstrated either partial or complete responses. At the latest follow-up, three patients remained alive and these were the patients in which CAR T cells could be detected for the longest duration, as compared to patients who did not have similar responses to CAR T cell therapy. The trial targeting L1CAM[44] utilized a first-generation construct transduced into CD8+ T cells. Six patients were treated with CAR T cells over one (one patient), two (four patients) or three (one patient) infusions, however efficacy was limited as only one patient had a partial response to CAR T cell therapy with the other five patients presenting progressive disease. The most prolonged response observed in this trial, as measured by time to relapsed disease, was 174 days. Importantly neither T cell expansion nor persistence were observed in any of the treated patients.

2.2.2 Challenges

Though the patients from the trial targeting GD2 were treated with first-generation CAR T cells, CAR T cells were detected for up to 192 weeks in a patient with a positive clinical outcome whereas CAR T cells were not detectable past 28 weeks in patients with poorer clinical outcomes.[43] Together these results demonstrate that there is a correlation between extended CAR T cell persistence and clinical efficacy thus indicating that efforts toward enhancing CAR T cell persistence would be beneficial. Additionally, targeting L1CAM was not effective as determined by time to relapse and the failure of CAR T cells to persist or expand.[44] Unlike other trials, however, this trial utilized a pure CD8+ population and the functional deficits observed may be attributed to a lack of "help" from CD4+ cells. Lastly, neither of the previously discussed trials employed preconditioning chemotherapy in their design, inclusion of which may have improved CAR T cell persistence and efficacy by creating immunological space for the adoptively transferred cells to engraft into.

2.3 Mesothelioma

2.3.1 Clinical efforts

Mesothelioma is a lethal malignancy arising from cells of the mesothelium, most commonly occurring in the pleural cavity,[45] and has been an area of interest for targeted immunotherapies, including CAR T cells. Cell surface glycoprotein mesothelin is upregulated on mesotheliomas among other cancers[46,47] and has successfully been targeted as a TAA for therapy of mesothelioma and pancreatic ductal adenocarcinoma (PDAC).[48–50] Three mesothelioma patients and one PDAC patient were treated with multiple intravenous or intratumoral infusions of CAR T cells expressing a mesothelin-targeted 4-1BBζ construct. Moderate antitumor efficacy, trafficking to the tumor site after intravenous administration, and persistence of CAR T cells were observed.[50] However, one patient enrolled on this trial developed anaphylaxis and cardiac arrest, highlighting the potential for patients to develop serious immune responses against CAR T cells expressing murine-derived components[51] and the potential for serious on-target, off-tumor toxicities as mesothelin is expressed by pericardial tissue. In a separate, ongoing clinical trial (NCT02414269), patients are being treated with intrapleural delivery of mesothelin-targeting CD28ζ CAR T cells followed by maintenance anti PD-1 antibody therapy. This combinatorial approach has demonstrated superior efficacy as compared to a regimen consisting of only CAR T cells and highlights concomitant blockade of immunosuppressive pathways as a key requirement for CAR T cell therapy of mesothelioma.

2.3.2 Challenges

Two key challenges observed in these trials are the development of immune responses against infused CAR T cells and the existence of immunosuppression within the TME. Anti-CAR immune responses present two problems that must be overcome. The first is the potential for serious allergic reactions directed against the CAR, which may be life threatening, and the second is that an anti-CAR response may abrogate CAR T cell persistence, potentially hindering clinical efficacy.

2.4 Ovarian cancer

2.4.1 Clinical efforts

Ovarian cancer is one of the leading causes of cancer related deaths in the world[52,53] and prognosis for these patients is poor despite combination therapy of surgery and chemotherapy.[54] CAR T cells for the therapy of this disease have been investigated in two clinical trials, each targeting a different antigen. The first trial employed first-generation CAR T cells targeted to α-folate receptor (α-FR), with concomitant IL-2,[55] administered to 14 patients presenting metastatic disease. Patients in this study did not demonstrate a response to CAR T cell therapy, a result that can be correlated to the lack of persistence of the infused cells. Furthermore, a selection of patients received infusions of CAR T cells that were radiolabeled with Indium-111 to facilitate in vivo tracking of the CAR T cells by positron emission tomography. Imaging demonstrated that a majority of the infused CAR T cells localized to the lungs of these patients rather than to disease sites. Lastly, the development of an anti-CAR immune response was observed in three patients, as the α-FR targeting scFv of this CAR was derived from a murine antibody.

Mucin 16 a highly glycosylated cell surface protein composed of a cleaved soluble domain, CA-125, and a retained portion MucCD,[56] and is commonly overexpressed by ovarian cancers.[57] CA-125 is a biomarker used in the detection of ovarian cancer,[58] therefore the retained portion of the protein is an attractive target for CAR T cell therapy. Accordingly a CAR T cell targeting this antigen was engineered and validated in established xenograft[59,60] and syngeneic[61] murine models of ovarian cancer. A clinical protocol employing a second-generation CAR T cell targeted to this antigen was designed[62] and clinical evaluation is underway (NCT00562640). A second clinical trial is evaluating an IL-12 secreting armored CAR T cell targeted to the same antigen (NCT02498912). Results from both have not yet been reported.

2.4.2 Challenges

Key challenges identified in the trial of α-FR targeting CAR T cells were a lack of persistence of infused CAR T cells, failure of CAR T cells to traffic to disease sites, and the development of anti-CAR immune responses in multiple patients. Ovarian cancer also presents a tumor microenvironment in the form of ascites,[63] a milieu composed of immunosuppressive cytokines including IL-10 and TGFβ[64] and inhibitory cell types such as TAMs and Tregs, which together function to impede CAR T cell function.[65–67] Overcoming these factors in crucial to improving the therapeutic efficacy of CAR T cells in the setting of ovarian cancer.

3. Overcoming challenges to CAR T cell therapy of solid tumors

For each of the previously discussed clinical trials, we have identified several potential problems that have limited the efficacy of the administered CAR T cells. Here we will describe the mechanisms that contribute to diminished CAR T cell efficacy in solid tumors in addition to providing methods to overcome these problems.

3.1 Antigen heterogeneity

B cell malignancies express lineage markers such as CD19 and CD22, which are ubiquitously expressed on all B cells thus making them amenable to serving as target antigens for CAR T cell therapy across multiple B cell malignancies. In contrast, solid tumors which not only do not express lineage markers, but also have an inherent heterogeneity of targetable antigens within each class of malignancy. This is demonstrated by clinical studies wherein EGFRvIII,[38] HER2,[68] and IL-13Rα2[36,69] have each been independently evaluated as a target antigen for CAR T cell therapy of GBM; while MucCD and α-FR[55] have been investigated in the context of ovarian cancer.

A major focus of preclinical studies in the field of CAR T cells for solid tumors is the identification of novel target antigens, and while antigens specific to particular tumor type are useful, they possess a limited therapeutic range. Antigens expressed by multiple tumor types, like cell lineage-specific markers, would enable targeting across malignancies. Pan-cancer is a term used to describe a target antigen that is expressed by numerous cancer types, and while rare, B7-H3,[70,71] mesothelin,[48–50] and HER2[68,72,73] are some of the antigens that have been identified as target antigens for CAR T cell therapy of multiple malignancies. It is important to note that these pan-cancer antigens are not true cell lineage markers and though a tumor may stain positively by immunohistochemistry or flow cytometry for an antigen, every cell in that lesion may not express the target antigen, leaving a population inherently resistant to CAR T cell therapy. Further complicating this heterogeneity is the potential for an abrogation of durable therapeutic efficacy resulting from a modulation of the targeted antigen's expression, wherein the targeted malignancy loses (antigen-negative relapse) or down-regulates (antigen-positive relapse) the targeted antigen. Antigen loss was observed upon comparing pre- and posttherapy biopsies of patients' tumors in five out of seven evaluated patients enrolled in the clinical evaluation of EGFRvIII targeted CAR T cells for GBM[38] and was also observed in the trial targeting IL-13Rα2 wherein relapsed disease sites were noted to display decreased levels of target antigen as compared to the primary lesions.[69]

Antigen loss can be come from either epitope loss due to splicing variations that result in an aberrant isoform of the targeted antigen[74,75] or from comprehensive loss of antigen expression.[75,76] Differing strategies must be employed to address these phenomena. Epitope loss was observed in the case of CD19 targeting CAR T cells[74,75] and a possible route to surmount this hurdle is to engineer CAR T cells to target different epitopes of the same target. However this has yet to be explored. In contrast, comprehensive loss of antigen expression cannot be overcome by targeting the same antigen, rather separate antigens must be targeted concomitantly or sequentially following confirmation of antigen loss.

One method to achieve combinatorial antigen targeting is to use CAR T cell products, each targeting separate antigens (Fig. 2iii). This is currently being evaluated in hematological malignancies by simultaneously targeting CD19 and CD20[79] or CD19 and CD123[80] or CD19 and CD22 (NCT03330691), and due to the lack of lineage-derived antigens may be a more efficacious strategy applied to solid tumors. Another method for targeting multiple antigens is the creation of multivalent CAR T cell products

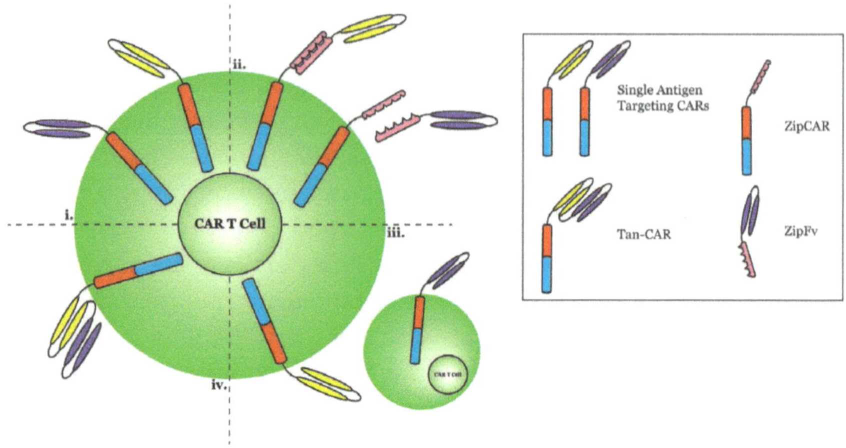

Fig. 2 Differing strategies exist to facilitate the targeting of multiple TAAs by CAR T cells, including (i) coexpression of multiple CARs on a single transduced T cell[77]; (ii) expression of a programmable CAR that allows for temporal control of the antigen(s) targeted by the CAR T cell[78]; (iii) utilization of multiple CAR T cells, each expressing a CAR against a separate antigen[79,80]; (iv) expression of a CAR containing two or more antigen binding domains, allowing for the recognition of multiple antigens by the same receptor.[81]

that allow for the recognition of multiple target antigens by a single transduced T cell (Fig. 2i). The latter can be accomplished by expressing two or more separate CARs in a single T cell, as was demonstrated in an orthotopic, patient-derived xenograft model of GBM.[77] T cells were transduced with a tricistronic construct encoding CARs targeting HER2, IL-13Rα2, and EphA2, and CAR T cells targeting all three antigens were more efficacious than CAR T cells targeting one or two antigens by a measure of delayed tumor growth and overall survival of tumor-bearing mice. CARs can also be engineered to contain multiple antigen recognition domains in a single receptor[79,81,82] (Fig. 2iv). The creation of a TanCAR, wherein an IL-13rα2 targeting mutein was fused to a HER2 targeting CD28z CAR construct, afforded a receptor with two antigen targeting domains expressed in tandem.[82] Evaluated in a murine orthotopic xenograft model of GBM, T cells transduced with the TanCAR construct were more efficacious than CAR T cells targeting either HER2 or IL-13rα2 alone, as measured by progression free survival and overall survival. Another method to achieve combinatorial antigen targeting is the use of programmable CARS,[78] wherein T cells are engineered with CAR constructs that have their antigen binding domains replaced with one half of a leucine zipper domain. Antigen recognition is facilitated by the coadministration of the antigen binding domain fused to a leucine zipper domain complimentary to that expressed on the T cells. This allows for temporal control of antigen targeting by facilitating sequential or concomitant targeting of multiple antigens through coadministration of differing binding domains with the CAR T cell (Fig. 2ii).

Loss of therapeutic efficacy of CAR T cells due to downregulation of target antigen expression was observed clinically in the context of CD22 targeting CAR T cells in the setting of leukemia and was recapitulated in a murine xenograft model.[76] This phenomenon was also demonstrated in another preclinical model using CAR T cells targeting neuroblastoma associated antigen anaplastic lymphoma kinase (ALK). In this model, T cell effector function increased when CAR T cells engaged cells expressing higher levels of ALK.[83] Downregulation of antigen expression, "antigen dimming," can be addressed by engineering an optimized CAR possessing increased binding affinity for the targeted antigen. However, this method increases the risk for on-target, off-tumor toxicity in cases where the target antigen is one overexpressed by tumor tissue but also expressed at low levels on healthy tissue.[84]

3.2 Suboptimal CAR T cell trafficking

One of the differences between the hematological malignancies and solid tumors is the access of CAR T cells to the site of tumors. In hematological malignancies, CAR T cells are infused intravenously and are instantly able to exert their activity at the tumor site, whereas in solid tumors, CAR T cells face challenges of migrating to and infiltrating into disease sites. Physical barriers, a hostile TME, and a lack of matched chemokine receptors on CAR T cells relative to chemokines secreted by cancer cells and stroma of solid tumors result in suboptimal migration of the infused CAR T cells to tumor sites.[85,86] Preclinical and clinical investigations have demonstrated that this suboptimal intratumoral accumulation and persistence of CAR T cells is diminished following their intravenous infusion and that initial trafficking of CAR T cells is usually to the spleen, lung and liver without specific accumulation in the sites of tumor.[55,87] Discussed below are some of the obstacles hindering efficient migration of CAR T cells to solid malignancies, and strategies to overcome these hurdles.

3.2.1 Mismatched chemokine receptors and associated chemokines

In cancer, trafficking of immune cells to the sites of tumors is directed and facilitated by chemokines such as CCL2, CXCL9, CXCL10, and CXCL11, which are secreted by tumor and tumor-associated cells.[85] The expression of cognate receptors to these chemokines, such as CXCR3 and CCL5, on activated T cells allows migration to tumor sites. This relationship between chemokines secreted by tumor cells and their respective receptors on T cells is pivotal in directing a T cell to tumor sites.[8] For CAR T cells to efficiently migrate, there must be a match between chemokine receptors expressed by CAR T cells and the chemokines secreted by the tumor and associated stroma. Unfortunately, there is rarely a proper alignment between the chemokine receptors and secreted chemokines, resulting in inefficient CAR T cell infiltration of solid tumors. For example, tumors producing small amounts of the CXCR3 ligands CXCL9, CXCL10, and CXCL11, or small amounts of CCR5 ligands CCL3 and CCL4, are incapable of attracting a therapeutically sufficient number of CXCR3^{high} CD8+ CAR T cells.[85] Additionally, secretion of CCL2 by tumors has been linked with enhanced trafficking of T cells into these tumors and analysis of tumor infiltrating T cells demonstrated high expression of the CCL2 receptor CCR2.[88] Furthermore, glioblastoma and other highly vascularized tumors secrete elevated levels of vascular endothelial growth factor (VEGF), which in addition to promoting angiogenesis, impedes secretion of chemoattractants such as CXCL10 and CXCL11 and downregulates the expression of adhesion molecules such as ICAM-1, ICAM-2, and VCAM-1, contributing to inefficient lymphocyte homing.[89,90] TGFβ in TME also inhibits T cells trafficking by downregulating adhesion molecules.[91]

One strategy to overcome insufficient trafficking is to engineer CAR T cells that coexpress better matched chemokine receptors. CAR T cells coexpressing chemokine receptor CCR2b enhanced intratumoral infiltration of CAR T cells and also augmented the antitumor activity against established xenograft models of malignant pleural mesothelioma tumors that secrete elevated levels of CCL2.[92,93] Alternatively, targeting downstream signaling cascades resulting from inhibitory signals can also be employed to indirectly increase the surface expression of chemokine receptors on CAR T cells. For example, blocking the localization of protein kinase A to the immune synapse increased the expression of CXCR3, the receptor for T cell migratory chemokine CXCL10. This strategy demonstrated increased in vitro migration of T cells in the presence of CXCL10 and demonstrated increased in vivo antitumor efficacy against syngeneic and xenograft murine models of mesothelioma.[94]

3.2.2 Structural hindrance

Solid tumors are surrounded by a dense peritumoral fibrous stromal framework, which inhibits access to and accumulation of lymphocytes within tumor. The shielding of tumor cells by a dense peritumoral stroma also functions to prevent proper interaction with lymphocytes, such as CAR T cells.[86] Cancer-associated fibroblasts (CAFs) constitute the structural network which facilitates tumor growth, persistence, invasion, and metastasis by reshaping the extracellular matrix (ECM). Heparanase is expressed by endogenous T cells and enhances T cell penetration through stroma by degrading ECM. However expression of heparinase on CAR T cells is frequently downregulated or lost during in vitro culturing, thus affecting trafficking of CAR T cells to tumor sites.[90,95] The chemokines CXCL12, VEGF, and PDGF are abundantly secreted by tumors and their associated stroma, and are tasked to recruit bone marrow derived endothelial cells resulting in the development of abnormal, tortuous, and leaky vascular networks that resist the infiltration of tumors by lymphocytes.[86]

For successful infiltration into tumor sites, CAR T cells must overcome the structural stromal networks shielding tumors. The ECM is an integral part of the stromal network and functions to inhibit the infiltration of CAR T cells to the tumor sites leading to subdued antitumor activity. One strategy to break down the stroma is to overexpress an ECM degrading enzyme on CAR T cells. In one study, an ECM degrading heparanase was coexpressed on CAR T cells, and preclinical evaluation of these engineered cells in xenograft models of neuroblastoma demonstrated superior antitumor activity as compared to non-secreting CAR T cells.[95] Strategies aimed at targeting stromal cells within TME may enhance the activity of infused CAR T cells by facilitating greater engagement of CAR T cells and tumor cells. CAR T cells targeting fibroblast activating protein (FAP), which is upregulated on tumor-associated fibroblasts in differing solid tumors, reduced the number of stromal cells, allowing administered CAR T cells to better control tumor growth in preclinical murine models of mesothelioma and lung cancer. CAR T cells engineered to target FAP in the tumor stroma also demonstrated augmented antitumor activity in multiple syngeneic murine models when given as a single agent or in combination with vaccination therapy.[96] Additionally, anti-FAP CAR T cells reduced extracellular matrix proteins and glycosaminoglycans resulting in diminished tumor vasculature, inhibition of tumor growth, and higher endogenous CD8+ activity in lung cancer xenograft and syngeneic murine pancreatic models.[97]

CAR T cells that target abnormal tumoral vasculature have also been investigated. Investigators developed CAR T cells targeting prostate specific membrane antigen (PSMA) expressed on endothelial cells, and anti-PSMA demonstrated regression of tumor burden in a preclinical model of murine ovarian cancer.[98] In a study employing a murine melanoma model, CAR T cells engineered to target both the tumor vasculature (VEGFR2) and melanoma specific antigens (gp100, TRP1, and TRP2) demonstrated synergistic antitumor activity resulting in enhanced survival of tumor-bearing mice.[99]

3.2.3 Use of oncolytic viruses

Several investigators have demonstrated the successful use of oncolytic viruses harboring chemotactic chemokines that attract CAR T cells to tumor sites.[100] The local administration of cytokines and oncolytic viruses has shown clinically promising results.[101,102] Combinatorial therapy of an oncolytic adenoviral vector expressing CCL5 and GD2 targeting CAR T cells was evaluated in a murine xenograft model of neuroblastoma, and demonstrated improved infiltration of CAR T cells into the tumor, affording improved survival of tumor-bearing mice.[103] Furthermore, combinatorial therapy of oncolytic viruses and CAR T cells did not demonstrate any adverse effect on T cells. Together these results suggest that administration of CAR T cells in combination with oncolytic viruses may increase the infiltration and homing of CAR T cells to solid tumors.

3.2.4 Regional administration of CAR T cells

Typical administration of CAR T cells is currently achieved through intravenous infusion, requiring infused CAR T cells to migrate to and then infiltrate disease sites. To overcome the problem of suboptimal intratumoral infiltration, several groups are exploring local administration of CAR T cells directly to tumor sites for the treatment of solid tumors. Investigators have compared the efficacy between intravenous and regional delivery of CAR T cells, and shown that CAR T cells delivered directly to the tumor site demonstrate superior persistence and enhanced antitumor activity as compared to peripherally infused CAR T cells.[104,105] CAR T cells delivered intratumorally have been observed to largely remain at the site of administration with limited dispersion into the systemic circulation, therefore regional delivery may also minimize the risks of on-target, off-tumor toxicities in the setting of localized disease.[84,87] Adusumilli and colleagues demonstrated regionally delivered (intrapleural) mesothelin-targeted CAR T cells to induce greater T cell derived antitumor efficacy with reduced T cell doses, increased CD8+ T cell proliferation and persistence, and enhanced production of cytokines in an orthotopic xenograft model of malignant pleural disease.[105] This study resulted in the initiation of a phase I clinical trial wherein CAR T cells targeting mesothelin are regionally delivered to patients presenting malignant pleural disease (NCT02414269). Glioblastoma (GBM) and other central nervous system (CNS) malignancies are attractive diseases for regional delivery of CAR T cells because tumor sites are protected by the blood-brain-barrier (BBB) resulting in poor infiltration. Regional delivery of CAR T cells and CAR NK cells has demonstrated promising results against primary and metastatic CNS diseases in preclinical models[106–108] and clinical evaluation.[36] Ongoing clinical trials exploring the regional delivery of CAR T cells include intratumoral administration for treatment of multiple solid tumors including hepatocellular carcinoma and triple-negative invasive breast carcinoma (NCT02587689), intracranial administration in the context of brain and central nervous system tumors (NCT00730613), intrapleural administration for malignant pleural mesothelioma (NCT02414269), and hepatic artery administration for the treatment of liver metastases (NCT01373047). Potential limitations of regional delivery include being technically challenging as compared to peripheral administration, and the lack of migration of administered CAR T cells from primary tumor sites to metastases.

3.3 Immunosuppressive tumor microenvironment

Unlike hematological malignancies, treatment of solid tumors with CAR T cells presents the infused cells with a significant challenge in the form of an immunosuppressive TME; components of which contribute to progression and metastasis of tumors.[109] The immunosuppressive TME presents numerous hurdles to CAR T cell therapy including physical barriers, metabolic inhibition, tumor-derived soluble factors, inhibitory cytokines, immunosuppressive cells, and the expression of coinhibitory receptors on activated T cells. Together these hostile conditions of the TME inhibit the ability of CAR T cells to infiltrate and engage targeted tumor antigens of solid tumors.[109,110] For CAR T cell therapy to be efficacious against solid tumors, administered CAR T cells must surmount these hurdles posed by the immunosuppressive TME. Modifications to CAR T cells have improved their antitumor efficacy against solid tumors in preclinical models and these modified CAR T cells have progressed to clinical evaluation. Discussed below are some of the strategies being employed to overcome the challenges posed by the TME.

3.3.1 Metabolic hurdles

After intravenous infusion, CAR T cells are faced with metabolic challenges in addition to the physical and structural hurdles discussed earlier.

The prevalence of hypoxia in TME has been shown to adversely affect the activation, proliferation, and antitumor efficacy of T cells.[111] Nutrient starvation within TME may present as a key inhibitory factor of CAR T cell function in solid tumors. Lack of sufficient glucose and other metabolites in TME has been shown to impede both T cell proliferation and production of crucial effector cytokines, including interleukin 2 and interferon-γ.[112,113] Low levels of amino acids such as tryptophan, arginine, and lysine can result in the shutdown of protein translation in effector T cells resulting in diminished survival.[114] The expression of IDO1 by tumor cells and MDSCs is induced by interferon-γ and catalyzes the conversion of tryptophan into kynurenine. This tryptophan depletion facilitates effector T cell inactivity and death, and induces the accumulation of Tregs.[115,116] Investigated in vivo, CD19 targeting CAR T cells failed to control the progression of CD19 and IDO expressing tumors in a murine xenograft model. However, when tumor-bearing mice were pretreated with cyclophosphamide and fludarabine, IDO was observed to be downregulated and CAR T cells were better able to control tumor progression.[117] This demonstrates another dimension for the use of preconditioning chemotherapy, wherein its use is not only to make space for the engraftment of CAR T cells, but also to affect tumors in a way that makes them less resistant to CAR T cell therapy.

3.3.2 Inhibitory cytokines and soluble factors within the tumor microenvironment

Cytokines involved in the suppression of inflammation at disease sites may contribute to diminished antitumor efficacy of adoptively transferred CAR T cells. TGFβ is one of the inhibitory cytokines produced most abundantly by tumors and tumor-associated cells. It is known to enhance matrix formation, promote epithelial-to-mesenchymal transition, and skew immune responses toward a Th2 phenotype, in addition to directly suppressing effector functions of T cells.[118,119] Elevated levels of inhibitory cytokines, such as IL-4 and IL-10, in the TME of solid tumors are correlated with an inhibition of T cell differentiation, proliferation, and antitumor effector functions.[110,120,121]

Soluble factors capable of suppressing effector T cell activity have been identified in serum, ascites, and tissue extracts from cancer patients. Tumors and tumor-associated macrophages (TAMs) are known to secrete prostaglandin E2 (PGE2), an immunosuppressive derivative of arachidonic acid, synthesized by inducible cyclooxygenase 2.[122,123] PGE2 has been reported to inhibit proliferation of T cells, suppress CD4 + T cell function and impede

CD8+ T cell differentiation.[124] Furthermore, PGE2 produced by both tumor and tumor-associated cells mediates angiogenesis and promotes immunosuppression through the recruitment of Tregs and MDSCs to disease sites.[125] Another soluble factor known to suppress antitumor immunity is adenosine. Tumors have been noted to be surrounded by a "purinergic halo" comprised of adenosyl phosphates derived from hypoxia-induced release of adenosine triphosphate (ATP). Extracellular ATP is hydrolyzed to its diphosphate (ADP) and monophosphate (AMP) forms by the activity of ectonucleotidase CD39. AMP is then hydrolyzed to adenosine by a second ectonucleotidase, CD73.[126] The resulting adenosine binds to the adenosine 2A (A_{2A}) G protein-coupled receptor (GPCR) on cytotoxic T cells and functions to antagonize TCR signaling.[127] This is observed as a decreased release of interferon-γ and associated proinflammatory effects of cytotoxic T cells.[28] The combination of PGE2 and adenosine was shown to suppress T cell function immunosuppression through independent GPCRs and activate protein kinase A (PKA) in a cyclic AMP-dependent manner, resulting in inhibition of TCR signaling.[128]

PGE2 and adenosine are known to suppress immune responses by Treg-mediated mechanisms in addition to activating PKA signaling pathways.[129] CAR T cells engineered to produce a small peptide called regulatory subunit I anchoring disruptor (RIAD), which disrupts the interaction of PKA and ezrin, inhibited the immunosuppressive actions of adenosine and PGE2, resulting in enhanced antitumor efficacy, increased levels of proinflammatory cytokines, and increased trafficking of infused T cells to the disease sites.[94] Prior work has also demonstrated that small molecule-based pharmacological modulation of the A_{2A} receptor can lead to increased interferon-γ secretion, which directly correlates with increased cytolytic activity of tumor-specific T cells.[28] This has been shown to have an additive effect in combination with PD-1 checkpoint blockade in a syngeneic murine model of breast cancer treated with HER2 targeting CAR T cells.[130]

TGFβ is known to suppresses antitumor immune responses while simultaneously creating a pro-tumor environment. One strategy being employed to counteract the effects of TGFβ is the use of anti-TGFβ antibodies to systemically block the TGFβ receptor. This demonstrated an enhanced antitumor response in a syngeneic murine thymoma model.[131] Additionally, CAR T cells engineered to express a dominant-negative TGFβ receptor resulted in augmented antitumor activity in a xenograft model of prostate cancer.[25] The coexpression of dominant-negative TGFβRII increased proliferation of T cells, elevated cytokine production and long-term in vivo persistence, and abrogated exhaustion. This impressive preclinical efficiency has led to phase I clinical trial of PSMA CAR T cells coexpressing TGFβRII for metastatic prostate cancer (NCT03089203).

3.3.3 Immunosuppressive cells

Immune regulatory cells capable of inhibiting antitumor immune responses are prevalent within the TME of solid malignancies and are a population comprised of Tregs, MDSCs, TAMs, and TANs.[123] TAMs and TANs are often observed to be skewed toward anti-inflammatory and pro-tumor M2 and N2 phenotypes, respectively. They function in combination with Tregs and MDSCs to produce immunosuppressive factors such as TGFβ, PGE2, arginase, and reactive oxygen and nitrogen species, which contribute to diminished CAR T cell efficacy.[132,133] TAMs have also been observed to express elevated levels of surface coinhibitory ligands (such as PD-L1, among others) that, upon binding to their cognate receptors on T cells, induce dysfunction (Fig. 3B). Tregs produce elevated levels of immunosuppressive cytokines such as TGFβ and IL-10. However, this population may also inhibit CAR T cell function by cell-cell contact wherein Tregs express coinhibitory ligands which interact with CAR T cells present in the tumor.[136]

Overcoming TAM mediated inhibition of CAR T cell function was demonstrated using IL-12 secreting armored CAR T cells in a syngeneic murine model of ovarian peritoneal carcinomatosis. In addition to prolonging overall survival of treated mice, armored CAR T cells were able to deplete TAMs in a FAS/FAS-ligand dependent manner, and conferred resistance to PD-1/PD-L1-mediated inhibition of CAR T cells.[61] IL-12 secreting armored CAR T cells have also demonstrated the capacity to resist Treg-mediated inhibition of CAR T cell function in a syngeneic murine model of hematological malignancy.[5] Together these results demonstrate a multifaceted role played by IL-12 in conferring resistance to immunosuppressive cell types in the context of CAR T cell therapy and use of IL-12 armored CAR T cells is currently under clinical investigation for the treatment of ovarian cancer (NCT02498912).

3.3.4 Upregulation of coinhibitory receptors on T cells

T cells have also been to shown to express activation-induced surface inhibitory receptors such as PD-1, LAG-3, TIM-3, and CTLA-4, which, upon binding to their cognate ligands, are known to antagonize effector function of T cells resulting in limited persistence and diminished antitumor response[122,137] (Fig. 3A). These inhibitory receptors have been shown to

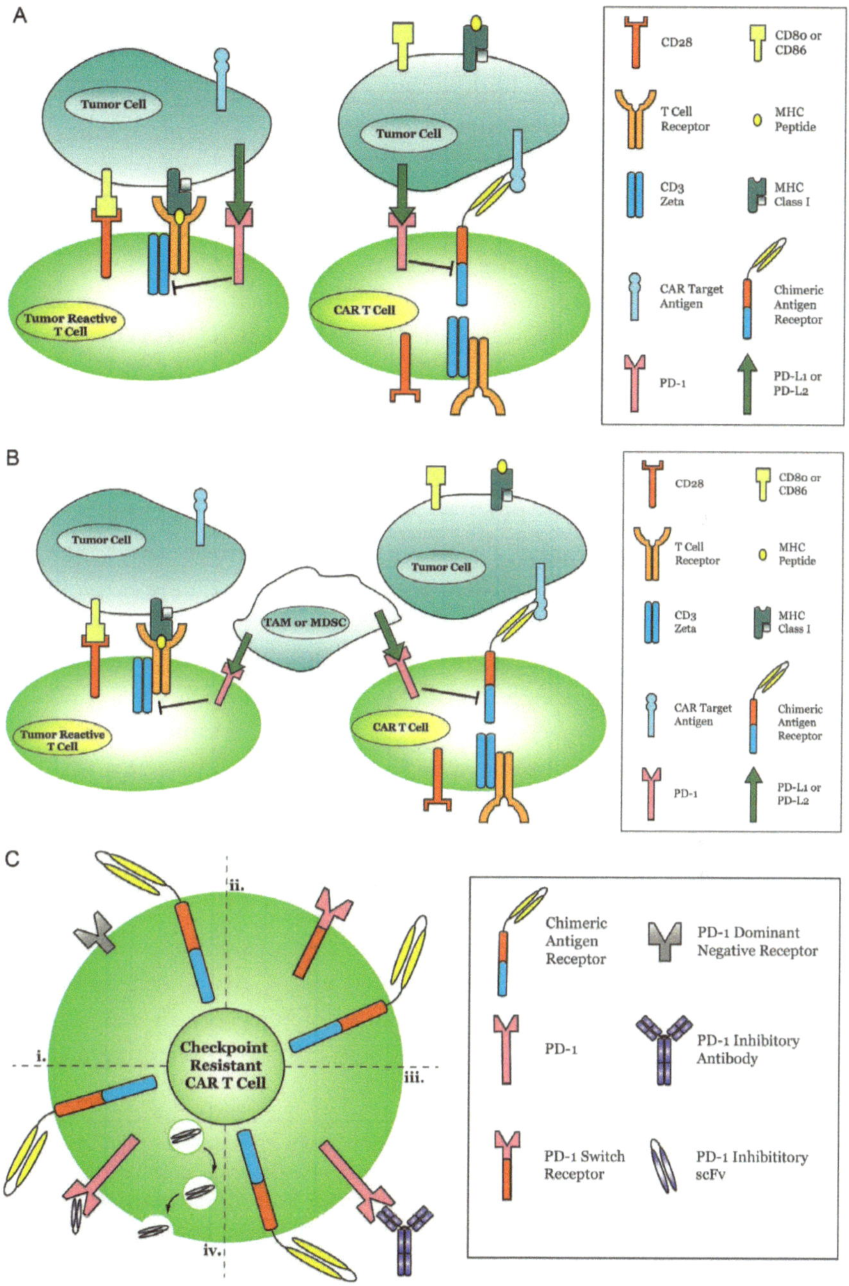

Fig. 3 (A) Tumor cells express coinhibitory ligands which impede T cell effector functions upon binding to their cognate receptors on T cells or CAR T cells. (B) T cell effector function may also be abrogated independent of tumor cell derived coinhibition.
(Continued)

be further upregulated on infused CAR T cells as well as on tumor infiltrating CARs, indicating that these cells are dysfunctional.[137] The expression of Fas (CD95) on T cells can also limit the activity of CAR T cells. The engagement of Fas receptor on activated T cells with Fas-ligands on tumors cells as well as tumor vasculature induces apoptosis of T cells leading to diminished T cell-mediated antitumor activity.[123]

Strategies to block the PD-1/PD-L1 interaction may enhance the antitumor efficacy of CAR T cells and include the use of PD-1 and PD-L1 blocking antibodies,[138] genetic removal of PD-1 from CAR T cells,[139] modification of CAR T cells to secrete PD-L1or PD-1 blocking scFvs[9], and coexpressing switch receptors on CAR T cells[134] (Fig. 3C). PD-1/PD-L1 axis blockade by PD-1 checkpoint antibody, intrinsic PD-1 shRNA blockade or PD-1 dominant-negative receptor expression augmented the antitumor efficacy of CAR T cells in a preclinical model of orthotopic malignant pleural disease.[26] Similarly, genetic removal of PD-1 from CAR T cells by a CRISPR/CAS9 system increased the efficacy of prostate stem cell antigen targeting CAR T cells in a xenograft model of prostate cancer model by rescuing dysfunctional T cells.[139] CAR T cells engineered to secrete an scFv that blocks PD-1 demonstrated improved antitumor activity in addition to enhancing the efficacy of bystander tumor-specific T cells in both hematologic and solid tumor preclinical mouse models. scFv secreting CAR T cells also demonstrated equal or greater antitumor efficacy than was observed in combination of CAR T cells and PD-1 checkpoint blockade antibody.[9] Separately, CAR T cells engineered to secrete a PD-1 blocking scFv demonstrated decreased expression of markers indicative of T cell dysfunction, and when applied to a xenograft model of lung cancer, scFv secreting CAR T cells demonstrated improved control of tumor growth and increased overall survival of tumor-bearing mice, as compared to groups treated with CAR T cells lacking the PD-1 blocking scFv.[140] In a murine model of renal carcinoma, carbonic anhydrase IX (CAIX) targeting CAR T cells capable of secreting a PD-L1

Fig. 3—Cont'd Immunosuppressive cell types such as TAMs or MDSCs or T_{regs} (not shown), are abundant in the TME, and may also express coinhibitory ligands which function to abrogate antitumor T cell effector function *in trans*. (C) Differing strategies exist to enhance CAR T cell function by removal of coinhibitory signaling, including, (i) coexpression of a PD-1 dominant-negative receptor by CAR T cells[26]; (ii) coexpression of a PD-1 switch receptor that allows for the switching of an intended inhibitory signal to one of costimulation[134]; (iii) concomitant CAR T cell therapy and antibody-based checkpoint blockade therapy[135]; and (iv) coexpression of secreted checkpoint receptor inhibiting scFv.[9]

targeting scFv resulted in a fivefold improvement in antitumor activity as compared to CAIX-targeting CAR T cells lacking this PD-L1 targeting scFv.[141] The CAIX-PD-L1 CAR T cells also demonstrated enhanced secretion of granzyme B, which induces apoptosis of the target cells. Therapeutic application of CAR T cells expressing PD-1 blocking antibody is currently being explored in several clinical trials for the treatment of various solid tumors (NCT02873390, NCT03182816, NCT03615313, and NCT03179007). Preliminary data has demonstrated a twofold expansion of CAR T cells in combination with PD-1 blockade as compared to CAR T cell monotherapy in patients with refractory/relapsed non-Hodgkin lymphoma.[135]

A PD-1 switch receptor is a synthetic receptor wherein the intracellular domain of PD-1 is replaced by that of a costimulatory receptor such as CD28, thus switching an intended inhibitory signal to one of costimulation. In a xenograft model of prostate cancer, coexpression of this switch receptor on CAR T cells demonstrated enhanced cytokine production and greater antitumor efficacy as compared to CAR T cells lacking the switch receptor.[134] Similarly, in a xenograft model of pancreatic cancer, expression of a chemokine switch receptor composed of the ectodomain of the IL-4 receptor fused to the intracellular signaling domain of the IL-7 receptor converts the immunosuppressive effects of IL-4 into an activating signaling cascade within CAR T cells, thereby enhancing antitumor efficiency, improving persistence, and increasing intratumoral expansion of CAR T cells.[142] Lastly, CAR T cells engineered to secrete IL-7 and IL-15 have demonstrated augmented antitumor activity with improved persistence in preclinical models of solid tumors.[143,144]

3.4 CAR T cell persistence

Persistence of adoptively transferred CAR T cells is crucial for long-term efficacy. However, as discussed earlier, CAR T cells administered to patients with solid tumors do not persist long term. Discussed below are some of the multiple factors that contribute to this phenomenon and solutions to these challenges.

The development of anti-CAR immune responses that could lead to a depletion of infused CAR T cells was demonstrated in multiple different clinical evaluations of CAR T cells containing murine-derived targeting domains.[51,55] To address this problem, the use of human-derived components will prevent the development of anti-CAR responses as the reactive epitopes will not be present in these fully human constructs,[145] and this strategy is currently under clinical evaluation (NCT02374333).

Another factor that limits the persistence of CAR T cells is the activation of intrinsic signaling pathways that result in abrogation of T cell activity after stimulation through the T cell receptor.[146] Activation of these intrinsic loops result in accumulation of transcription factors such as Ikaros, kinases such as diacylglycerol kinase that phosphorylates and inactivates diacylglycerol, an essential secondary messenger for signal transduction cascades that are crucial for T cell activation, phosphatases such as SHP1 phosphatase that dephosphorylate TCR-associated kinases such as LCK and ZAP70 resulting in reduced T cell activity, and ubiquitin ligases such as CblB that ubiquitylate TCR intracellular domains affording a reduction in T cell activity.[123,147] Building from this, a preclinical study was conducted wherein the signaling strength of a CD19 targeting second-generation CAR was modulated by silencing selected immunoreceptor tyrosine-based activation motifs (ITAMs) of the CD3ζ domain.[148] Selectively silencing two ITAMs of this domain afforded increased persistence of CAR T cells in a murine xenograft model of leukemia and warrants further investigation in the context of solid tumors.

3.5 Site-specific CAR integration into the T cell genome

One final method to improve the efficacy of CAR T cells for the treatment of solid tumors is to precisely direct the integration of the CAR to a specific site within the genome. Retroviral or lentiviral transduction of CAR constructs into T cells typically results in nonspecific integration of the CAR into the genome. Recognition of specific sites of integration in the genomes of T cells could be utilized to improve CAR T cells expansion and persistence. A study employing a CRISPR/CAS9 system to direct targeted CAR integration to the T cell receptor α constant (TRAC) locus demonstrated uniform CAR expression on T cells, augmented antitumor activity, and high persistence, as compared to retrovirally generated CAR T cells.[149] Targeted integration of CAR to the TRAC locus avoided tonic signaling with improved persistence and delayed differentiation and exhaustion of effector T cells.

Separately, a case study reported that clonal expansion of single CAR T cell resulted in the remission of disease in a patient with chronic lymphocytic leukemia (CLL).[150] Subsequent analysis demonstrated that there was a random insertion of the CAR into the TET2 gene locus and that this patient also had a hypomorphic mutation of the second TET2 allele, resulting in biallelic disruption of the TET2 gene. This disruption, which afforded an altered differentiation program skewed toward a central memory phenotype of the T cells with enhanced antitumor activity, led to a positive clinical response. CAR T cells engineered with the use of transcription activator-like effector nucleases (TALEN) mediated gene editing of TRAC and CD52 gene loci produced efficient antitumor activity and improved persistence.[151] Likewise, another study demonstrated high antitumor activity of the universal CAR T cells by editing PD-1 and CTLA-4 loci.[139]

These innovative strategies could be employed to improve expansion of CAR T cells and help to overcome the hurdles faced by CAR T cells within solid tumors. The site-specific delivery of CAR into the T cell genome demonstrates a promising method to generate phenotypically efficient T cell subtypes with enhanced antitumor activity and persistence. Further research is warranted to elucidate additional genomic sites amenable to targeted insertion of the CAR, validate the feasibility of this technology on a clinical manufacturing scale, and to address safety concerns associated with targeted delivery of CARs into the T cell genome.

4. Concluding remarks

While CAR T cell therapy has yielded impressive outcomes for patients presenting with hematological malignancies, it has not had similar successes with solid tumors. The limitations of CAR T cell therapy for patients with solid tumors have been attributed to heterogenous expression of targeted antigens, inefficient trafficking to disease sites, and suboptimal persistence and expansion of administered CAR T cells due to a hostile tumor microenvironment. Several strategies are being evaluated to enhance the efficacy of CAR T cell therapy for solid tumors.

To date, clinical trials for CAR T cell therapy in patients with solid tumors have demonstrated the feasibility and tolerability of the approach. Compared to the traditional anti-cancer therapies of chemotherapy and radiation, CAR T cell therapy has the prospect of improved tumor targeting and life-long tumor immune surveillance. Focusing on the elucidation of ideal target antigens for specific tumors with minimal on-target off-tumor effects may pave the way for improved antitumor activity with abrogated toxicity. Strategies incorporating advanced CAR designs such as TanCARs, programmable SUPRA CARs, SynNotch CARs, and "AND" CARs have shown reduced off-target toxicity in preclinical studies.[78,152–155]

The issue of suboptimal trafficking of CAR T cells to disease sites can be countered with regional delivery, coexpression of rationally matched chemokine receptors, and combination therapy with oncolytic viruses. These strategies will continue to be explored in the clinical setting to evaluate whether preclinical data translate to clinically meaningful responses. The problem of suboptimal persistence and expansion of CAR T cells in the hostile TME has been addressed in preclinical models employing CAR T cells engineered to secrete proinflammatory cytokines such as IL-12[5,6,61] and IL-18.[7] Rejuvenation of endogenous immune cells by the effects of these cytokines represents another method of bolstering the efficacy of CAR T cells.

References

1. Zinkernagel RM, Doherty PC. Restriction of in vitro T cell-mediated cytotoxicity in lymphocytic choriomeningitis within a syngeneic or semiallogeneic system. *Nature*. 1974;248(5450):701–702. https://doi.org/10.1038/248701a0.
2. Dembić Z, Haas W, Weiss S, et al. Transfer of specificity by murine α and β T-cell receptor genes. *Nature*. 1986;320(6059):232–238. https://doi.org/10.1038/320232a0.
3. Boomer JS, Green JM. An enigmatic tail of CD28 signaling. *Cold Spring Harb Perspect Biol*. 2010;2(8):a002436. https://doi.org/10.1101/cshperspect.a002436.
4. Kuhn NF, Purdon TJ, van Leeuwen DG, et al. CD40 ligand-modified chimeric antigen receptor T cells enhance antitumor function by eliciting an endogenous antitumor response. *Cancer Cell*. 2019;35(3):473–488.e6. https://doi.org/10.1016/j.ccell.2019.02.006.
5. Pegram HJ, Lee JC, Hayman EG, et al. Tumor-targeted T cells modified to secrete IL-12 eradicate systemic tumors without need for prior conditioning. *Blood*. 2012;119(18):4133–4141. https://doi.org/10.1182/blood-2011-12-400044.
6. Pegram HJ, Purdon TJ, van Leeuwen DG, et al. IL-12-secreting CD19-targeted cord blood-derived T cells for the immunotherapy of B-cell acute lymphoblastic leukemia. *Leukemia*. 2015;29(2):415–422. https://doi.org/10.1038/leu.2014.215.
7. Avanzi MP, Yeku O, Li X, et al. Engineered tumor-targeted T cells mediate enhanced anti-tumor efficacy both directly and through activation of the endogenous immune system. *Cell Rep*. 2018;23:2130–2141.
8. Calzascia T, Masson F, Di Berardino-Besson W, et al. Homing phenotypes of tumor-specific CD8 T cells are predetermined at the tumor site by crosspresenting APCs. *Immunity*. 2005;22(2):175–184. https://doi.org/10.1016/j.immuni.2004.12.008.
9. Rafiq S, Yeku OO, Jackson HJ, et al. Targeted delivery of a PD-1-blocking scFV by CAR-T cells enhances anti-tumor efficacy in vivo. *Nat Biotechnol*. 2018;36(9):847–858. https://doi.org/10.1038/nbt.4195.
10. Sadelain M, Brentjens R, Rivière I. The basic principles of chimeric antigen receptor design. *Cancer Discov*. 2013;3(4):388–398. https://doi.org/10.1158/2159-8290.CD-12-0548.
11. Kahlon KS, Kahlon KS, Brown C, et al. Specific recognition and killing of glioblastoma multiforme by 13-zetakine redirected cytolytic T cells. *Cancer Res*. 2004;64(24):9160–9166. https://doi.org/10.1158/0008-5472.CAN-04-0454.
12. Eshhar Z, Waks T, Grosst G, Schindler DG. Specific activation and targeting of cytotoxic lymphocytes through chimeric single chains consisting of antibody-binding domains and the y or C subunits of the immunoglobulin and T-cell receptors. *Immunology*. 1993;90:720–724. https://www.ncbi.nlm.nih.gov/pmc/articles/PMC45737/pdf/pnas01100-0366.pdf. Accessed 21 March 2018.

13. Jenkins MK, Chen CA, Jung G, Mueller DL, Schwartz RH. Inhibition of antigen-specific proliferation of Type 1 murine T cell clones after stimulation with immobilized anti-CD3 monoclonal antibody. *J Immunol.* 1990;144:16–22. http://www.ncbi.nlm.nih.gov/pubmed/2153162.
14. Gong MC, Latouche J-B, Krause A, Heston WDW, Bander NH, Sadelain M. Cancer patient T cells genetically targeted to prostate-specific membrane antigen specifically lyse prostate cancer cells and release cytokines in response to prostate-specific membrane antigen. *Neoplasia.* 2002;1:123–127. https://doi.org/10.1038/sj.neo.7900018.
15. Brentjens RJ, Latouche JB, Santos E, et al. Eradication of systemic B-cell tumors by genetically targeted human T lymphocytes co-stimulated by CD80 and interleukin-15. *Nat Med.* 2003;9(3):279–286. https://doi.org/10.1038/nm827.
16. Maher J, Brentjens RJ, Gunset G, Rivière I, Sadelain M. Human T-lymphocyte cytotoxicity and proliferation directed by a single chimeric TCRζ/CD28 receptor. *Nat Biotechnol.* 2002;20(1):70–75. https://doi.org/10.1038/nbt0102-70.
17. Imai C, Mihara K, Andreansky M, et al. Chimeric receptors with 4-1BB signaling capacity provoke potent cytotoxicity against acute lymphoblastic leukemia. *Leukemia.* 2004;18:676–684. https://doi.org/10.1038/sj.leu.2403302.
18. Park JH, Rivière I, Gonen M, et al. Long-term follow-up of CD19 CAR therapy in acute lymphoblastic leukemia. *N Engl J Med.* 2018;378(5):449–459. https://doi.org/10.1056/NEJMoa1709919.
19. Maude SL, Frey N, Shaw PA, et al. Chimeric antigen receptor T cells for sustained remissions in leukemia. *N Engl J Med.* 2014;371(16):1507–1517. https://doi.org/10.1056/NEJMoa1407222.
20. Neelapu SS, Locke FL, Bartlett NL, et al. Axicabtagene ciloleucel CAR T-cell therapy in refractory large B-cell lymphoma. *N Engl J Med.* 2017;377(26):2531–2544. https://doi.org/10.1056/NEJMoa1707447.
21. Zhao Y, Wang QJ, Yang S, et al. A herceptin-based chimeric antigen receptor with modified signaling domains leads to enhanced survival of transduced T lymphocytes and antitumor activity. *J Immunol.* 2009;183(9):5563–5574. https://doi.org/10.4049/jimmunol.0900447.
22. Zhong XS, Matsushita M, Plotkin J, Riviere I, Sadelain M. Chimeric antigen receptors combining 4-1BB and CD28 signaling domains augment PI 3 kinase/AKT/Bcl-X L activation and CD8 T cell-mediated tumor eradication. *Mol Ther.* 2010;18(2):413–420. https://doi.org/10.1038/mt.2009.210.
23. Wilkie S, Picco G, Foster J, et al. Retargeting of human T cells to tumor-associated MUC1: the evolution of a chimeric antigen receptor. *J Immunol.* 2008;180(7):4901–4909. https://doi.org/10.4049/jimmunol.180.7.4901.
24. Carpenito C, Milone MC, Hassan R, et al. Control of large, established tumor xenografts with genetically retargeted human T cells containing CD28 and CD137 domains. *Proc Natl Acad Sci U S A.* 2009;106:3360–3365. https://doi.org/10.1073/pnas.0813101106.
25. Kloss CC, Lee J, Zhang A, et al. Dominant-negative TGF-β receptor enhances PSMA-targeted human CAR T cell proliferation and augments prostate cancer eradication. *Mol Ther.* 2018;26(7):1855–1866. https://doi.org/10.1016/j.ymthe.2018.05.003.
26. Cherkassky L, Morello A, Villena-Vargas J, et al. Human CAR T cells with cell-intrinsic PD-1 checkpoint blockade resist tumor-mediated inhibition. *J Clin Investig.* 2016;126(8):3130–3144. https://doi.org/10.1172/JCI83092.
27. Mailankody S, Htut M, Lee KP, et al. JCARH125, anti-BCMA CAR T-cell therapy for relapsed/refractory multiple myeloma: initial proof of concept results from a phase 1/2 multicenter study (EVOLVE). In: *Blood.* American Society of Hematology; 2018; vol 132:Abstract 957. https://doi.org/10.1182/blood-2018-99-113548.

28. Beavis PA, Henderson MA, Giuffrida L, et al. Targeting the adenosine 2A receptor enhances chimeric antigen receptor T cell efficacy. *J Clin Investig.* 2017;127(3):929–941. https://doi.org/10.1172/JCI89455.
29. Dong H, Strome SE, Salomao DR, et al. Tumor-associated B7-H1 promotes T-cell apoptosis: a potential mechanism of immune evasion. *Nat Med.* 2002;8(8):793–800. https://doi.org/10.1038/nm730.
30. Topalian SL, Drake CG, Pardoll DM. Targeting the PD-1/B7-H1(PD-L1) pathway to activate anti-tumor immunity. *Curr Opin Immunol.* 2012;24(2):207–212. https://doi.org/10.1016/j.coi.2011.12.009.
31. Ansell SM, Lesokhin AM, Borrello I, et al. PD-1 blockade with nivolumab in relapsed or refractory Hodgkin's lymphoma. *N Engl J Med.* 2015;3724372(22):311–319. https://doi.org/10.1056/NEJMoa1411087.
32. Weller M, Wick W, Aldape K, et al. Glioma. *Nat Rev Dis Primers.* 2015;1:15017. https://doi.org/10.1038/nrdp.2015.17.
33. Brown CE, Warden CD, Starr R, et al. Glioma IL13Rα2 is associated with mesenchymal signature gene expression and poor patient prognosis. *PLoS One.* 2013;8(10): e77769. https://doi.org/10.1371/journal.pone.0077769.
34. Brown CE, Starr R, Aguilar B, et al. Stem-like tumor-initiating cells isolated from IL13Rα2 expressing gliomas are targeted and killed by IL13-zetakine-redirected T cells. *Clin Cancer Res.* 2012;18(8):2199–2209. https://doi.org/10.1158/1078-0432.CCR-11-1669.
35. Brown CE, Badie B, Barish ME, et al. Bioactivity and safety of IL13Rα2-redirected chimeric antigen receptor CD8+ T cells in patients with recurrent glioblastoma. *Clin Cancer Res.* 2015;21(18):4062–4072. https://doi.org/10.1158/1078-0432.CCR-15-0428.
36. Brown CE, Alizadeh D, Starr R, et al. Regression of glioblastoma after chimeric antigen receptor T-cell therapy. *N Engl J Med.* 2016;375(26):2561–2569. https://doi.org/10.1056/NEJMoa1610497.
37. Wong AJ, Ruppert JM, Bigner SH, et al. Structural alterations of the epidermal growth factor receptor gene in human gliomas. *Proc Natl Acad Sci U S A.* 1992;89(7):2965–2969. https://doi.org/10.1073/pnas.89.7.2965.
38. O'Rourke DM, Nasrallah MP, Desai A, et al. A single dose of peripherally infused EGFRvIII-directed CAR T cells mediates antigen loss and induces adaptive resistance in patients with recurrent glioblastoma. *Sci Transl Med.* 2017;9(399). https://doi.org/10.1126/scitranslmed.aaa0984.
39. Van Den Bent MJ, Gao Y, Kerkhof M, et al. Changes in the EGFR amplification and EGFRvIII expression between paired primary and recurrent glioblastomas. *Neuro-Oncology.* 2015;17(7):935–941. https://doi.org/10.1093/neuonc/nov013.
40. Org Felsberg J, Hentschel B, Kaulich K, et al. Cancer therapy: clinical epidermal growth factor receptor variant III (EGFRvIII) positivity in EGFR-amplified glioblastomas: prognostic role and comparison between primary and recurrent tumors. *Clin Cancer Res.* 2017;23:6846–6855. https://doi.org/10.1158/1078-0432.CCR-17-0890.
41. Matthay KK, Maris JM, Schleiermacher G, et al. Neuroblastoma. *Nat Publ Group.* 2016;2:16078. https://doi.org/10.1038/nrdp.2016.78.
42. Pule MA, Savoldo B, Myers GD, et al. Virus-specific T cells engineered to coexpress tumor-specific receptors: persistence and antitumor activity in individuals with neuroblastoma. *Nat Med.* 2008;14:1264–1270. https://doi.org/10.1038/nm.1882.
43. Louis CU, Savoldo B, Dotti G, et al. Antitumor activity and long-term fate of chimeric antigen receptor-positive T cells in patients with neuroblastoma. *Blood.* 2011;118:6050–6056. https://doi.org/10.1182/blood-2011.

44. Park JR, DiGiusto DL, Slovak M, et al. Adoptive transfer of chimeric antigen receptor re-directed cytolytic T lymphocyte clones in patients with neuroblastoma. *Mol Ther.* 2007;15(4):825–833. https://doi.org/10.1038/sj.mt.6300104.
45. Robinson BW, Musk AW, Lake RA. Malignant mesothelioma. *Lancet.* 2005;366(9483):397–408. https://doi.org/10.1016/S0140-6736(05)67025-0.
46. Chang K, Pastan I. Molecular cloning of mesothelin, a differentiation antigen present on mesothelium, mesotheliomas, and ovarian cancers. *Proc Natl Acad Sci.* 2002;93(1):136–140. https://doi.org/10.1073/pnas.93.1.136.
47. Argani P, Iacobuzio-Donahue C, Ryu B, et al. Mesothelin is overexpressed in the vast majority of ductal adenocarcinomas of the pancreas: identification of a new pancreatic cancer marker by serial analysis of gene expression (SAGE). *Clin Cancer Res.* 2001;7:3862–3868. www.ncbi.nlm.nih.gov/SAGE. Accessed 31 March 2019.
48. Kelly RJ, Sharon E, Pastan I, Hassan R. Mesothelin-targeted agents in clinical trials and in preclinical development. *Mol Cancer Ther.* 2012;11(3):517–525. https://doi.org/10.1158/1535-7163.mct-11-0454.
49. Hassan R, Miller AC, Sharon E, et al. Major cancer regressions in mesothelioma after treatment with an anti-mesothelin immunotoxin and immune suppression. *Sci Transl Med.* 2013;5(208):208ra147. https://doi.org/10.1126/scitranslmed.3006941.
50. Beatty GL, Haas AR, Maus MV, et al. Mesothelin-specific chimeric antigen receptor mRNA-engineered T cells induce antitumor activity in solid malignancies. *Cancer Immunol Res.* 2014;2(2):112–120. https://doi.org/10.1158/2326-6066.CIR-13-0170.
51. Maus MV, Haas AR, Beatty GL, et al. Cancer immunology miniatures T cells expressing chimeric antigen receptors can cause anaphylaxis in humans. *Cancer Immunol Res.* 2013;1:26–31. https://doi.org/10.1158/2326-6066.CIR-13-0006.
52. Singh AP, Senapati S, Ponnusamy MP, et al. Clinical potential of mucins in diagnosis, prognosis, and therapy of ovarian cancer. *Lancet Oncol.* 2008;9(11):1076–1085. https://doi.org/10.1016/S1470-2045(08)70277-8.
53. Sun CC, Ramirez PT, Bodurka DC. Quality of life for patients with epithelial ovarian cancer. *Nat Clin Pract Oncol.* 2007;4(1):18–29. https://doi.org/10.1038/ncponc0693.
54. Matulonis UA, Sood AK, Fallowfield L, Howitt BE, Sehouli J, Karlan BY. Ovarian cancer. *Nat Rev Dis Primers.* 2016;2(1). 16061https://doi.org/10.1038/nrdp.2016.61.
55. Kershaw MH, Westwood JA, Parker LL, et al. A phase I study on adoptive immunotherapy using gene-modified T cells for ovarian cancer. *Clin Cancer Res.* 2006;12(20):6106–6115. https://doi.org/10.1158/1078-0432.CCR-06-1183.
56. O'Brien TJ, Beard JB, Underwood LJ, Dennis RA, Santin AD, York L. The CA 125 gene: an extracellular superstructure dominated by repeat sequences. *Tumor Biol.* 2001;22(6):348–366. https://doi.org/10.1159/000050638.
57. Badgwell D, Bast RC. Early Detection of Ovarian Cancer. vol. 23. IOS Press; 2007. https://www.ncbi.nlm.nih.gov/pmc/articles/PMC3851959/pdf/DM23-05-309382.pdf. Accessed 5 June 2019.
58. Krivak TC, Tian C, Rose GS, Armstrong DK, Maxwell GL. A gynecologic oncology group study of serum CA-125 levels in patients with stage III optimally debulked ovarian cancer treated with intraperitoneal compared to intravenous chemotherapy: an analysis of patients enrolled in GOG 172. *Gynecol Oncol.* 2009;115(1):81 85. https://doi.org/10.1016/j.ygyno.2009.06.021.
59. Chekmasova AA, Rao TD, Nikhamin Y, et al. Successful eradication of established peritoneal ovarian tumors in SCID-Beige mice following adoptive transfer of T cells genetically targeted to the MUC16 antigen. *Clin Cancer Res.* 2010;16(14):3594–3606. https://doi.org/10.1158/1078-0432.CCR-10-0192.

60. Koneru M, Purdon TJ, Spriggs D, Koneru S, Brentjens RJ. IL-12 secreting tumor-targeted chimeric antigen receptor T cells eradicate ovarian tumors *in vivo*. *Oncoimmunology*. 2015;4(3):e994446. https://doi.org/10.4161/2162402X.2014.994446.
61. Yeku OO, Purdon TJ, Koneru M, Spriggs D, Brentjens RJ. Armored CAR T cells enhance antitumor efficacy and overcome the tumor microenvironment. *Sci Rep*. 2017;7(1):10541. https://doi.org/10.1038/s41598-017-10940-8.
62. Koneru M, O'Cearbhaill R, Pendharkar S, Spriggs DR, Brentjens RJ. A phase I clinical trial of adoptive T cell therapy using IL-12 secreting MUC-16ecto directed chimeric antigen receptors for recurrent ovarian cancer. *J Transl Med*. 2015;13(1):102. https://doi.org/10.1186/s12967-015-0460-x.
63. Kim S, Kim B, Song YS. Ascites modulates cancer cell behavior, contributing to tumor heterogeneity in ovarian cancer. *Cancer Sci*. 2016;107(9):1173–1178. https://doi.org/10.1111/cas.12987.
64. Chen LL, Ye F, Lü WG, Yu Y, Chen HZ, Xie X. Evaluation of immune inhibitory cytokine profiles in epithelial ovarian carcinoma. *J Obstet Gynaecol Res*. 2009;35(2):212–218. https://doi.org/10.1111/j.1447-0756.2008.00935.x.
65. Sinha P, Clements VK, Bunt SK, Albelda SM, Ostrand-Rosenberg S. Cross-talk between myeloid-derived suppressor cells and macrophages subverts tumor immunity toward a type 2 response. *J Immunol*. 2007;179(2):977–983. https://doi.org/10.4049/jimmunol.179.2.977.
66. Bak SP, Alonso A, Turk MJ, Berwin B. Murine ovarian cancer vascular leukocytes require arginase-1 activity for T cell suppression. *Mol Immunol*. 2008;46(2):258–268. https://doi.org/10.1016/j.molimm.2008.08.266.
67. Jon G, Quatromoni EE. Tumor-associated macrophages and lung cancer. *Am J Transl Res*. 2012;4(4):376–389. http://www.ajtr.org/files/AJTR1209004.pdf. Accessed 5 June 2019.
68. Ahmed N, Brawley V, Hegde M, et al. HER2-specific chimeric antigen receptor-modified virus-specific T cells for progressive glioblastoma: a phase 1 dose-escalation trial. *JAMA Oncol*. 2017;3(8):1094–1101. https://doi.org/10.1001/jamaoncol.2017.0184.
69. Brown CE, Badie B, Barish ME, et al. Bioactivity and safety of IL13R 2-redirected chimeric antigen receptor CD8+ T cells in patients with recurrent glioblastoma. *Clin Cancer Res*. 2015;21(18):4062–4072. https://doi.org/10.1158/1078-0432.CCR-15-0428.
70. Majzner RG, Theruvath JL, Nellan A, et al. CAR T cells targeting B7-H3, a pan-cancer antigen, demonstrate potent preclinical activity against pediatric solid tumors and brain tumors. *Clin Cancer Res*. 2019;25(8):2560–2574. https://doi.org/10.1158/1078-0432.ccr-18-0432.
71. Du H, Hirabayashi K, Ahn S, et al. Antitumor responses in the absence of toxicity in solid tumors by targeting B7-H3 via chimeric antigen receptor T cells. *Cancer Cell*. 2019;35(2):221–237.e8. https://doi.org/10.1016/J.CCELL.2019.01.002.
72. Morgan RA, Yang JC, Kitano M, Dudley ME, Laurencot CM, Rosenberg SA. Case report of a serious adverse event following the administration of T cells transduced with a chimeric antigen receptor recognizing ERBB2. *Mol Ther*. 2010;18:843–851. https://doi.org/10.1038/mt.2010.24.
73. van Schalkwyk MCI, Papa SE, Jeannon J-P, Urbano TG, Spicer JF, Maher J. Design of a phase I clinical trial to evaluate intratumoral delivery of ErbB-targeted chimeric antigen receptor T-cells in locally advanced or recurrent head and neck cancer. *Hum Gene Ther Clin Dev*. 2013;24(3):134–142. https://doi.org/10.1089/humc.2013.144.
74. Fischer J, Paret C, El Malki K, et al. CD19 isoforms enabling resistance to CART-19 immunotherapy are expressed in B-ALL patients at initial diagnosis. *J Immunother (Hagerstown, Md 1997)*. 2017;40(5):187–195. https://doi.org/10.1097/CJI.0000000000000169.

75. Sotillo E, Barrett DM, Black KL, et al. Convergence of acquired mutations and alternative splicing of CD19 enables resistance to CART-19 immunotherapy. *Cancer Discov*. 2015;5:1282–1295. https://doi.org/10.1158/2159-8290.CD-15-1020.
76. Fry TJ, Shah NN, Orentas RJ, et al. CD22-targeted CAR T cells induce remission in B-ALL that is naive or resistant to CD19-targeted CAR immunotherapy. *Nat Med*. 2018;24(1):20–28. https://doi.org/10.1038/nm.4441.
77. Bielamowicz K, Fousek K, Byrd TT, et al. Trivalent CAR T cells overcome interpatient antigenic variability in glioblastoma. *Neuro-Oncology*. 2018;20(4):506–518. https://doi.org/10.1093/neuonc/nox182.
78. Cho JH, Collins JJ, Wong WW. Universal chimeric antigen receptors for multiplexed and logical control of T cell responses. *Cell*. 2018;173:1426–1438. https://doi.org/10.1016/j.cell.2018.03.038.
79. Schneider D, Xiong Y, Wu D, et al. A tandem CD19/CD20 CAR lentiviral vector drives on-target and off-target antigen modulation in leukemia cell lines. *J Immunother Cancer*. 2017;5(1):42. https://doi.org/10.1186/s40425-017-0246-1.
80. Ruella M, Barrett DM, Kenderian SS, et al. Dual CD19 and CD123 targeting prevents antigen-loss relapses after CD19-directed immunotherapies. *J Clin Investig*. 2016;126(10):3814–3826. https://doi.org/10.1172/JCI87366.
81. Qin H, Ramakrishna S, Nguyen S, et al. Preclinical development of bivalent chimeric antigen receptors targeting both CD19 and CD22. *Mol Ther Oncolytics*. 2018;11:127–137. https://doi.org/10.1016/J.OMTO.2018.10.006.
82. Hegde M, Mukherjee M, Grada Z, et al. Tandem CAR T cells targeting HER2 and IL13Rα2 mitigate tumor antigen escape. *J Clin Investig*. 2016;126(8):3036–3052. https://doi.org/10.1172/JCI83416.
83. Walker AJ, Majzner RG, Zhang L, et al. Tumor antigen and receptor densities regulate efficacy of a chimeric antigen receptor targeting anaplastic lymphoma kinase. *Mol Ther*. 2017;25:2189–2201. https://doi.org/10.1016/j.ymthe.2017.06.008.
84. van der Stegen SJC, Davies DM, Wilkie S, et al. Preclinical in vivo modeling of cytokine release syndrome induced by ErbB-retargeted human T cells: identifying a window of therapeutic opportunity? *J Immunol*. 2013;191(9):4589–4598. https://doi.org/10.4049/jimmunol.1301523.
85. Harlin H, Meng Y, Peterson AC, et al. Chemokine expression in melanoma metastases associated with CD8 + T-Cell recruitment. *Cancer Res*. 2009;69(7):3077–3085. https://doi.org/10.1158/0008-5472.CAN-08-2281.
86. Kakarla S, Song XT, Gottschalk S. Cancer-associated fibroblasts as targets for immunotherapy. *Immunotherapy*. 2012;4(11):1129–1138. https://doi.org/10.2217/imt.12.112.
87. Parente-Pereira AC, Burnet J, Ellison D, et al. Trafficking of CAR-engineered human T cells following regional or systemic adoptive transfer in SCID beige mice. *J Clin Immunol*. 2011;31(4):710–718. https://doi.org/10.1007/s10875-011-9532-8.
88. Brown CE, Vishwanath RP, Aguilar B, et al. Tumor-derived chemokine MCP-1/CCL2 is sufficient for mediating tumor tropism of adoptively transferred T cells. *J Immunol*. 2007;179(5):3332–3341. https://doi.org/10.4049/jimmunol.179.5.3332.
89. Huang H, Langenkamp E, Georganaki M, et al. VEGF suppresses T-lymphocyte infiltration in the tumor microenvironment through inhibition of NF-κB-induced endothelial activation. *FASEB J*. 2015;29(1):227–238. https://doi.org/10.1096/fj.14-250985.
90. Slaney CY, Kershaw MH, Darcy PK. Trafficking of T cells into tumors. *Cancer Res*. 2014;74(24):7168–7174. https://doi.org/10.1158/0008-5472.CAN-14-2458.
91. Lohr J, Ratliff T, Huppertz A, et al. Effector T-cell infiltration positively impacts survival of glioblastoma patients and is impaired by tumor-derived TGF-β. *Clin Cancer Res*. 2011;17(13):4296–4308. https://doi.org/10.1158/1078-0432.CCR-10-2557.

92. Craddock JA, Lu A, Bear A, et al. Enhanced tumor trafficking of GD2 chimeric antigen receptor T cells by expression of the chemokine receptor CCR2b. *J Immunother (Hagerstown, Md 1997)*. 2010;33(8):780–788. https://doi.org/10.1097/CJI.0b013e3181ee6675.
93. Moon EK, Carpenito C, Sun J, et al. Expression of a functional CCR2 receptor enhances tumor localization and tumor eradication by retargeted human t cells expressing a mesothelin-specific chimeric antibody receptor. *Clin Cancer Res*. 2011;17:4719–4730. https://doi.org/10.1158/1078-0432.CCR-11-0351.
94. Newick K, O'Brien S, Sun J, et al. Augmentation of CAR T-cell trafficking and antitumor efficacy by blocking protein kinase a localization. *Cancer Immunol Res*. 2016;4(6):541–551. https://doi.org/10.1158/2326-6066.CIR-15-0263.
95. Caruana I, Savoldo B, Hoyos V, et al. Heparanase promotes tumor infiltration and antitumor activity of CAR-redirected T lymphocytes. *Nat Med*. 2015;21(5):524–529. https://doi.org/10.1038/nm.3833.
96. Wang L-CS, Lo A, Scholler J, et al. Targeting fibroblast activation protein in tumor stroma with chimeric antigen receptor T cells can inhibit tumor growth and augment host immunity without severe toxicity. *Cancer Immunol Res*. 2014;2:154–166. https://doi.org/10.1158/2326-6066.CIR-13-0027.
97. Lo A, Wang LCS, Scholler J, et al. Tumor-promoting desmoplasia is disrupted by depleting FAP-expressing stromal cells. *Cancer Res*. 2015;75(14):2800–2810. https://doi.org/10.1158/0008-5472.CAN-14-3041.
98. Santoro SP, Kim S, Motz GT, et al. T cells bearing a chimeric antigen receptor against prostate-specific membrane antigen mediate vascular disruption and result in tumor regression. *Cancer Immunol Res*. 2015;3(1):68–84. https://doi.org/10.1158/2326-6066.cir-14-0192.
99. Chinnasamy D, Tran E, Yu Z, Morgan RA, Restifo NP, Rosenberg SA. Simultaneous targeting of tumor antigens and the tumor vasculature using T lymphocyte transfer synergize to induce regression of established tumors in mice. *Cancer Res*. 2013;73(11):3371–3380. https://doi.org/10.1158/0008-5472.CAN-12-3913.
100. Wong RJ, Patel SG, Kim S-H, et al. *Cytokine Gene Transfer Enhances Herpes Oncolytic Therapy in Murine Squamous Cell Carcinoma*. vol. 12. Mary Ann Liebert; 2001. www.liebertpub.com. Accessed 2 June 2019.
101. van Herpen CM, van der Laak JA, de Vries IJ, et al. Intratumoral recombinant human interleukin-12 administration in head and neck squamous cell carcinoma patients modifies locoregional lymph node architecture and induces natural killer cell infiltration in the primary tumor. *Clin Cancer Res*. 2005;11(5):1899–1909.
102. Carpenter SG, Carson J, Fong Y. Regional liver therapy using oncolytic virus to target hepatic colorectal metastases. *Semin Oncol*. 2010;37(2):160–169. https://doi.org/10.1053/j.seminoncol.2010.03.001.
103. Nishio N, Diaconu I, Liu H, et al. Armed oncolytic virus enhances immune functions of chimeric antigen receptor-modified T cells in solid tumors. *Cancer Res*. 2014;74(18):5195–5205. https://doi.org/10.1158/0008-5472.CAN-14-0697.
104. Katz SC, Point GR, Cunetta M, et al. Regional CAR-T cell infusions for peritoneal carcinomatosis are superior to systemic delivery. *Cancer Gene Ther*. 2016;23(5):142–148. https://doi.org/10.1038/cgt.2016.14.
105. Adusumilli PS, Cherkassky L, Villena-Vargas J, et al. Regional delivery of mesothelin-targeted CAR T cell therapy generates potent and long-lasting CD4-dependent tumor immunity. *Sci Transl Med*. 2014;6(261):261ra151. https://doi.org/10.1126/scitranslmed.3010162.
106. Choi BD, Suryadevara CM, Gedeon PC, et al. Intracerebral delivery of a third generation EGFRvIII-specific chimeric antigen receptor is efficacious against human glioma. *J Clin Neurosci*. 2014;21(1):189–190. https://doi.org/10.1016/j.jocn.2013.03.012.

107. Han J, Chu J, Keung Chan W, et al. CAR-engineered NK cells targeting wild-type EGFR and EGFRvIII enhance killing of glioblastoma and patient-derived glioblastoma stem cells. *Sci Rep*. 2015;5(1). 11483https://doi.org/10.1038/srep11483.
108. Priceman SJ, Tilakawardane D, Jeang B, et al. Regional delivery of chimeric antigen receptor-engineered T cells effectively targets HER2 + breast cancer metastasis to the brain. *Clin Cancer Res*. 2018;24(1):95–105. https://doi.org/10.1158/1078-0432.CCR-17-2041.
109. Swartz MA, Iida N, Roberts EW, et al. Tumor microenvironment complexity: emerging roles in cancer therapy. *Cancer Res*. 2012;72(10):2473–2480. https://doi.org/10.1158/0008-5472.CAN-12-0122.
110. Newick K, Moon E, Albelda SM. Chimeric antigen receptor T-cell therapy for solid tumors. *Mol Ther Oncolytics*. 2016;3 16006https://doi.org/10.1038/mto.2016.6.
111. Zhang Y, Ertl HCJ. Starved and asphyxiated: how can CD8+ T cells within a tumor microenvironment prevent tumor progression. *Front Immunol*. 2016;7:32. https://doi.org/10.3389/fimmu.2016.00032.
112. Fischer K, Hoffmann P, Voelkl S, et al. Inhibitory effect of tumor cell-derived lactic acid on human T cells. *Blood*. 2007;109:3812–3819. https://doi.org/10.1182/blood-2006-07.
113. Jacobs SR, Herman CE, MacIver NJ, et al. Glucose uptake is limiting in T cell activation and requires CD28-mediated Akt-dependent and independent pathways. *J Immunol*. 2008;180(7):4476–4486. https://doi.org/10.4049/jimmunol.180.7.4476.
114. Howie D, Waldmann H, Cobbold SP. Nutrient sensing via mTOR in T cells maintains a tolerogenic microenvironment. *Front Immunol*. 2014;5:409. https://doi.org/10.3389/fimmu.2014.00409.
115. Platten M, Wick W, Van Den Eynde BJ. Tryptophan catabolism in cancer: beyond IDO and tryptophan depletion. *Cancer Res*. 2012;72(21):5435–5440. https://doi.org/10.1158/0008-5472.CAN-12-0569.
116. Munn DH, Mellor AL. IDO in the tumor microenvironment: inflammation, counter-regulation, and tolerance. *Trends Immunol*. 2016;37(3):193–207. https://doi.org/10.1016/j.it.2016.01.002.
117. Ninomiya S, Narala N, Huye L, et al. Tumor indoleamine 2,3-dioxygenase (IDO) inhibits CD19-CAR T cells and is downregulated by lymphodepleting drugs. *Blood*. 2015;125(25):3905–3916. https://doi.org/10.1182/blood-2015-01-621474.
118. Oh SA, Li MO. TGF-β: guardian of T cell function. *J Immunol*. 2013;191(8):3973–3979. https://doi.org/10.4049/jimmunol.1301843.
119. Massagué J. TGFβ in Cancer. *Cell*. 2008;134(2):215–230. https://doi.org/10.1016/J.CELL.2008.07.001.
120. Kaminska B, Wesolowska A, Danilkiewicz M. TGF beta signalling and its role in tumour pathogenesis. *Acta Biochim Pol*. 2005;52(2):329–337. www.actabp.pl. Accessed 6 June 2019.
121. Hishii M, Nitta T, Ishida H, et al. Human glioma-derived interleukin-10 inhibits antitumor immune responses in vitro. *Neurosurgery*. 1995;37(6):1160–1167. https://doi.org/10.1227/00006123-199512000-00016.
122. Yu S, Li A, Liu Q, et al. Chimeric antigen receptor T cells: a novel therapy for solid tumors. *J Hematol Oncol*. 2017;10(1):78. https://doi.org/10.1186/s13045-017-0444-9.
123. Newick K, O'Brien S, Moon E, Albelda SM. CAR T cell therapy for solid tumors. *Annu Rev Med*. 2017;68(1):139–152. https://doi.org/10.1146/annurev-med-062315-120245.
124. Goodwin JS. Suppression of human T-cell mitogenesis by prostaglandin. Existence of a prostaglandin-producing suppressor cell. *J Exp Med*. 2004;146(6):1719–1734. https://doi.org/10.1084/jem.146.6.1719.

125. Kobayashi K, Omori K, Murata T. Role of prostaglandins in tumor microenvironment. *Cancer Metastasis Rev*. 2018;37(2–3):347–354. https://doi.org/10.1007/s10555-018-9740-2.
126. Beavis PA, Stagg J, Darcy PK, Smyth MJ. CD73: a potent suppressor of antitumor immune responses. *Trends Immunol*. 2012;33(5):231–237. https://doi.org/10.1016/j.it.2012.02.009.
127. Linnemann C, Schildberg FA, Schurich A, et al. Adenosine regulates CD8 T-cell priming by inhibition of membrane-proximal T-cell receptor signalling. *Immunology*. 2009;128. https://doi.org/10.1111/j.1365-2567.2009.03075.x.
128. Su Y, Huang X, Raskovalova T, et al. Cooperation of adenosine and prostaglandin E2 (PGE2) in amplification of cAMP-PKA signaling and immunosuppression. *Cancer Immunol Immunother*. 2008;57(11):1611–1623. https://doi.org/10.1007/s00262-008-0494-5.
129. Mandapathil M, Szczepanski MJ, Szajnik M, et al. Adenosine and prostaglandin e2 cooperate in the suppression of immune responses mediated by adaptive regulatory T cells. *J Biol Chem*. 2010;285(36):27571–27580. https://doi.org/10.1074/jbc.M110.127100.
130. Beavis PA, Milenkovski N, Henderson MA, et al. Adenosine receptor 2A blockade increases the efficacy of anti-PD-1 through enhanced antitumor T-cell responses. *Cancer Immunol Res*. 2015;3(5):506–517. https://doi.org/10.1158/2326-6066.CIR-14-0211.
131. Wallace A, Kapoor V, Sun J, et al. Transforming growth factor-β receptor blockade augments the effectiveness of adoptive T-cell therapy of established solid cancers. *Clin Cancer Res*. 2008;14(12):3966–3974. https://doi.org/10.1158/1078-0432.CCR-08-0356.
132. Gabrilovich DI, Nagaraj S. Myeloid-derived suppressor cells as regulators of the immune system. *Nat Rev Immunol*. 2009;9(3):162–174. https://doi.org/10.1038/nri2506.
133. Mussai F, Egan S, Hunter S, et al. Neuroblastoma arginase activity creates an immunosuppressive microenvironment that impairs autologous and engineered immunity. *Cancer Res*. 2015;75(15):3043–3053. https://doi.org/10.1158/0008-5472.CAN-14-3443.
134. Liu X, Ranganathan R, Jiang S, et al. A chimeric switch-receptor targeting PD1 augments the efficacy of second-generation CAR T cells in advanced solid tumors. *Cancer Res*. 2016;76(6):1578–1590. https://doi.org/10.1158/0008-5472.CAN-15-2524.
135. Locke FL, Neelapu SS, Bartlett NL, et al. Phase 1 results of ZUMA-1: a multicenter study of KTE-C19 anti-CD19 CAR T cell therapy in refractory aggressive lymphoma. *Mol Ther*. 2017;25(1):285–295. https://doi.org/10.1016/j.ymthe.2016.10.020.
136. Sakaguchi S, Miyara M, Costantino CM, Hafler DA. FOXP3 + regulatory T cells in the human immune system. *Nat Rev Immunol*. 2010;10(7):490–500. https://doi.org/10.1038/nri2785.
137. Moon EK, Wang L-C, Dolfi DV, et al. Cancer therapy: preclinical multifactorial T-cell hypofunction that is reversible can limit the efficacy of chimeric antigen receptor-transduced human T cells in solid tumors. *Clin Cancer Res*. 2014;20(16):4262–4273. https://doi.org/10.1158/1078-0432.CCR-13-2627.
138. Chong EA, Melenhorst JJ, Lacey SF, et al. PD-1 blockade modulates chimeric antigen receptor (CAR)-modified T cells: refueling the CAR. *Blood*. 2017;129(8):1039–1041. https://doi.org/10.1182/blood-2016-09-738245.
139. Ren J, Liu X, Fang C, Jiang S, June CH, Zhao Y. Multiplex genome editing to generate universal CAR T cells resistant to PD1 inhibition. *Clin Cancer Res*. 2017;23(9):2255–2266. https://doi.org/10.1158/1078-0432.CCR-16-1300.
140. Li S, Siriwon N, Zhang X, et al. Enhanced cancer immunotherapy by chimeric antigen receptor-modified T cells engineered to secrete checkpoint inhibitors. *Clin Cancer Res*. 2017;23(22):6982–6992. https://doi.org/10.1158/1078-0432.CCR-17-0867.

141. Suarez ER, Chang D-K, Sun J, et al. Chimeric antigen receptor T cells secreting anti-PD-L1 antibodies more effectively regress renal cell carcinoma in a humanized mouse model. *Oncotarget*. 2016;7(23):34341–34355. https://doi.org/10.18632/oncotarget.9114.
142. Mohammed S, Sukumaran S, Bajgain P, et al. Improving chimeric antigen receptor-modified T cell function by reversing the immunosuppressive tumor microenvironment of pancreatic cancer. *Mol Ther*. 2017;25(1):249–258. https://doi.org/10.1016/j.ymthe.2016.10.016.
143. Shum T, Omer B, Tashiro H, et al. Constitutive signaling from an engineered IL7 receptor promotes durable tumor elimination by tumor-redirected T cells. *Cancer Discov*. 2017;7(11):1238–1247. https://doi.org/10.1158/2159-8290.CD-17-0538.
144. Hurton LV, Singh H, Najjar AM, et al. Tethered IL-15 augments antitumor activity and promotes a stem-cell memory subset in tumor-specific T cells. *Proc Natl Acad Sci*. 2016;113(48):E7788–E7797. https://doi.org/10.1073/pnas.1610544113.
145. Maude SL, Barrett DM, Rheingold SR, et al. Efficacy of humanized CD19-targeted chimeric antigen receptor (CAR)-modified T cells in children and young adults with relapsed/refractory acute lymphoblastic leukemia. *Blood*. 2016;128(22):217. http://www.bloodjournal.org/content/128/22/217. Accessed 6 June 2019.
146. Wang LCS, Riese MJ, Moon EK, Albelda SM. Overcoming intrinsic inhibitory pathways to augment the antineoplastic activity of adoptively transferred T cells: re-tuning your CAR before hitting a rocky road. *Oncoimmunology*. 2013;2(11). e26492https://doi.org/10.4161/onci.26492.
147. Riese MJ, Wang LCS, Moon EK, et al. Enhanced effector responses in activated CD8+ T cells deficient in diacylglycerol kinases. *Cancer Res*. 2013;73(12):3566–3577. https://doi.org/10.1158/0008-5472.CAN-12-3874.
148. Feucht J, Sun J, Eyquem J, et al. Calibration of CAR activation potential directs alternative T cell fates and therapeutic potency. *Nat Med*. 2019;25(1):82–88. https://doi.org/10.1038/s41591-018-0290-5.
149. Eyquem J, Mansilla-Soto J, Giavridis T, et al. Targeting a CAR to the TRAC locus with CRISPR/Cas9 enhances tumour rejection. *Nature*. 2017;543(7643):113–117. https://doi.org/10.1038/nature21405.
150. Fraietta JA, Nobles CL, Sammons MA, et al. Disruption of TET2 promotes the therapeutic efficacy of CD19-targeted T cells. *Nature*. 2018;558(7709):307–312. https://doi.org/10.1038/s41586-018-0178-z.
151. Qasim W, Zhan H, Samarasinghe S, et al. Molecular remission of infant B-ALL after infusion of universal TALEN gene-edited CAR T cells. *Sci Transl Med*. 2017;9. https://doi.org/10.1126/scitranslmed.aaj2013.
152. Hegde M, Orange JS, Ahmed N. Tandem CAR T cells targeting HER2 and IL13Ra a 2 mitigate tumor antigen escape The Journal of Clinical Investigation. *J Clin Invest*. 2016;126(8). https://doi.org/10.1172/JCI83416.
153. Morsut L, Roybal KT, Xiong X, et al. Engineering customized cell sensing and response behaviors using synthetic notch receptors. *Cell*. 2016;164(4):780–791. https://doi.org/10.1016/j.cell.2016.01.012.
154. Roybal KT, Williams JZ, Morsut L, et al. Engineering T cells with customized therapeutic response programs using synthetic notch receptors. *Cell*. 2016;167(2):419–432.e16. https://doi.org/10.1016/j.cell.2016.09.011.
155. Roybal KT, Rupp LJ, Morsut L, et al. Precision tumor recognition by T cells with combinatorial antigen-sensing circuits. *Cell*. 2016;164(4):770–779. https://doi.org/10.1016/j.cell.2016.01.011.

Index

Note: Page numbers followed by "*f*" indicate figures, and "*t*" indicate tables.

A

Abnormal splicing, 33–34
Acetyl-coenzyme A (AcCoA), 267
Activation-induced cell death (ACID), 294–296
Acute lymphoblastic leukemia (ALL), 294–296
Acute myeloid leukemia (AML), 105–106, 110–112*t*, 124–125
Adaptive T-cell therapy, 206–208
Ad death protein (ADP), 221–222
Adenosine, 311
Adenovirus death protein (ADP), 268–269
Adoptive cell transfer (ACT), 153–154, 219, 225–226
Aldefluor, 131*f*, 132–133
Aldehyde dehydrogenase (ALDH), 129–130, 132–133
Angiogenesis, 81–83
Anti-apoptosis mechanisms, 130
Antibody-based targeting, 160–162
Anticancer immunotherapy, 153
Antigen.garnish tool, 36*t*, 40
Antigen heterogeneity, 302–305
Antigenicity of cancer, 141–142
Antigen presenting cell (APC), 74, 146–147, 195, 294
Antigen processing machinery (APM), 143–146
Anti-tumor immunity, 79, 218–219
Anti-tumor T-cell responses, 28
Anti-tumor vaccines, 159
ATP-binding cassette (ABC) transporter, 129–130
Autoimmunity, 18

B

β_2-microglobulin, 68–69
Bispecific T-cell engagers (BiTE), 161, 249–251
B16 melanoma, 8–10
Bovine Herpesvirus-4 (BoHV-4), 159–160

C

Calreticulin, 77–78
Cancer, 62. See also *specific cancer*
- immunology, 2–4
- immunosurveillance and immunoediting in pancreatic cancer, 190–191
- immunotherapy, 190

Cancer-associated fibroblasts (CAFs), 86–87, 262, 306–307
Cancer stem cells (CSCs), 120–121
- ALDH activity, 132–133
- antibody-based targeting, 160–162
- cell surface markers, 132
- characteristics, 125*f*
- clinical implications, 135–137, 136*f*
- clonal evolution model, 121–122, 123*f*
- concepts, 125–126
- dendritic cells, 155–160, 158–159*f*
- epithelial-to-mesenchymal transition, 128–129
- generation of floating tridimensional cell clones, 133–134
- hierarchical model, 121–122, 123*f*
- historical perspectives, 122–125
- immune checkpoint inhibitors, 163–164
- immune evasion, 143–153
- innate and innate-like immune response, 164–170
- mesenchymal-to-epithelial transition, 129
- metastasis, 129
- methods for identification, isolation and characterization, 131–135, 131*f*
- recapitulating tumor heterogeneity, 135
- resistance to conventional therapies, 129–130
- self-renewal ability, 134–135
- side population, 133
- signaling pathways, 126–128
- sphere-generation assay, 133–134
- stem-like state, 128
- T cell-based therapy, 162–163
- tumor antigens expression, 138–143

Cancer stem cells (CSCs) (*Continued*)
in tumor immunoediting, 174*f*
tumor-initiating ability, 134
tumorspheres, 133–134
vaccination, 155–160, 158–159*f*
Cancer/testis (CT) antigens, 141
Candidate antigens, 142–143
Carcinoembryonic antigen (CEA), 192–194, 207–208
Carcinogenesis, 4, 16–17
CAR T cell therapy. *See* Chimeric antigen receptor (CAR) T cell therapy
Caspase recruitment domain-containing protein 9 (CARD9), 203–204
Cationic antimicrobial peptides (CMPs), 268–269
CD47 protein, 77–78
Cell surface markers, 132
Cervical cancer, 107–108
Cervical intraepithelial neoplasia (CIN), 107–108, 110–112*t*
Checkpoint inhibitors, 13–16, 163–164
Chemokine matching, 252–254
Chemokines
and CAR T-cell therapy, 243–249
mismatched receptors, 305–306
pro-T-cell cytokines and, 255
structural hindrance, 306–308
Chemotherapy, 129–130
Chimeric antigen receptor (CAR) T cell therapy, 162–163, 206–210, 207*t*, 218–219, 257*f*
advantage, 228
antigen heterogeneity, 302–305
barriers, 231–232
carriage of oncolytic virus, 251–252
cell persistence, 315–316
cell trafficking, 305–309
chemokines and, 243–249
clinical data, 230–231
cytokines and, 240–243
design, 229–230
development of, 227–234
genome, 316–317
glioblastoma, 297–299
immune checkpoint molecules, 256–261
immunosuppressive cells, 312
immunosuppressive tumor microenvironment, 309–315
inhibitory cytokines, 310–312
mesothelioma, 300–301
metabolic hurdles, 309–310
neuroblastoma, 299–300
ovarian cancer, 301–302
regional administration of, 308–309
safety concerns, 232–234
for solid tumors, 297–302
soluble factors, 310–312
upregulation of coinhibitory receptors, 312–315
Chromosomal fusions, 33
Circular RNA, 34
Cisplatin, 12–14
Classical immunotherapy, 195–196, 197–198*t*
Clonal evolution model, 121–122, 123*f*
Colon cancer, 106–107, 110–112*t*
Colony stimulating factor-1 (CSF-1), 200–203
Conventional DCs (cDCs), 192–194
COX2 gene, 84
Coxsackie and adenovirus receptor (CAR) expression, 267–268
Coxsackievirus B3 (CVB3), 236–237
Cripto-1 protein, 159–160
Cytokine-induced killer (CIK) cells, 169–170
Cytokines
cancer stem cells, 151
and chimeric antigen receptor T-cell therapy, 240–243
and DC, 239–240
immunosuppressive microenvironment, 83–85
inhibitory, 310–312
Cytolytic T cell (CTL), 9–10
Cytosine deaminase (CD), 268–269
Cytosolic proteins, 67
Cytotoxic drugs, 129–130
Cytotoxic molecules, 268–269
Cytotoxic T lymphocyte (CTL), 68, 71–72, 79–80
Cytotoxic T-lymphocyte antigen 4 (CTLA4), 85, 146–148

D

Damage-associated molecular patterns (DAMPs), 235–236
Darwinian selection, 64–65, 121, 137

Death receptor signaling, 78–80
Dectin-1, 203–204
Dendritic cells (DCs)
cytokines and, 239–240
in pancreatic cancer, 191–194, 193*t*
vaccines, 155–160, 158–159*f*
Detoxification, from reactive oxygen species, 130
Differential agretopicity index (DAI), 40
DNAJB8 protein, 142
Double-edged sword, 83

E

E-cadherin, 128
Embryogenesis
development in, 124
epithelial-to-mesenchymal transition during, 128
Endoplasmic reticulum (ER), 37–38, 67
Endoplasmic reticulum aminopeptidase (ERAP) enzymes, 37–38
Endothelial cells, tumor blood vessels, 81–83
Epidermal growth factor receptor variant three (EGFRvIII), 297–298
Epidisco pipeline, 44–45
Epi-Seq processes, 45–46
Epithelial cell, 4–5
Epithelial cell adhesion molecule (EpCAM), 141–142
Epithelial-to-mesenchymal transition, 128–129
European Medicines Agency (EMA), 162
Exome sequencing, 29
Exosomes, 78, 80–81
Extracellular matrix (ECM), 306–307

F

Fas ligand (FasL), 78–79
Fibroblast activation protein (FAP), 205, 307
Floating tridimensional cell clones, 133–134
Fluorescence-activated cell sorting techniques, 124
Fluorescent imaging system, 135–136
Focal adhesion kinase (FAK), 204–205

G

Galectin-9, 203–204
Gamma delta (γδ) T cells, 167–169
Germline, 34–35
Glioblastoma multiforme (GBM), 143–144, 146–147, 151, 297–299
Glucocorticoid-induced TNFR-related protein (GITR), 262–263
Graft-versus-leukemia effect, 294
Granulocyte–macrophage colony-stimulating factor (GM-CSF), 85–86

H

Hedgehog (Hh) signaling pathways, 127
Hellstrom paradox, 2–3
Hematopoietic stem cell markers, 132
Heparanase, 306–308
Hepatic metastasis, 110–112*t*
Hepatocellular carcinoma (HCC), 108, 110–112*t*
Hepatoma-associated hepatitis C virus (HCV), 140
Hierarchical model, cancer stem cells, 121–122, 123*f*
High-risk human papillomavirus (hr-HPV), 4
Human endogenous retroviruses (hERVs), 34
Human epidermal growth factor receptor 2 (HER2), 208
Human leukocyte antigen (HLA), 26, 66–67
Human telomerase reverse transcriptase (hTERT) promoter, 221–222

I

Immune checkpoint inhibitors, 163–164, 218–219
in pancreatic cancer, 195–196, 197–198*t*
targeting of, 80
Immune checkpoints
cancer stem cells, 139, 146–148
molecules, 256–261
Immune co-stimulatory molecules, 255–256
Immune Epitope Database (IEDB), 36
Immune evasion, cancer stem cells
extrinsic mechanisms, 149–153
immune checkpoints, 146–148
immune response and targeted immunotherapies, 153–170
membrane-bound and soluble factors, 148–149
MHC-I and APM downregulation, 143–146

Immune privileged tumors, 89–90
Immune system, 62–63, 66–68
Immunodominant antigen, 71–72
Immunoediting
cancer stem cells in, 174*f*
concept of, 65
stages of, 64–65, 65*f*
Immunogenic cell death (ICD), 235–238
Immunogenicity, 28
Immunohistochemistry (IHC), 299
Immunology of cancer stem cells, 137–138, 170–171, 173–175
immune evasion
extrinsic mechanisms, 149–153
immune checkpoints, 146–148
membrane-bound and soluble factors, 148–149
MHC-I and APM downregulation, 143–146
immune response and targeted immunotherapy
antibody-based targeting, 160–162
dendritic cells, 155–160, 158–159*f*
immune checkpoint inhibitors, 163–164
innate and innate-like immune response, 164–170
T cell-based therapy, 162–163
vaccination, 155–160, 158–159*f*
tumor antigens expression, 138–143
Immunomodulatory mAbs
with cisplatin, 12–14
injection of, 17
therapeutic efficacy of, 8–12
tumor-bearing mice treating with, 5–7
in vitro evaluation of, 15
Immunoreceptor tyrosine-based activation motifs (ITAMs), 229–230
Immunosuppressed animal models, 63
Immunosuppressive cells, 312
metabolic pathways, 265–268
molecules delivery targeting, 261–264
Immunosuppressive microenvironment
clinical relevance, 86–87
cytokines, 83–85
recruitment of inhibitory immune cells, 85–86
tumor microenvironment, 81–83
immunosuppressive cells, 312
inhibitory cytokines, 310–312
metabolic hurdles, 309–310
in pancreatic cancer, 196–199
soluble factors, 310–312
upregulation of coinhibitory receptors, 312–315
Immunosurveillance, 62–65
Immunotherapy, 2–3, 6, 14
cancer stem cells, 136–137
with natural killer cells, 191–194, 193*t*
Indel variant calling, 30–33
Indoleamine 2,3-dioxygenase (IDO), 84–85
Indoleamine 2,3-dioxygenase 1 (IDO1), 4–5
Integrated model, 122
Intercellular adhesion molecule-1 (ICAM-1), 79–80, 222–223
Interferon (IFN)-α, 143–146
Interferon γ (IFN-γ), 143–146, 190–191
Interleukin 10 (IL-10), 84, 86, 241–242
Interleukin-13 receptor alpha 2 (IL-13rα2), 297–298
Intratumor heterogeneity, 120–122, 123*f*
Invasive cervical cancer (ICC), 4
In vitro evaluation, of immunomodulatory mAbs, 15
Ipilimumab monotherapy, 196–199

K

Killer inhibitory receptors (KIRs), 67

L

Leukemia-initiating cells (L-IC), 124
Leukemia stem cells, 132
Leukocyte function antigen-1 (LFA-1), 79–80
Lewis Lung cancer (LLC), 103
Loss of antigenicity
clinical relevance, 72
defects in antigen processing, 70–71
immune system recognition of transformed cells, 66–68
MHC class I, 68–70
tumor-specific antigen, 71–72
Loss of immunogenicity
clinical relevance, 80–81
death receptor signaling, 78–80

inhibitory signals, 76–78
reduced costimulation, 76–78
regulation of immune cell activity, 73–75
sensitivity loss, 78–80
upregulation of inhibitory immune checkpoints, 75–76, 77*f*
Low-molecular-mass polypeptides (LMPs), 67
Lung cancer, 105, 110–112*t*
Lung metastasis, 110–112*t*
Lymphocyte activation gene-3 (LAG-3), 76
Lymphoid cells, 4–5
Lymphoma, 106

M

Macrophages, tumor-associated, 199–204
Major histocompatibility complex (MHC)
alleles, 41
MHC I, 66–68
antigen processing, 37–38
binding prediction, 35–37
immune evasion of cancer stem cells, 143–146
loss of antigenicity, 68–70
MHC II binding prediction, 38–39
Matrix metalloproteinase 8 (MMP-8), 270
Melanoma antigen recognized by T cells (MART)-1, 140–141
Melanoma-associated antigen gene (MAGE) family, 141
Melanoma cell lines, 71–72, 73*f*
Mesenchymal-to-epithelial transition (MET), 129
Mesothelin, 206–207
Mesothelioma, 300–301
Microsatellite, 69–70
MixMHC2pred tool, 36*t*, 39
Monoclonal antibodies (mAbs), 71–72
with cisplatin, 12–14
injection of, 17
therapeutic efficacy of, 8–12
tumor-bearing mice treating with, 5–7
in vitro evaluation of, 15
Mouse colon cancer model, 12–13
Mouse mammary tumor virus (MMTV), 127
Mucin 16 (MUC16), 206–207
Mucin (MUC)-1, 141–142
MuPeXI workflow, 43–44
Myeloid cells, 264
Myeloid-derived suppressor cells (MDSCs), 85–87, 103, 149–152

N

Natural killer (NK) cells, 67, 164–167, 191–194, 193*t*
Neoantigens, 26–28, 27*f*, 138
from high-throughput sequencing data, 29–35
prediction/prioritization tools, 36*t*
prediction tools, 28
prediction workflows, 41–46, 42*t*
protein changes result in, 35–41
sequencing platform, 29
sources, 28–29, 31–32*t*
T-cell receptors, 46, 47*t*
tumor antigens, 139
NeoEpiScope workflow, 45
Neoplastic cells, 3
NetChop tool, 36*t*, 37–38
NetMHCpan tool, 36–37, 36*t*
NetMHCIIpan tool, 36*t*, 39
Neuroblastoma, 299–300
Neutrophil extracellular traps (NETs), 152
Next-generation sequencing, 28
Non-transformed cells, microenvironment, 81–83
Notch signaling pathway, 127–128
Nucleotide excision repair (NER), 130
Nucleotide oligomerization domain, leucine rich repeat and pyrin domain-containing 3 (NLRP3) signaling, 203–204

O

Olfactory receptor family 7 subfamily C member 1 (OLF7C1), 142
Oncogenic signaling pathways, 75–76
Oncolytic virus (OV), 70–72, 218–219, 257*f*
bispecific T-cell engagers, 249–251
CAR T-cell therapy carriage of, 251–252
clearance, 226
COX2/PGE2 pathway, 265
cytokines, 239–243
cytotoxic molecules, 268–269
diversity, 220–221

Oncolytic virus (OV) (*Continued*)
immunogenic cell death, 235–238
local inflammation, 234–235
molecules delivery, 269–273
options for administration, 225–226
with payloads, 239–273
pro-apoptotic molecules, 268–269
recombinant, 224–225
tumor tropism, 221–224
use of, 308
vaccination strategies, 238–239
OpenVax pipeline, 44–45
Optitype tool, 35, 36*t*
Ovarian cancer, 103–104, 110–112*t*, 301–302

P

Pancreatic cancer, 190
adaptive T-cell therapy for, 206–208
cancer immunosurveillance and immunoediting in, 190–191
chimeric antigen receptor T cells, 206–210, 207*t*
classical immunotherapy with immune checkpoint inhibitors in, 195–196, 197–198*t*
dendritic cells in, 191–194, 193*t*
immunosuppressive tumor microenvironment in, 196–199
reprogramming microenvironment by targeting regulators, 204–206
stroma/microenvironment in, 201–202*t*
tumor-associated macrophages for, 199–204
Pancreatic ductal adenocarcinoma (PDAC), 86–87, 300–301
Passive immunotherapy, 141, 153–154
Pattern-recognition receptor (PRR), 203–204
Peptide antigens, 74
Peripheral blood mononuclear cells (PBMCs), 148–149
Phagocytosis, 77–78
Phenotypic heterogeneity, 122, 144
Post-mitotic leukemia cells, 124
Pro-apoptotic molecules, 268–269
Programmed cell death 1 (PDCD/PD1), 75–76, 146–148, 195–196, 235–236
Programmed death-ligand 1 (PD-L1), 75–76, 80, 146–148
Prostaglandin E2 (PGE2), 310–311
Pro-T-cell cytokines, 255
Proteasome, 67
pVACseq tool, 42–43

R

Radiotherapy, 129–130
Reference genomes, 29–30
Regressors, 10
Regulatory subunit I anchoring disruptor (RIAD), 311
Regulatory T cells (Tregs), 102
acute myelocytic leukemia, 105–106
in cancer immunology and immunotherapy, 110–112*t*
cervical cancer, 107–108
cervical intraepithelial neoplasia, 107–108
colon cancer, 106–107
hepatocellular carcinoma, 108
lung cancer, 105
lymphoma, 106
ovarian cancer, 103–104
RNA
circular, 34
editing, 34

S

Self-renewal ability, 134–135
Sensitivity loss, 78–80
Seq2hla tool, 35, 36*t*
Sézary syndrome (SS), 106, 110–112*t*
Side population (SP), 133
Signal transducer and activator of transcription 3 (STAT3), 146–147, 151
Single chain fragment variable (scFv), 294
Single nucleotide variations (SNVs), 26, 28–30
Sipuleucel-T, 157
Slow-cycling leukemia cells, 124
Solid tumors, 231–232
Somatic variants, 34–35
Sonic hedgehog (Shh), 127, 205–206
Spleens, 8–10
Splicing, abnormal, 33–34
Stem cells, 135
Stem-like state, cancer stem cells, 128
Stromal cells, 204–206
Structural variants, 33

T

TANK-binding kinase 1 (TBK-1), 239–240
Targeting regulators, 204–206
T cell-based therapy, 162–163
T cell immunoglobulin-3 (TIM-3), 76
T cell-mediated immune response, 141
T cell receptor (TCR), 40, 46, 47*t*, 67, 74, 153–154
Temozolomide, 135–136
TGF-β signaling, 83, 103–104, 149
Therapeutic efficacy, of immunomodulatory mAbs, 8–12
Th2-type inflammation
  promotes tumor growth, 4–5
  tumor-bearing mice, treating, 5–7
TIminer tool, 44
Transcriptomic meta-analysis, 148
Transformed cells, 66–68
Tumor antigens, 138–143
  candidate antigens, 142–143
  neoantigens, 139
  tumor-associated antigens, 140–142
  viral antigens, 140
Tumor-associated antigens (TAAs), 64–65, 78, 138, 140–142, 294
Tumor-associated macrophages (TAMs), 86, 150–151, 196–204
Tumor-bearing mice, 5–7
Tumor-derived exosomes, 78, 80–81
Tumor-draining lymph nodes (TLNs), 6, 10
Tumor growth, Th2-type inflammation promotes, 4–5
Tumor heterogeneity, 135
Tumor immunoediting, 174*f*
Tumor infiltrating lymphocytes (TILs), 64, 75–76
Tumor-initiating cells, 125–126, 134
Tumor microenvironment (TME), 81–83, 296–297
  immunosuppressive cells, 312
  inhibitory cytokines, 310–312
  metabolic hurdles, 309–310
  in pancreatic cancer, 196–199
  soluble factors, 310–312
  upregulation of coinhibitory receptors, 312–315
Tumor necrosis factor (TNF), 78–79, 102–103
Tumor necrosis factor receptor type I (TNFR1), 102–103
Tumor necrosis factor receptor type II (TNFR2), 102–103
  acute myelocytic leukemia, 105–106
  in cancer immunology and immunotherapy, 110–112*t*
  cervical cancer, 107–108
  cervical intraepithelial neoplasia, 107–108
  colon cancer, 106–107
  hepatocellular carcinoma, 108
  lung cancer, 105
  lymphoma, 106
  ovarian cancer, 103–104
Tumor progression, 4, 16–17
Tumor-specific antigens (TSAs)
  cancer stem cells, 138
  loss of antigenicity, 71–72
Tumor-specific monoclonal antibodies (mAbs), 153
Tumorspheres, 133–134, 159–160

V

Vaccination
  anti-tumor, 159
  cancer stem cells, 155–160, 158–159*f*
  dendritic cells, 155–160, 158–159*f*
  immunology of cancer stem cells, 155–160, 158–159*f*
  oncolytic virus, 238–239
Vaccine-induced T-cell responses, 28
Vaccinia, 223
Variant allelic fraction (VAF), 42–43
Variant call format (VCF) file, 41
Vascular endothelial growth factor (VEGF), 81–83, 305–306
Viral antigens, 140
Virus-like particle (VLP), 159–160

W

Whole exome sequencing (WES), 29
Whole-genome sequencing (WGS), 29
Wnt signaling pathways, 127

X

Xenograft models, 135–136

Kolibri 360 GmbH
Ferdinand-Jühlke-Straße 7
99095 Erfurt, Deutschland
produktsicherheit@kolibri360.de